T0177344

Energy Harvesting Communications

Energy Harvesting Communications

Principles and Theories

Yunfei Chen
University of Warwick
Coventry, UK

Registered Offices
John Wiley & Sons, Inc., 111 River Street, Hoboken, NJ 07030, USA
John Wiley & Sons Ltd, The Atrium, Southern Gate, Chichester, West Sussex, PO19 8SQ, UK

Editorial Office
The Atrium, Southern Gate, Chichester, West Sussex, PO19 8SQ, UK

For details of our global editorial offices, customer services, and more information about Wiley products visit us at www.wiley.com.

Wiley also publishes its books in a variety of electronic formats and by print-on-demand. Some content that appears in standard print versions of this book may not be available in other formats.

Library of Congress Cataloging-in-Publication Data

Names: Chen, Yunfei, 1976- author.
Title: Energy harvesting communications : principles and theories / Yunfei
   Chen, University of Warwick, Coventry, UK.
Description: First edition. | Hoboken, NJ : John Wiley & Sons, Inc., [2019] |
   Includes bibliographical references and index. |
Identifiers: LCCN 2018038943 (print) | LCCN 2018039730 (ebook) | ISBN
   9781119383055 (Adobe PDF) | ISBN 9781119383086 (ePub) | ISBN 9781119383000
   (hardcover)
Subjects: LCSH: Wireless communication systems–Power supply. | Energy
   harvesting. | Microharvesters (Electronics)
Classification: LCC TK5103.17 (ebook) | LCC TK5103.17 .C44 2019 (print) | DDC
   621.382/32–dc23
LC record available at https://lccn.loc.gov/2018038943

Cover Design: Wiley
Cover Images: © Verticalarray/Shutterstock; © Iscatel/Shutterstock

Set in 10/12pt WarnockPro by SPi Global, Chennai, India
Printed in Singapore by C.O.S. Printers Pte Ltd

10 9 8 7 6 5 4 3 2 1

To my parents

Contents

Preface

Wireless communication has provided unprecedented convenience for our daily lives over the past few decades. This convenience largely comes from the replacement of data cables with wireless interconnections. However, the remaining power cables or batteries used in conventional wireless systems are still restricting their mobility or lifetime and hence, are preventing them from being deployed in more and wider applications. Meanwhile, wireless power has emerged as a recent innovation to substitute the power cable. Many breakthroughs have been made. The innovation is further boosted by green communications that aims to meet the governmental targets for emission reduction by harvesting solar energy, wind energy, and other renewable sources.

Given these recent development, it is reasonable to adopt wireless power or energy harvesting in communications so that the last cable in wireless systems can be removed to exploit the full potential of wireless communications. This leads to energy harvesting wireless communications, which is the topic and the motivation of this book.

The use of harvested energy brings big challenges to system designs. First and most importantly, the energy source becomes random or dynamic. This leads to fundamental changes to wireless system designs. Secondly, energy harvesting changes the characteristics of signals and channels utilized in system designs. On the other hand, the use of harvested energy also creates great opportunities. It allows perpetual and sustainable operations of wireless systems. Many conventional wireless systems can be upgraded by adding the energy harvesting functionality to improve their sustainability. For example, sensor networks can use energy harvesting to prolong their lifetime. Cognitive radios can exchange energy for transmission opportunities. Relaying networks can encourage more idle nodes to be involved in relaying by offering them wireless power.

Chapters 1–5 focus on the challenges brought by energy harvesting in wireless communications. Chapters 6–8 focus on different wireless applications enhanced by energy harvesting. Specifically, this book will follow the flow of energy from the energy source, to the energy harvester, to the wireless device, and then to the application. Chapter 1 gives a brief introduction of energy harvesting wireless communications. Chapter 2 discusses different energy sources harvested for wireless communications. Chapter 3 examines the energy harvesters used for two widely used sources. Chapters 4 and 5 study the physical layer and upper layer of the energy harvesting wireless device, respectively. Chapters 6–8 investigate wireless powered communications, energy harvesting cognitive radios and energy harvesting relaying as applications. The whole book focuses on principles and theories rather than systems and implementations.

I would like to acknowledge the contributions made by my students, Freeha Azmat, Youwen Wang, and Idris Adenopo, on the modeling parts, and the useful discussions with my visitors, Yan Gao and Zhibin Xie, on energy harvesting cognitive radios and interference. I am also very grateful to some of my collaborators, Professor Ning Cao, Dr Nan Zhao, Dr Hee-Seok Oh, Dr Gaofeng Pan, Dr Wei Feng, Dr Zhutian Yang, and Dr Mohamed-Slim Alouini, for their support. I would like to sincerely thank my PhD adviser, Dr Norman C. Beaulieu, who showed me the path to an academic career. Last but not least, I would like to give a special thank you to my wife for supporting me through out this book project.

I apologize in advance for any errors that may have occurred and welcome any comments and suggestions for further improvements.

Yunfei Chen
Coventry, UK, May 2018

Acronyms

3G	Third Generation
4G	Fourth Generation
5G	Fifth Generation
AC	Alternating Current
AF	Amplify-and-Forward
AP	Access Point
AWGN	Additive White Gaussian Noise
BCR	Bit Correct Rate
BER	Bit Error Rate
BFSK	Binary Frequency Shift Keying
BPSK	Binary Phase Shift Keying
CDF	Cumulative Distribution Function
CDMA	Code Division Multiple Access
CR	Cognitive Radio
CSMA/CA	Carrier Sensing Multiple Access with Collision Avoidance
D2D	Device-to-Device
DF	Decode-and-Forward
DPS	Dynamic Power Splitting
DPSK	Differential Phase Shift Keying
EA-MAC	Energy Adaptive Media Access Control
EME	Average Energy to Minimum Eigenvalue
EWMA	Exponentially Weighted Moving Average
FCC	Federal Communications Commission
GSM	Global System for Mobile Communications
HAP	Hybrid Access Point
HE-MAC	Harvest-then-Transmit Media Access Control
LOS	Line-Of-Sight
MAC	Media Access Control
MB	Moment-Based
ME	Maximum Eigenvalue
MIMO	Multiple-Input-Multiple-Output
MME	Maximum-to-Minimum Eigenvalue
ML	Maximum Likelihood
MSE	Mean Squared Error
NLOS	Non-Line-Of-Sight

NOMA	Non-Orthogonal Multiple Access
NRMSE	Normalized Root Mean Squared Error
OFDM	Orthogonal-Frequency-Division-Multiplexing
OFDMA	Orthogonal Frequency Division Multiple Access
OMA	Orthogonal Multiple Access
OSI	Open Systems Interconnection
PAPR	Peak-to-Average-Power Ratio
PB	Power Beacon
PDF	Probability Density Function
PHY	Physical
PS	Power Splitting
PSK	Phase Shift Keying
PU	Primary User
PV	Photovoltaic
QoS	Quality of Service
RC	Resistor-Capacitor
RD	Relay to Destination
RF	Radio Frequency
RFID	Radio Frequency Identification
ROC	Receiver Operating Characteristics
SIR	Signal-to-Interference Ratio
SINR	Signal-to-Interference-plus-Noise Ratio
SNR	Signal-to-Noise Ratio
SR	Source to Relay
STC	Standard Testing Conditions
SWIPT	Simultaneous Wireless Information and Power Transfer
TCP/IP	Transmission Control Protocol/Internet Protocol
TDD	Time-Division-Duplex
TDMA	Time Division Multiple Access
TS	Time Switching
QoS	Quality of Service
WCMA	Weather Conditioned Moving Average
WPC	Wireless Powered Communications
WSN	Wireless Sensor Network

1

Introduction

1.1 Background

Energy harvesting wireless communication is one of the most recent advances in communications techniques. It refers to any communications systems that use devices powered by energy from either the ambient environment or a dedicated power transmitter in a cable-less or battery-free way. This new method of communications has two main benefits. First, conventional communications devices rely either on batteries or fixed mains connections for energy supply. However, all batteries have a limited lifetime, while mains connections are not flexible. Energy harvesting provides a promising solution to perpetual and flexible operations of communications devices. Specifically, wireless communication replaces the data cable with wireless interconnection, while energy harvesting aims to replace the power cable with harvested energy, the very last cable in wireless communications. Together, energy harvesting wireless communication provides unprecedented convenience for our daily life. Secondly, energy efficiency is a key issue in wireless communications systems. This is particularly important for conventional wireless devices that rely on batteries, such as sensor nodes or mobile phones. Due to this importance, many studies have been conducted to improve the energy efficiency of wireless communications, including green communications. In fact, most studies on modern communications systems are about their energy efficiency or spectral efficiency. For energy harvesting wireless communications, since the devices are powered by the ambient environment or power transmitter, this problem is less severe for the communication devices. Thus, the main benefits of energy harvesting are the increased convenience and the improved energy efficiency in communications.

These two benefits are achieved at certain costs and have generated several issues. In energy harvesting wireless communications, although the energy supply becomes wireless and endless to provide convenience and efficiency, respectively, the quality of the energy supply drops. For the energy harvested from the ambient environment, such as the sun or wind, there is great uncertainty in the energy availability. This is due to the unpredictable or uncontrollable changes in the ambient environment. For the energy harvested from the power transmitter, there is less uncertainty but to allow energy harvesting and information delivery at the same device, one has to either use two separate sets of equipment or share the same set of equipment. If energy harvesting and information delivery use separate equipment, the hardware cost increases. For example, two radio frequency (RF) fronts may be needed. If both share the same

Energy Harvesting Communications: Principles and Theories, First Edition. Yunfei Chen.
© 2019 John Wiley & Sons Ltd. Published 2019 by John Wiley & Sons Ltd.

equipment, the coordination between energy harvesting and information delivery becomes complicated. The system throughput may also decrease.

These issues have significant impact on wireless system designs, as power and bandwidth are two precious resources for communications. Hence, most research works in energy harvesting wireless communications focus on these two issues. For example, when the energy supply is insufficient, the wireless node may not be able to transmit or receive data when it wants to. Also, when energy harvesting and information delivery are performed in the same frequency band using the same transceiver, the original time interval for data transmission may have to be divided into two parts for the best coordination between energy harvesting and information delivery. In the physical layer, these will affect signal transmission, signal detection, and signal estimation. In the upper layer, these will affect user scheduling, channel assignment, message control, and message routing. These issues will also change the ways in which many recently proposed wireless systems operate. Thus, the effect of energy harvesting and new designs based on energy harvesting will be investigated for both legacy systems and recently proposed systems in this book.

1.2 Relationship with Green Communications

Green communications is a concept proposed recently to tackle the energy efficiency problem of communications as well as to reduce the CO_2 footprint of communications devices. According to the report from the International Telecommunication Union, information and communications technologies account for 2% of all CO_2 emissions. Among them, mobile calls contribute 125 million tonnes of CO_2 emissions every year. These figures are steadily increasing due to the fact that more and more communications systems are being deployed. Thus, under the pressure from governments to achieve emission reduction targets by 2020 and 2050, telecommunications carriers will have to take action. Moreover, this action will eventually reduce their energy bills as well. For example, by shutting down or properly scheduling base stations in the evenings when the traffic is low, one of the methods proposed for green communications, the carriers can reduce their operational costs.

Energy harvesting can be one way of implementing green communications by powering wireless devices using environmentally friendly energy sources, such as solar power and wind power. This can replace or reduce the battery power and the mains power to save energy. However, energy harvesting does not necessarily save energy. For example,

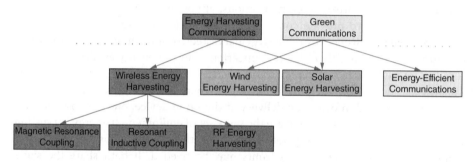

Figure 1.1 Relationship between energy harvesting communications and green communications.

in wireless energy harvesting using power transmitters, a large amount of energy will be wasted during the transmission loss in order to provide convenience at the remote node. Thus, energy harvesting communications and green communications are related to each other but do not belong to each other. Figure 1.1 shows the relationship between energy harvesting wireless communications and green communications.

1.3 Potential Applications

Energy harvesting wireless communication has a lot of interesting applications. One main application is wireless sensor networks (Sudevalayam and Kulkarni 2011). Many sensor nodes are designed for low-power and low-data-rate scenarios, which are very suitable for energy harvesting communications. These applications mainly use energy harvested from the ambient energy sources, such as the sun and the wind. Thus, they also fall into the category of green communications. In other applications, wireless energy harvesting from a dedicated wireless power transmitter can also be used, such as cellular communications (Huang and Lau 2014). Among these applications, the fifth-generation (5G) mobile communications system is a good use case.

1.3.1 Energy Harvesting for 5G

It is noted that 5G is actually a general backbone network that aims to support the Internet of Things, vehicular communications and other applications, in addition to cellular communications. In this sense, it is an enabling technology for wireless sensor networks too. In this section, we mainly focus on its cellular communications function.

For example, in Liu et al. (2015c), an integrated energy and spectrum harvesting mechanism for 5G networks has been proposed. Spectrum harvesting refers to cognitive radio operations for spectrum opportunities, while energy harvesting uses the ambient energy opportunities to support short-distance communications. A multi-tier network was considered and the effects of spectrum and energy harvesting on device-to-device (D2D) communications, Femtocell, Picocell and Macrocell operations have been discussed. It was shown that the aggregate network throughput of such a network can be greatly improved due to spectrum and energy harvesting.

In Hossain and Hasan (2015), a general overview on the 5G cellular network was given. Several key enabling technologies and research challenges have been discussed. Among them, the importance of using energy harvesting to improve the energy efficiency of 5G systems has been investigated. It was suggested that, for 5G services that do not have strict requirements on reliability or quality of service (QoS), ambient sources can be used for harvesting, similar to Liu et al. (2015a). However, for 5G services that require QoS, dedicated RF power transmitters can be used so that energy is always available when needed. A similar discussion is found in Buzzi et al. (2016) with more recent reviews.

In Ding et al. (2017b), energy harvesting was combined with non-orthogonal multiple access (NOMA) to improve energy efficiency and spectrum efficiency in 5G networks. In particular, the NOMA users could be powered by energy harvesting to relay the information in 5G networks. Similarly, in Khan et al. (2016), energy harvesting was combined with millimeter waves, another important 5G technique. In this case, 5G devices tried to harvest energy from the millimeter waves.

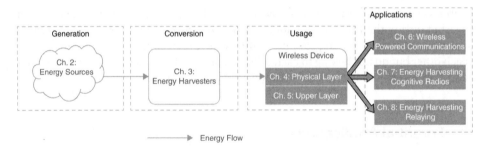

Figure 1.2 Organization of chapters in the book.

There are many other studies on the application of energy harvesting in 5G mobile communications. Since this is not the focus of the book, we have only given a very brief discussion here. Also, owing to the importance of energy harvesting, there is a huge investment in the integration of energy harvesting into existing systems around the world. Due to the enormous number of projects funded in this area, we do not discuss them here.

1.4 Outline of Chapters

In the following chapters, different aspects of energy harvesting wireless communications will be discussed. In Chapter 2, we will discuss different sources of energy that can be harvested and used for communications. Empirical and mathematical models will be examined. This deals with the energy source in energy harvesting wireless communications. In Chapter 3, we will study the relevant energy harvesters for different sources. Their principles and theories will be discussed. This deals with the energy conversion in energy harvesting wireless communications. In Chapter 4 and Chapter 5, the effect of energy harvesting and new techniques based on energy harvesting will be investigated for the physical layer and the upper layer at the wireless device, respectively. This deals with the energy usage in energy harvesting wireless communications. Finally, from Chapter 6 to Chapter 8, the application of energy harvesting in recently proposed systems, including wireless powered systems, cognitive radio systems and relaying systems, will be studied. These deal with the application of energy harvesting wireless communications. Figure 1.2 shows a diagram of how these chapters are organized, related, and what kind of problems they deal with.

2

Energy Sources

2.1 Introduction

There are many different types of energy sources available for harvesting in wireless communications. Depending on their characteristics, they can be categorized as follows:

- *Uncontrollable and unpredictable*: These energy sources cannot be controlled to generate the amount of energy required at a specific time in a specific location. Also, they do not follow commonly used predictive models or implementation of such predictive models is too complicated for relevant applications. An example of such an energy source is mechanical vibration. A piezoelectric or electrostatic energy harvester can convert the vibrational energy into electricity but it may be hard to predict when or where the vibration will occur and it is even harder to generate it intentionally (Mitcheson et al. 2008).
- *Uncontrollable but predictable*: These energy sources cannot be controlled to generate the energy when and where it is needed. However, their generation follows certain patterns that have been well studied and are relatively predictable with acceptable errors. For example, solar energy is mainly determined by solar activities and weather conditions. It is hard to control the movement of the sun or the weather to achieve the level of solar energy desired but solar energy has strong diurnal and seasonal cycles that can be predicted (Bergozini et al. 2010). This prediction can be further improved by incorporating weather data in the forecast.
- *Controllable and partially predictable*: These energy sources can be controlled to produce the amount of energy required at a specific time in a specific location by the wireless device. In other words, these energy sources are controlled by the communications system. Wireless power is a good example of energy sources in this category. In wireless powered communications systems, a radio frequency (RF) signal can be sent by the power transmitter to the remote wireless device for electricity. Also, in an indoor environment, the indoor light can be controlled for the wireless device to harvest its energy using a photovoltaic cell (Wang et al. 2010). These energy sources are only partially predictable, because their behaviors are not fully deterministic. For example, channel fading may change the received wireless power randomly. Obstacles in the room may change the illumination too.

Figure 2.1 shows some commonly used energy sources in energy harvesting wireless communications. They have different characteristics. For example, the solar energy can

Energy Harvesting Communications: Principles and Theories, First Edition. Yunfei Chen.
© 2019 John Wiley & Sons Ltd. Published 2019 by John Wiley & Sons Ltd.

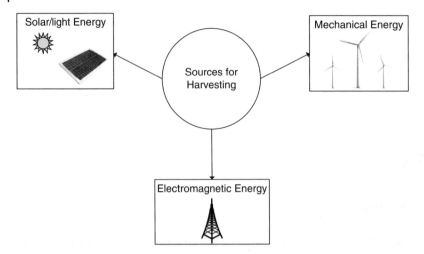

Figure 2.1 Some commonly used energy sources for energy harvesting wireless communications.

only be used when or where it is sunny. The wind energy can only be used when or where it is windy. The electromagnetic energy can only be used when radio transmissions are not blocked. Hence, not all the energy sources can be used in all wireless communications systems due to size, mobility or power limitations of the wireless device. It is important to choose the appropriate source for harvesting in the designs of energy harvesting wireless communications systems. In the next section, we will discuss some of these energy sources, their characteristics, and their applications.

2.2 Types of Sources

In this section, we briefly discuss some commonly used energy sources in energy harvesting wireless communications. These sources can be mainly divided into three categories: mechanical energy, solar/light energy, and electromagnetic energy. All of them need to be converted into electricity using transducers.

2.2.1 Mechanical Energy

Mechanical energy is commonly available in our daily life. Many devices can be used to convert vibration, motion, stress, pressure or strain into electricity. Their main principle is the conversion of mechanical energy from the displacement and oscillation of a spring-mounted mass component into electricity. Based on the randomness of the mechanical energy source, they can be categorized into three types: random vibration energy; steady flow energy; and intermittent motion energy.

The random vibration energy is often seen in built environments, such as bridges, buildings and train tracks (Roundy et al. 2003). They follow certain amplitudes and frequencies but may be random due to the random occurrence of events. The vibration energy can be extracted from these sources but the amount of energy extracted depends on the amplitude and frequency of vibration. In some cases, the presence of the energy harvesting device may also affect the vibration due to the harvester's own weight, as vibration is normally generated by the movement of a mass on a supporting frame and the harvester could add weight to the mass.

The steady flow energy comes from fluid flow, such as air or water, through pipes, or the continuous motion around a shaft, etc. Wind power is one of the most important examples of this energy. It uses the wind turbine to convert the air flow energy into electricity. Another example is the use of blood flow in vessels and breathing in human subjects to generate energy for body sensors that can monitor human body temperature or blood pressure (Mitcheson et al. 2008). The air flow and the blood flow are relatively stable so that the energy harvested from these flows is more deterministic.

The intermittent motion energy falls between vibration and flow. These energy sources come from cyclical motion in the natural environment but the energy can only be harvested during a short period of the cycle. For example, a sensor monitoring the surface of a road can harvest energy from vehicles passing over it but this energy is only available periodically. Also, energy can be harvested from human walking through shoes but only when the foot steps on the ground.

These three types of mechanical energy have different levels of randomness, leading to different levels of predictability. Based on the transduction method, the mechanical energy sources can also be categorized into three types: electromagnetic; electrostatic; and piezoelectric.

In the electromagnetic method, a magnet is used with a metal coil based on the law of induction. This method produces electricity by moving the coil through the magnetic field created by the stationary magnet (Moghe et al. 2009). When the coil moves or the distance between the magnet and the coil changes due to mechanical motion or vibration, an alternating current (AC) will be generated in the coil, which can be used to power up the wireless devices. This motion can be either controllable or uncontrollable. The advantage of this method is that no contact between coil and magnet is required and the electricity generated can be used directly. However, it is hard to integrate the electromagnetic device with the sensor circuit due to its size.

In the electrostatic method, the mechanical motion or vibration is used to change the distance between two electrodes of a capacitor against an electrical field (He et al. 2009). This will change the capacitance of a variable capacitor. The variable capacitor is made of two plates, one fixed and one moving. It needs to be initially charged. When vibration or motion separates the two plates, the vibration or motion is transformed into electricity due to the capacitance change, as the voltage across the capacitor will also change to generate a current in the circuit for use. This method allows the integration of the harvesting device into the sensor circuit.

In the piezoelectric method, a layer of piezoelectric material is used on top of the wireless device so that mechanical strain on the surface of the wireless device will be converted into electricity (Mitcheson et al. 2008). It uses a cantilever structure with a mass attached to a piezoelectric beam that has contacts on both sides of the piezoelectric material. The strain creates charge separation across the device to generate a voltage proportional to the stress applied. In some cases, the amount of energy harvested from this method is small and therefore, it may need to be combined with other methods. Also, the piezoelectric materials are breakable.

Different motion, vibration and strain sources have different power densities. For example, for a wind turbine operating at a wind speed between 2 m/s and 9 m/s, it can generate a power of about 100 mW (Ramasur and Hancke 2012). The blood flow can generate a power of 1 μW, while the running shoes can generate a power of several milliwatts (Paradiso and Starner 2005; Mitcheson et al. 2008). Finger typing can generate a power of 7 mW, while lower limb movement could generate a power of 67 W (Mitcheson et al. 2008). Also, these energy sources have different applications. For example, for a wireless sensor, it is unlikely to use a wind turbine or any harvesters based on the electromagnetic method due to their bulky sizes. On the other hand, the piezoelectric method is well suited for the sensor networks due to their size but only for low-power applications due to the limited power.

2.2.2 Solar/Light Energy

Light is perhaps one of the most commonly used sources of energy for harvesting. The photons from the light source can be converted into electricity using photovoltaic cells. The photovoltaic cell has two types of semiconductor materials and their area of contact forms a PN junction. When the photons arrive from the light source, the photovoltaic cell will release electrons to produce electricity.

For outdoor applications, solar energy is a very reliable source for self-powered devices. It has been used in many wireless networks to replace batteries by providing an almost unlimited energy supply (Sitka et al. 2004). In most of these applications, a solar panel is used to convert the radiation from the sun into electricity. This method has been well established with relatively low cost and high efficiency over a wide range of wavelengths. Also, the energy level provided by a solar panel is very close to the nominal energy required by wireless devices. Specifically, the solar power density is around 1370 W/m^2 when it arrives at the Earth and after attenuation and conversion, the available power density is still around 2 W/m^2. However, one main disadvantage of solar energy is its heavy reliance on the weather, time, and the operating environment. For example, in the evenings when the sun goes down, there is hardly any solar energy to harvest. Also, for indoor applications, solar energy may not be available. In this case, it must be complemented by other energy sources. In general, solar energy is uncontrollable but can be predicted in standard conditions. In most cases, it can provide more power than any other energy sources and thus is suitable for power-consuming energy-harvesting communications applications.

For indoor applications, illumination from indoor lights is another source of energy. Its radiation is typically at the level of 1 W/m^2 and given an efficiency of 15%, the converted electricity could be at the level of 0.15 W/m^2. Typical values range between 10 $\mu W/cm^2$ and 100 $\mu W/cm^2$ (Wang et al. 2010). This is much smaller than the power density of the Sun. Indoor light is relatively controllable compared with the Sun but still varies depending on obstacles, distances, and operations.

Another type of energy that also uses the thermal effect is thermal energy (Leonov 2013). It uses the thermoelectric effect by converting the temperature difference between two metals or semiconductors of different materials into electricity. This is also called the Seebeck effect. Such temperature difference naturally occurs in human bodies or in certain machines. The amount of power converted depends on the thermoelectric properties of the materials and the temperature difference but in general is on the order of 10 $\mu W/cm^2$ to 1 mW/cm^2 (Leonov 2013). This is suitable for wearable sensors, such as fitness bands and smart watches.

2.2.3 Electromagnetic Energy

The electromagnetic energy in this subsection mainly refers to RF energy. The advantage of RF energy over solar energy is that it can work under most conditions: indoor or outdoor, day or evening, sunny or cloudy. It can be as controllable as the light illumination but can also be as unpredictable as vibration. RF energy sources can be divided into two main categories: near field; and far field (Lu et al. 2015). The near field applications include magnetic resonance or inductive coupling. They are often used to charge devices in a wireless way over a very short distance. Due to the short distance, their efficiency can be higher than 80% but this efficiency decreases quickly with distance. This method is completely controllable and predictable. However, for wireless communications systems, this short distance may not be realistic. Hence, energy harvesting wireless communications often use the far field method that can harvest energy over a distance of more than 10 m.

The sources of RF energy in the far field method can be from the ambient environment, such as radiations from the cellular base station, TV transmitter or WiFi router. It can also be from dedicated power transmitters. One unique advantage of RF energy is that most wireless systems are implemented using radio waves too and hence, information delivery can be combined with energy transfer in the same system and sometimes by the same signal.

The level of power from a global system for mobile communications (GSM) base station is around -40 dBm/cm^2. Studies show that other ambient sources, such as TV, third generation (3G) and WiFi produce even weaker power. For example, a 3G base station generates a power density of around -50 dBm/cm^2, while WiFi signals provide a power density of around -70 dBm/cm^2 (Pinuela et al. 2013). Hence, although there are many different ambient RF energy sources, in general their power densities are very low, as their power densities decay quickly with distance. Consequently, these ambient RF sources can only be used for low-power applications, such as radio frequency identification (RFID) or sensor networks, or the wireless device must be very close to the sources. To harvest enough energy for more power-consuming wireless operations, either a large antenna or a wide-band antenna need to be used. Alternatively, dedicated sources of RF energy are required, as in wireless powered communications systems, at additional cost.

There are other types of energy available for harvesting. For example, the pyroelectric effect of materials can be used to generate electricity. Biomedical substances can be used to generate biochemical energy. Acoustic waves can be converted into electricity using transducers or resonators too. Alternatively, all the above energy sources can be combined. Table 2.1 gives an overview of the amount of energy available from different sources.

2.3 Predictive Models of Sources

The amount of energy from most energy sources varies with time. This time-variance leads to uncertainty in the energy supply for wireless communications. Thus, it is very useful to have accurate models that describe this energy uncertainty, because wireless communications systems can use these models to make critical decisions on the usage of energy.

Table 2.1 Typical amount of energy from different sources.

Source	Typical amount of energy
Solar	100 mW/cm^2 (sunny)
Indoor light	0.01 ∼ 0.1 mW/cm^2
Wind	0.38 mW/cm^3 (at 5 m/s)
Piezoelectric	0.2 ∼ 0.4 mW/cm^3
Electrostatic	0.05 ∼ 0.1 mW/cm^3
Ambient RF	0.2 nW/cm^2 ∼ 1μW/cm^2

In this section, we will discuss some important modeling studies on the amount of energy provided by various sources. The data used to derive these models were collected by using a certain measuring equipment or energy harvester. Thus, there is a conversion loss from the available energy at the source as the input of the equipment to the measured or harvested energy as the output of the equipment. Nevertheless, since the same equipment or energy harvester is used to collect all data, this conversion loss can be considered as constant so that the behaviors of the data before collection and after collection are approximately the same to justify the usefulness of the derived models. In the next chapter, we will discuss models that describe the conversion loss or the efficiency of the energy harvester. Using the energy source model here and the energy harvester model in the next chapter, one can predict how much energy is available for wireless communications. We will also discuss models of the harvested power directly in the next chapter. In the following, we will focus on the solar energy models and the ambient RF energy models, as they are the two most widely used energy sources in wireless communications systems.

2.3.1 Solar Energy Modeling

As discussed before, solar energy is not controllable but due to its clear diurnal and seasonal patterns, it is predictable. However, its prediction is highly dependent on the weather conditions.

The simplest model for solar energy prediction is the exponentially weighted moving average (EWMA) model (Cox 1961; Kansal et al. 2007). It divides the data at different time slots on different days into a matrix, where the columns of the matrix could represent different time slots on the same day while the rows of the matrix could represent different days. It uses a weighting factor of ρ to predict the solar energy in the next time slot by linearly combining the current measurement and the previously predicted solar energy. The weighting factor decreases with time to lay a higher emphasis on the measurements taken at a time closer to the time to be predicted. Mathematically, the EWMA predictor is given by

$$\hat{E}(t+1) = \rho E(t) + (1-\rho)\hat{E}(t) \tag{2.1}$$

where $\hat{E}(t+1)$ is the predicted value in the next time slot $t+1$, $E(t)$ is the measurement in the current time slot t, and $\hat{E}(t)$ is the predicted value in the current time slot t.

The exponential weighting can be seen by replacing $\hat{E}(t)$ with $\rho E(t-1) + (1-\rho)\hat{E}(t-1)$ to give

$$\hat{E}(t+1) = \rho E(t) + (1-\rho)\rho E(t-1) + (1-\rho)^2 \hat{E}(t-1) \tag{2.2}$$

where the predicted value at time slot $t-1$ has a smaller weighting factor of $(1-\rho)^2$ since $0 < \rho < 1$. Using (2.1) in solar energy prediction, one has

$$\hat{E}_t(d+1) = \rho E_t(d) + (1-\rho)\hat{E}_t(d) \tag{2.3}$$

where d represents the day, t represents the time slot on that day, $\hat{E}_t(d+1)$ is the predicted value at time t on the $(d+1)$th day, $E_t(d)$ is the measured value at time slot t on the dth day, and $\hat{E}_t(d)$ is the predicted value at time slot t on the dth day. If a matrix is used, $E_t(d)$ is the element on the dth row and tth column of the measurement matrix. From (2.3), the prediction for each time slot is calculated by taking into account the predicted and measured values at the same time slot on the previous day.

The EWMA model works well when the weather condition is stable over a few days or does not change at all. However, if the weather does change, its accuracy will decrease. For example, if the weather keeps switching between sunny and cloudy on different days, using the predicted and measured values on the current day will not help the prediction for the next day much. In this case, the weather conditioned moving average (WCMA) model can be used (Piorno et al. 2009). The WCMA model again divides data into a matrix with rows representing days and columns representing time slots. However, unlike EWMA, it uses the average of the measured values on a few previous days, not just one previous day. Specifically, one has

$$\hat{E}_{t+1}(d) = \rho E_t(d) + (1-\rho)\frac{\sum_{i=1}^{D} E_{t+1}(i)}{D} \tag{2.4}$$

where $\hat{E}_{t+1}(d)$ is the value predicted for the next time slot $t+1$ on the dth day, $E_t(d)$ is the measured value at the current time slot t on the dth day, D is the number of previous days used, and $E_{t+1}(i)$ is the measured value at time slot $t+1$ on the ith day, $i = 1, 2, \cdots, D$. This model can improve the accuracy over the EWMA model by using an average of the values at the same time slots on previous days instead of one single predicted value on the previous day. However, if there is a cloudy day followed by many sunny days or vice versa, this method will cause errors. To reduce the error, the measurements in the previous time slots on the same day can also be used, to replace a single measurement at the current time slot on the same day assuming that the weather conditions are at least stable for the whole day. In this case, the WCMA model can be modified as

$$\hat{E}_{t+1}(d) = \rho E_t(d) + (1-\rho)G(K,d,t)\frac{\sum_{i=1}^{D} E_{t+1}(i)}{D} \tag{2.5}$$

where

$$G(K,d,t) = \frac{\mathbf{V} \cdot \mathbf{P}}{\mathbf{1} \cdot \mathbf{P}} \tag{2.6}$$

K is the number of time slots in the past on the same day, $\mathbf{V}$ is a $K \times 1$ vector with the kth element being $v_k = \frac{E_{t-k+1}(d)}{1/D\sum_{i=1}^{D} E_{t-k+1}(i)}$, $\mathbf{P}$ is a $K \times 1$ vector with the kth element being $p_k = \frac{k-1}{K}$, $\cdot$ is the dot product of two vectors, and $k = 1, 2, \cdots, K$. One sees that there is an additional weighting factor $G(K,d,t)$ to consider the variance over different time slots.

Another simple predictor uses a 2-D linear filter. In this case, the predicted value is calculated as

$$\hat{E}_{t+1}(d) = a_1 E_{t+1}(d-1) + a_2 E_t(d) + a_3 E_t(d-1) \tag{2.7}$$

where the parameters of a_1, a_2 and a_3 can be optimized using training data. There are other more complicated models, such as adaptive management or neural networks. They are not discussed here. In general, WCMA is better than EWMA. A detailed comparison of these models can be found in Bergozini et al. (2010).

2.3.2 Ambient RF Energy Modeling

Another widely used energy source is the ambient RF energy. Example measurements of such ambient RF energy at different time instants in different frequencies are shown in Figure 2.2. This plot shows the measurements taken from the GSM uplink from 880 to 915 MHz in the UK (Azmat et al. 2016). One can see that the input power of the energy harvester or the available power from this band changes with both time and frequency. It is close to zero in many cases but at certain time instants and frequencies, it could be as large as a few milliwatts. Thus, if a wireless device harvests energy from this band, it will be useful to have some model that describes the change of power at different time instants and frequencies so that the device knows when and where to harvest. To simplify the problem, a wide-band harvester can also be used so that all input power at different frequencies can be collected. In this case, all the components at different frequencies will be added to give a simpler measurement plot similar to Figure 2.3, where the measurements become a function of time only. Figure 2.3 measures the power in

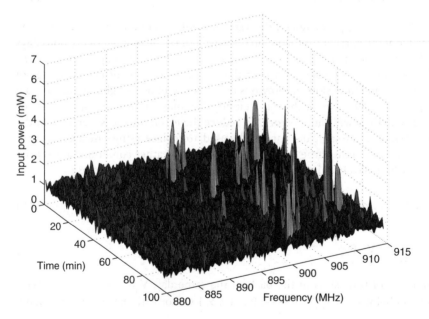

Figure 2.2 Example measurements of ambient RF energy at different time instants and frequencies in the GSM band from 880 to 915 MHz.

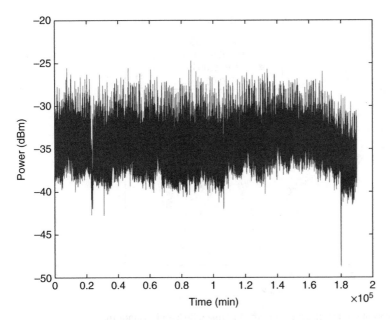

Figure 2.3 Time series of measured power in the 3G band from 1805 to 1880 MHz in the UK.

the 3G downlink from 1805 to 1880 MHz for four months in the UK. There are certain patterns embedded in the data that can be extracted.

Using these measurements and time series analysis methods, the relationship between the available ambient power and the time variable can be modeled. For example, machine learning algorithms can be used. The linear regression method states that the available power can be modeled as

$$\hat{E}(t) = \sum_{i=1}^{I} a_i E_i(t) + a_0 \tag{2.8}$$

where $a_1, \cdots, a_I$ are the parameters to be obtained via training, a_0 represents some random error or disturbance, and $E_i(t)$ are the features used. For polynomial regression, the available power can be predicted as

$$\hat{E}(t) = \sum_{i=1}^{I} a_i [E(t)]^i + a_0 \tag{2.9}$$

to a power of order I. There are other machine learning algorithms, such as support vector regression and random forest. All of them split the data into two parts, one part for training to obtain the optimal parameters and the other part for testing to calculate the prediction error. Define the normalized root mean squared error (NRMSE) as $NRMSE = \dfrac{\sqrt{\frac{1}{I}\sum_{t=1}^{I}(E(t)-\hat{E}(t))^2}}{\frac{1}{I}\sum_{i=1}^{I}E(t)}$. Figure 2.4 compares the prediction errors of different machine learning algorithms. In this case, the power measurements at different minutes are combined into hours and the combined hourly data are used to predict the RF energy in the next hours. One sees that the random forest algorithm has the highest

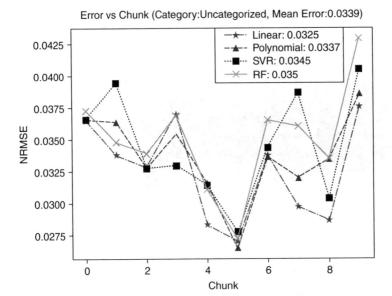

Figure 2.4 Comparison of prediction errors for different machine learning algorithms.

error, while the linear regression has the lowest error, in most cases. The average error is 0.0339, or about 3% error.

The predicted energy can be used to control the wireless device so that it knows when it should start to harvest the energy, as most energy harvesters have an activation energy below which they cannot operate and hence it is only meaningful to harvest energy above the activation level. Thus, there may be two types of errors. If the predicted energy is too high but the actual energy is below the activation level, the energy harvester will start but harvest no energy. Hence, it will waste existing energy. If the predicted energy is too low but the actual energy is above the activation level, the energy harvester will miss an energy opportunity. Figure 2.5 describes this situation, where over-estimation leads to wasted existing energy while under-estimation leads to missed energy. The error rate is 0.108 so that the total efficiency using linear regression is around 89.2%.

In addition to regression methods, the wavelets method can also be used to model the available ambient RF energy (Chen and Oh 2016). For example, the Daubechies D-2n wavelet can be used. For a time series $\mathbf{E}$ of length N, the level 1 Daubechies D-2n transform is given by

$$\mathbf{E} \rightarrow [\mathbf{m}^{(1)}|\mathbf{k}^{(1)}] \tag{2.10}$$

where $\mathbf{E} = [E(1)\ E(2)\ \cdots E(N)]$ is the time series, $\mathbf{m}^{(1)} = [m_1^{(1)}\ m_2^{(1)}\ \cdots m_{N/2}^{(1)}]$ is the first trend sub-signal whose ith element is given by $m_i^{(1)} = \mathbf{E} \cdot \mathbf{V}_i^{(1)}$, $\mathbf{k}^{(1)} = [k_1^{(1)}\ k_2^{(1)}\ \cdots k_{N/2}^{(1)}]$ is the first fluctuation sub-signal whose jth element is given by $k_j^{(1)} = \mathbf{E} \cdot \mathbf{U}_j^{(1)}$, $\mathbf{V}_i^{(1)}$ is the level 1 scaling signal and $\mathbf{U}_j^{(1)}$ is the level 1 wavelet that will be explained later. In practice, multiple levels are often used in order to obtain as many details about the time series as possible. Thus, $\mathbf{m}^{(1)}$ can be used as if it were the signal to obtain level 2 decomposition as

$$\mathbf{m}^{(1)} \rightarrow [\mathbf{m}^{(2)}|\mathbf{k}^{(2)}] \tag{2.11}$$

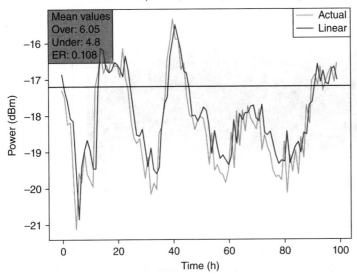

Figure 2.5 Effect of prediction error on wasted and missed energy using linear regression.

where $\mathbf{m}^{(2)}$ is the level 2 trend sub-signal (or low-frequency component) and $\mathbf{k}^{(2)}$ is the level 2 fluctuation sub-signal (or high-frequency component). They are defined in a similar way to $\mathbf{m}^{(1)}$ and $\mathbf{k}^{(1)}$, except that their lengths are $N/4$ now. Thus, one has

$$\mathbf{E} \rightarrow [\mathbf{m}^{(2)}|\mathbf{k}^{(2)}|\mathbf{k}^{(1)}]. \tag{2.12}$$

This process can continue until L levels are used. A level selection at this point is necessary to decide how many high-frequency components should be included to reconstruct the signal with the low-frequency component. Once this is decided, if two levels are selected, the reconstructed signal would be

$$\hat{\mathbf{E}} = \sum_{i=1}^{N/4} m_i^{(2)} \mathbf{V}_i^{(2)} + \sum_{j=1}^{N/4} k_j^{(2)} \mathbf{U}_j^{(2)} + \sum_{j=1}^{N/2} k_j^{(1)} \mathbf{U}_j^{(1)} \tag{2.13}$$

and so on. There will be some difference between the original signal and the reconstructed signal. The choice of levels actually determines the amount of this difference. An important observation here is that, using these reconstruction equations, one can have an analytical relationship between the power measurement and the time. Thus, they provide a statistical model between the measured energy and the time.

The scaling signal and the wavelet are explained as follows. Specifically, one has $\mathbf{V}_i^{(l)} = \sum_{k=1}^{2n} c_k V_{2i-2+k}^{(l-1)}$ and $\mathbf{U}_j^{(l)} = \sum_{k=1}^{2n} d_k V_{2j-2+k}^{(l-1)}$, where $l = 1, 2, \cdots, L$ so that they can be iteratively derived with the initial conditions being the natural basis of V_i^0 whose ith element is 1 and all the other elements are zero, and c_k and d_k are the scaling and wavelet coefficients, respectively, and are constants that are predefined. Figure 2.6 compares the measured energy and the predicted energy using the Daubechies D-$2n$ wavelet with vanishing moments $n = 10$. One can see that the predicted value can track the actual measurement very well. Other wavelets methods can also be used to model or predict the ambient RF energy.

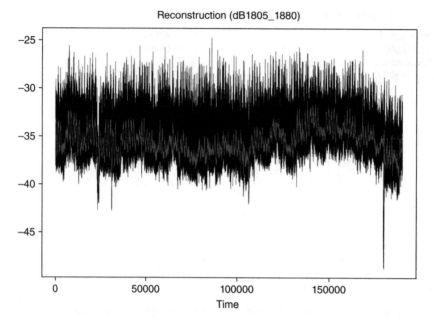

Figure 2.6 Comparison of predicted and measured power using the wavelets method for the band from 1805 to 1880 MHz in the UK (black outside represents measured power and light grey inside represents predicted power).

2.4 Summary

In this chapter, we have discussed different sources of energy available for harvesting to be used in wireless communications systems. This is the start of all energy harvesting wireless communications systems.

First, it has been shown that different sources have different characteristics and hence they can only be used for certain applications. For example, for mobile applications, solar energy may not be convenient due to the bulky size of a solar panel. However, for fixed nodes, such as a wireless sensor network access point, solar energy may be a good choice by providing adequate energy. On the other hand, RF energy may be convenient for radio communications systems due to its excellent integration with the RF circuits. However, it may not provide enough energy for operations.

Secondly, different types of energy are generated based on different principles. Some of them are green and renewable and hence they can be used for green communications. Others may not be green or renewable and are only used for convenience. It is important to consider the different amount of energy available from different sources so that the right source can be chosen for the considered application.

Thirdly, in many cases, the energy source is not controllable but predictable due to certain patterns or behaviors. The prediction models of the energy sources are useful in offline system planning or online real-time adaptation. Different methods can be used to model the available energy at the source. Since the measured energy is a time series, many time series analysis methods, such as regression and wavelets, can be used.

This chapter has discussed the energy source. In the next chapter, we will discuss the energy harvester, which acts as a transducer between different types of sources and electricity. It is also an interface between the natural environment and the communication circuit in the wireless device.

3

Energy Harvesters

3.1 Introduction

The energy harvester is the key component of an energy harvesting wireless communications system. Firstly, the harvester is the interface between the energy source in the natural environment and the circuit in the communications device. Thus, its design has a significant impact on the performance of the communications device. Any changes in the natural environment could be passed on to the communications device via the harvester. Secondly, the harvester supplies energy to the wireless device. Hence, the design of the wireless device also relies heavily on the characteristics of the energy harvester for efficient operation. For example, data transmission is only possible when enough energy is harvested. This necessitates the study of the conversion efficiency of the energy harvester.

In the previous chapter, we discussed some commonly used energy sources in energy harvesting wireless communications. These different energy sources will require different energy harvesters. These energy harvesters serve as transducers to convert non-electrical energy into electricity.

In this chapter, we will first discuss the principles and efficiencies of different energy harvesters. We will focus on two main harvesters that have been widely used in energy harvesting wireless communications systems: the radio frequency (RF) energy harvester and the photovoltaic (PV) panel, which convert the RF power and solar/light power into electricity, respectively.

Following the discussion of these two energy harvesters, we will examine the overall models that include both energy source and energy harvester. These models give a direct prediction of the amount and the arrival of energy available for communications without the need for any separate energy source models and energy harvester models. They have different complexity and accuracy, and thus they are suitable for different applications.

Finally, the battery and supercapacitor used as energy storage for some energy harvesting wireless communications systems will be discussed.

3.2 Photovoltaic Panels

Solar/light power is perhaps the most commonly used energy source. It can provide power for most wireless systems operating either outdoors, indoors with windows or in rooms with artificial light sources. In recent years, more and more solar-powered devices have been developed in our daily life, such as traffic display boards, sensor nodes, and roof panels. The solar power density at the outer layer of the Earth is around 137 mW/cm², which is available for satellites and other outer space communications devices. However, as the sunlight passes through the atmosphere, attenuation is incurred such that devices operating on the surface of the Earth can only expect a density of around 100 mW/cm². On cloudy days, this can be further reduced to 10 mW/cm². However, even for rooms with windows, a power density of around 0.5 mW/cm² is possible.

The efficiency of the PV cell normally ranges from 5 to 20%. This means that there is an energy loss of about 80–95%. In this case, if it is a sunny day, only 5–20 mW will be available for communications when using a PV cell with an area of 1 cm². The loss mainly comes from two sources: intrinsic loss; and extrinsic loss. The intrinsic loss includes photons in the solar radiation that are not absorbed by the PV cell due to energies lower than the band gap or photons that are lost to the lattice of the PV cell as heat. The extrinsic loss includes electrical loss, such as surface recombination, series resistance and shunt resistance, and optical loss, such as shading, incomplete absorption and surface reflection (Beeby and White 2010). In some cases, if the PV cell is encapsulated as a module, further loss may have incurred. All these losses add up to 80–95% of the incident light power. Thus, mathematically, the efficiency of the PV cell can be defined as

$$\eta = \frac{P}{P_i} = \frac{VI}{P_i} \tag{3.1}$$

where P_i is the incident light power at the input of the PV cell, for example, $P_i = 100$ mW on a sunny day with an area of 1 cm², P is the electrical power at the output of the PV cell and $P = VI$ (where V is the output voltage and I is the output current). For most PV cell studies, the relationship between V and I or the relationship between P and V are of interest, as the value of P_i depends on the operating environment only. Next, we will discuss the relationship between V and I or the relationship between P and V.

3.2.1 Principles

There are many different types of PV cells using different materials and properties. A typical silicon PV cell consists of a PN junction formed in a wafer of silicon, whose top and bottom are covered by metals to extract power from the PV cell. The upper metal contact is usually made of several separate metal fingers to allow the sun/light to arrive at the cell, while the lower metal contact is usually a continuous layer of metal. Also, the upper layer is often coated with some thin-film anti-reflective material or micro-scale texturing to minimize the optical loss due to surface reflection. Figure 3.1 shows a diagram of a typical PV cell and its components.

The principle of a PV cell is based on photoconductivity. When a photon from the light is absorbed by a semiconductor, if its energy is higher than the band gap energy of that semiconductor, an electron in the valence band will be excited into the conduction band to form an extra electron–hole pair. In this case, the conductivity of the semiconductor can be increased but after a very short period of time, the excited electron will lose its energy and recombine with a hole so that no useful electricity can be generated.

The idea of a PV cell is to physically separate electrons in the conduction band from holes in the valence band before they recombine. Specifically, a PN junction is formed by

Figure 3.1 Diagram of a typical PV cell.

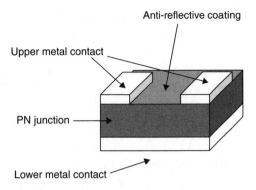

pushing a P-type semiconductor and a N-type semiconductor with different impurities together. The P-type semiconductor will gather a lot of holes, while the N-type semiconductor will gather a lot of electrons. This generates an electrostatic field and a built-in voltage across the junction. When photons fall on the PN junction, a net current will flow between P and N or a useful voltage will be generated. When a load is connected to them, power can be extracted from this current or voltage. Effectively, the PV cell is an unbiased diode, while the arrival of photons generates a photon current that adds to the drift current of this diode. Based on this equivalence, the relationship between V and I, or the $I - V$ curve can be expressed as

$$I = I_{ph} - I_0(e^{\frac{qV}{kT}} - 1) \tag{3.2}$$

where I is the output current of the PV cell, I_{ph} is the photo current generated by the incident light, I_0 is the reverse saturated current of the diode, q is the charge of an electron, k is the Boltzmann constant, T is the operating temperature of the PV cell in Kelvin, and V is the output voltage. In (3.2), the ideal diode equation has been used. Thus, a solar panel can be represented by a photon current generator connected to a diode. A diagram of such an ideal solar panel is shown in Figure 3.2.

Using (3.2), since $P = VI$, the $P - V$ curve can be derived as

$$P = [I_{ph} - I_0(e^{\frac{qV}{kT}} - 1)]V \tag{3.3}$$

Figure 3.2 Comparison of three one-diode models.

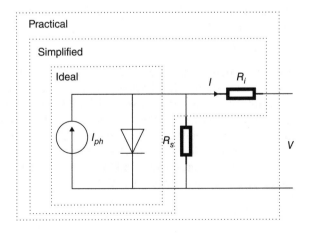

and the $P - I$ curve can be derived as

$$P = \frac{kTI}{q} \ln\left(1 + \frac{I_{ph} - I}{I_0}\right). \tag{3.4}$$

One sees from (3.4) that the output power P increases with I but decreases with $I_{ph} - I$. Hence, an optimal output power may exist. The optimal value of I that maximizes the output power is determined by

$$\ln\left(1 + \frac{I_{ph} - I}{I_0}\right) = \frac{I}{I_0 + I_{ph} - I}. \tag{3.5}$$

Using (3.3) and (3.4), the efficiency of the PV cell is calculated from (3.1) as

$$\eta = \frac{\left[I_{ph} - I_0\left(e^{\frac{qV}{kT}} - 1\right)\right] V}{P_i} = \frac{\frac{kTI}{q} \ln\left(1 + \frac{I_{ph} - I}{I_0}\right)}{P_i}. \tag{3.6}$$

Since the incident light power P_i is fixed, maximization of the efficiency is equivalent to the maximization of the output electrical power. Thus, maximum power position tracking techniques can be used to tune the parameters of the PV cell so that it can work at the output current determined by (3.5) to generate the maximum electrical power for best performance.

Note that, when the PV cell operates in the field, encapsulation may be needed to protect the cell from any damage, water, etc. It also allows the connection of multiple cells in series or parallel to increase the output voltage or the output current so that the generated output power is at a useful level. Note also that there has been a huge amount of work conducted on the device technology for PV cells. They can be categorized into first generation devices with high efficiency but high cost based on silicon wafers, second generation devices with low efficiency but low cost based on the deposit of semiconductor thin films onto inexpensive substrates such as glass or plastic, and third generation devices with even higher efficiency using multi-junctions (Beeby and White 2010). These are not discussed in detail here.

In the following, we will discuss different models of the PV cell that describe the relationship between I and V. Using them, the efficiency of the PV cell can be maximized.

3.2.2 Models

The simplest model for the PV cell is based on the ideal diode equation. It is given as (Xiao et al. 2013)

$$I = I_{ph} - I_0(e^{\frac{qV}{kAT}} - 1). \tag{3.7}$$

The only difference between (3.2) and (3.7) is that (3.7) has an additional parameter A, which is a constant called the diode ideality factor to be determined. All the other parameters are defined as before. The model in (3.7) only has three parameters to determine: I_{ph}, I_0 and A. Hence, it is the simplest model. However, this model ignores the leakage current of the diode as well as the internal resistance of the PV cell. Thus, it cannot describe the practical $I - V$ or $P - V$ curves.

To take the leakage current and the internal resistance into account, a practical model can be shown as (Villalva et al. 2009)

$$I = I_{ph} - I_0(e^{\frac{V+IR_i}{V_tA}} - 1) - \frac{V + IR_i}{R_s} \tag{3.8}$$

where R_i is the resistance in series with the diode representing the internal resistance, R_s is the shunt resistance in parallel with the diode representing the leakage current, $V_t = \frac{kT}{q}$ is the thermal voltage, and all other symbols are defined as before. This model has two extra parameters to determine: R_i and R_s. Thus, it is more complicated than (3.7). However, it takes all the important factors of a PV cell into account and hence it is more accurate.

To strike a balance between complexity and accuracy, a simplified model can also be used as (Xiao et al. 2004)

$$I = I_{ph} - I_0(e^{\frac{V+IR_i}{V_t A}} - 1) \tag{3.9}$$

where the leakage current is ignored, as it is quite small compared with the photo current and the diode current. This simplified model has one extra parameter, R_i, compared with the ideal model in (3.7), but one less parameter, R_s, compared with (3.8). Figure 3.2 compares these three one-diode models in terms of their equivalent circuits.

All the above models use one diode in their equivalent circuits. To increase the accuracy of the model further, some also proposed the use of more diodes. For example, in Babu and Gurjar (2014) and Gow and Manning (1999), two diodes were used to model the PV cell. In Nishioka et al. (2007), three diodes were used to increase the accuracy further. The larger the number of diodes used, the more accurate the model will be but the more complicated the computation will be. For this reason, existing studies hardly use any models with more than two diodes. Thus, in the following, we only discuss the two-diode models. Similar to the one-diode models, the ideal, practical and simplified two-diode models are given as

$$I = I_{ph} - I_1(e^{\frac{qV}{kA_1 T}} - 1) - I_2(e^{\frac{qV}{kA_2 T}} - 1), \tag{3.10}$$

$$I = I_{ph} - I_1(e^{\frac{V+IR_i}{V_t A_1}} - 1) - I_2(e^{\frac{V+IR_i}{V_t A_2}} - 1) - \frac{V+IR_i}{R_s}, \tag{3.11}$$

$$I = I_{ph} - I_1(e^{\frac{V+IR_i}{V_t A_1}} - 1) - I_2(e^{\frac{V+IR_i}{V_t A_2}} - 1) \tag{3.12}$$

respectively, where I_1 and I_2 are the reverse saturation currents of the two diodes, A_1 and A_2 are the ideality factors of the two diodes, and other symbols are defined as before. Figure 3.3 compares the three two-diode models. One sees that two extra parameters are added to each model, compared with the corresponding one-diode models. This increases the computational complexity dramatically. A comparison between these one-diode models and two-diode models can be found in Hasan and Parida (2016).

The establishment of the models in (3.7)–(3.12) is only a first step. The more important part of the modeling work is to determine the parameters of these models, as manufacturers normally do not provide the values of these parameters when the PV cells are sold. To do this, one could use experimental methods to obtain several pairs of values of V and I and then use these data to fit the curves so that the parameters of the models can be calculated. However, this method is material-specific and is required for each PV cell used. Thus, it is not efficient. On the other hand, manufacturers do provide certain data sheets for the PV cells sold. For example, all PV cell data sheets give the values of the open-circuit voltage V_{oc}, the short-circuit current I_{sc}, the voltage and the current that can achieve the maximum power V_{mp} and I_{mp}, respectively, the experimental maximum power P_{max}, the temperature coefficient for the open circuit K_v and the temperature

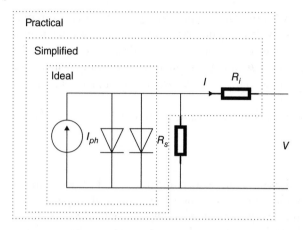

Figure 3.3 Comparison of three two-diode models.

coefficient for the short circuit K_s, under the standard testing conditions (STC) with a temperature of 300 K and a solar radiance of 1000 W/m². Thus, it will be useful if we can use these values to calculate the parameters of the models in (3.7)–(3.12). Next, we will use the model in (3.7) as an example to discuss this method. Note that, for some models since there are more parameters to determine than the values given in the data sheet, additional methods have to be used to obtain additional equations for the additional unknown parameters.

In the STC, when the PV cell is in short-circuit, the output voltage becomes 0. Thus,

$$I_{sc} = I_{ph}. \tag{3.13}$$

When the PV cell is in open-circuit, the output current becomes 0. Thus,

$$0 = I_{ph} - I_0(e^{\frac{qV_{oc}}{kAT_s}} - 1) \tag{3.14}$$

where T_s=300 K is the standard testing temperature. This gives the reverse saturation current as

$$I_0 = \frac{I_{sc}}{e^{\frac{qV_{oc}}{kAT_s}} - 1}. \tag{3.15}$$

Finally, the maximum power is achieved at V_{mp} and I_{mp}. This gives

$$I_{mp} = I_{ph} - I_0(e^{\frac{qV_{mp}}{kAT_s}} - 1). \tag{3.16}$$

Using (3.13)–(3.16), one has

$$\frac{e^{\frac{qV_{mp}}{kAT_s}} - 1}{e^{\frac{qV_{oc}}{kAT_s}} - 1} = \frac{I_{sc} - I_{mp}}{I_{sc}} \tag{3.17}$$

which can be solved to determine the value of the ideality factor A. Once A is obtained, it can be used in (3.16) to calculate I_0 and be used in (3.14) to calculate I_{ph}. Thus, we are able to calculate the parameters of I_{ph}, I_0 and A for the model in (3.7). In some cases, the PV cell may not operate in the STC. Thus, the parameters need to be updated according to the actual operating temperature T and radiance E. In this case, the short-circuit current can be updated to give

$$I_{ph}(T,E) = I_{sc}(T,E) = \frac{E}{E_s}I_{sc}[1 + K_s(T - T_s)] \tag{3.18}$$

where E and T are the operating radiance and temperature, respectively, E_s and T_s are the standard testing radiance and temperature, respectively, and K_s is the current temperature coefficient. Similarly, the open-circuit voltage can be updated to give

$$I_0(T,E) = \frac{I_{sc}(T,E)}{e^{\frac{qV_{oc}(T,E)}{kAT}} - 1} \qquad (3.19)$$

where $V_{oc}(T,E) = V_{oc}[1 + K_v(T - T_s)][1 + K_e(E - E_s)]$ and K_e can be estimated as $K_e = \frac{(V_{0.1} - V_{oc})}{V_{oc}(1-0.8)}$. The ideality factor is determined by the material and does not change with the operating conditions. Thus, it can still be derived from (3.17). Using (3.18), (3.19), and (3.17), the model parameters in (3.7) can be determined for different temperature and solar radiance.

As mentioned before, the ideal one-diode model is the simplest model. For other models, the calculation of the parameters will be more complicated, especially for R_i and R_s. For example, curve-fitting or iteration may be required. Details of these calculations can be found in Villalva et al. (2009) and Gow and Manning (1999).

The above models describe the $I - V$ curve of the PV cell, which gives the efficiency of the energy harvester. However, due to the large variety of device technologies for PV cells, these models may not apply to all PV cells. To be more specific, they are more accurate for the silicon-based PV cells than for other PV cells. Solar/light power can provide a modest level of energy for most wireless communications systems. However, it also has two main limitations. First, it requires the device be placed in an appropriate location with as much sun/light as possible. This is to allow enough energy to be harvested. Secondly, the size of the PV cell is restricted by the applications. For example, a PV cell, used for a laptop should not be larger than the laptop, and the roof panel should not be larger than the house. This restricts the amount of energy harvested for different applications. Thus, the illumination condition, the area of the PV cell, and the overall power usage of the wireless device should be designed jointly for different applications. Next, we discuss the RF energy harvester.

3.3 Radio Frequency Energy Harvester

The RF energy harvester converts the ambient or dedicated RF power into electricity. From the hardware's point of view, it is perhaps the most suitable energy source for energy harvesting wireless communications, as most wireless systems use radio waves for information delivery too. It is even possible to use the same radio signal for both information and energy as simultaneous wireless information and power transfer. This will be discussed in Chapter 6.

Compared with batteries that require either recharging or replacement, RF energy harvesting minimizes the operational or maintenance costs. Thus, it is a useful supplemental energy source to the battery. In some harsh environments or remote places, RF energy harvesting may also be indispensable. For example, for a wireless implant in human bodies, it will be painful to replace the battery using surgeries. For wireless sensors monitoring a high-temperature turbine or nuclear reactor, it will be hard to replace the battery. It is also inconvenient to use solar power or other energy sources in these applications. Thus, RF energy harvesting has its unique advantages in wireless communications.

The amount of energy available from RF energy harvesting depends on the RF energy source. For ambient sources, such as TV signals, cellular signals and WiFi signals, as discussed in the previous chapter, the available power density is usually between 0.2 nW/cm^2 and 1 $\mu W/cm^2$. This amount is small and hence such sources are mainly used for low-power sensor networks. For dedicated sources, if near-field transfer is used, the available power density can be on the order of watts or tens of watts. This is sufficient to power up most wireless devices. On the other hand, if far-field transfer is used, the available power density varies depending on the distance. For example, with a 1 W transmitter located at a distance of 10 m, the received power could be on the order of milliwatts. The propagation environment is another important factor that affects the available energy. A suburban environment will be more friendly to RF energy harvesting than a densely urban environment.

In addition to the source and the propagation environment, the performance or the design of the energy harvester has significant impact on the harvested energy too. This can be measured by the conversion efficiency. The conversion efficiency is defined as

$$\eta = \frac{P_o}{P_i} \tag{3.20}$$

where P_i is the input power at the RF energy harvester and P_o is the output power of the RF energy harvester. For near-field power transfer, the efficiency could be larger than 80%. For ambient sources or far-field dedicated transfer, the efficiency can be between 5% and 60%. Thus, a good design is important for RF energy harvesting. In the next subsection, we will discuss the main components of the RF energy harvester. Then, we will discuss the efficiencies of RF energy harvesting.

3.3.1 Principles

The main components of the RF energy harvester include the antenna, matching network, and rectifier. Thus, in many works, it is also called a "rectenna". Figure 3.4 shows a diagram of a typical RF energy harvester. In some RF energy harvesters, there will also be a voltage regulator and a supercapacitor or battery following the rectifier. They are used for power management required by certain wireless applications. We will focus on the antenna, matching network and rectifier only.

The antenna is the first part of the harvester that captures the electromagnetic waves from the air. It is also the front end that serves as the interface between the natural environment and the energy harvester. Its efficiency is the key to the whole harvester. This efficiency depends on the frequency and the bandwidth it operates with. To increase the efficiency, multiple antennas can be used to form an array. Also, antennas operating at multiple frequencies can be used to collect as much RF energy as possible. Beamforming

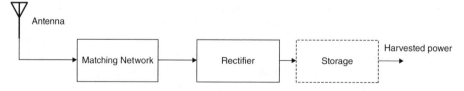

Figure 3.4 Diagram of a typical RF energy harvester.

and directional antennas can increase the efficiency further. In general, the antenna can be represented by a voltage source with a series impedance.

The overall impedance can be expressed as

$$Z_{ant} = R_{ant} + jX_{ant} \tag{3.21}$$

where the resistance R_{ant} is related to the material of the antenna as well as the antenna radiation, while X_{ant} is related to the antenna structure. For example, X_{ant} is inductive for a loop antenna and capacitive for a patch antenna. Typical values of the impedance are 300 Ω for a closed dipole antenna, 75 Ω for an open dipole antenna, and 50 Ω for other antennas.

The matching network is the second part of the harvester and lies between the antenna and the rectifier. According to the electromagnetic theories, the transmission loss is minimum or the power transfer efficiency is maximum if the impedance of the antenna equals the load connected to it. Thus, the matching network can be considered as part of the rectifier to adjust the overall impedance of the matching network and the rectifier to match that of the antenna's. This is achieved by using coils and inductors. Specifically, the matching network can be constructed using a transformer, a resistor-capacitor (RC) network or a parallel coil. Figure 3.5 shows the three different types of matching network.

In general, a RC network and a parallel coil are easier and cheaper to implement than a transformer. Thus, they are more widely used in sensor networks. Between them, a parallel coil is more commonly used for a high-impedance antenna, while a RC network is more commonly used for a low-impedance antenna.

The rectifier is the last part but the most important part of the harvester to rectify the captured alternating current into direct current for use in wireless communications. It mainly consists of diodes and capacitors. The commonly used diodes are the Schottky diodes.

Figure 3.5 Three different matching networks: (a) transformer; (b) parallel coil; and (c) RC network.

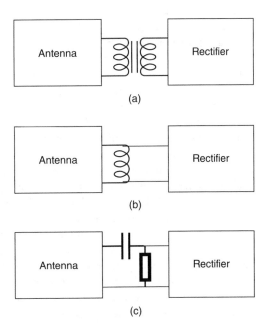

(a)

(b)

(c)

One popular rectifier is the modified Dickson multiplier. It can rectify the RF signal and increase the output voltage for different applications. In many cases, the rectifiers are used in a number of stages to increase the output voltage or system efficiency. Each stage normally consists of two diodes and two capacitors. When the input power is low, the output voltage is relatively independent of the stage number, while the system efficiency decreases with the stage number. For a large input power, both the output voltage and the system efficiency increases with the stage number.

Both the antenna and the rectifier have efficiencies but the overall efficiency from the input of the antenna to the output of the rectifier is of more interest. Next, we examine this overall efficiency as a function of several important system parameters.

3.3.2 Efficiencies

There has been a considerable amount of work conducted on rectenna designs (Nintanavongsa et al. 2012; Stoopman et al. 2013). To compare these designs, the conversion efficiency of the RF energy harvester is a key performance measure, which is defined in (3.20). Most of these studies have given the experimental values of their efficiencies.

These harvesters can be mainly classified as harvesters for low input power and harvesters for high input power.

Stoopman et al. (2013) designed a RF energy harvester with a high sensitivity of −26.3 dBm. It works at a frequency of 868 MHz with a long distance of 25 m, when the source transmits a power of 1.78 W. Its peak efficiency is 22%. Further improvements were made in Stoopman et al. (2014), where the sensitivity was increased to −27 dBm and the range was increased to 27 m. Also, the peak efficiency went up to 37%. Stoopman et al. (2014) designed a fully passive RFID tag operating at a range of 3 m. The peak efficiency is 37%. In Scorcioni et al. (2013), an energy harvester with a peak efficiency of 60% and a sensitivity of −21 dBm was achieved at 868 MHz. The efficiency is above 30% when the input power is between 0.05 mW and 1.5 mW. Such a wide range of input power is very useful. Sun et al. (2013) designed a dual-band harvester at the GSM1800 band and 3G band to harvest energy from multiple bands. Tests showed that it has a peak efficiency of 51%, and its normal operating efficiency is between 16% and 43%. In Le et al. (2008), a 906 MHz energy harvester based on a 36-stage rectifier was designed with a peak efficiency of 60% and a sensitivity of −22.5 dBm. It can operate at a distance of 42 m from a 4 W power source but the efficiency drops quickly when the input power is higher than 0.2 mW. In Kotani et al. (2009), another GSM band harvester was designed that can achieve a peak efficiency of 67.5%. Further tests also showed that this harvester could be used for a 500 MHz TV band to achieve a peak efficiency of 80%. All these designs operate at low input power and are suitable for long-range harvesting applications. Their peak efficiencies are normally achieved at an input power of less than 0.2 mW.

Masuch et al. (2012) designed a 2.4 GHz energy harvester with a peak efficiency of 22.7% achieved at 0.5 dBm and a sensitivity of −10 dBm. In Scorcioni et al. (2012a), another energy harvester at a frequency of 868 MHz was designed for RFID and remote powering applications. This design aimed to maximize the range of the input power that provides high efficiency. An efficiency higher than 40% for a dynamic range of 14 dB in the input power was obtained. This design has a peak efficiency of 60% achieved at 0.5 mW. Nintanavongsa et al. (2012) proposed a dual-rectifier energy harvester. The harvester can achieve an efficiency above 30% for up to 30 mW. It was tuned to 915 MHz but can be modified to other frequencies too. It achieves a peak efficiency of 72% at an input of 4 mW. These harvesters normally achieve their peak efficiencies at an input power of more than 0.2 mW.

For all these harvesters, it has been shown in Chen et al. (2016b) that their efficiency or input–output relationship can be heuristically modeled as

$$P_o = \frac{p_2 P_i^2 + p_1 P_i + p_0}{q_3 P_i^3 + q_2 P_i^2 + q_1 P_i + q_0} P_i \tag{3.22}$$

where the parameters $p_0, p_1, p_2, q_0, q_1, q_2$, and q_3 are different for different harvesters and can be determined by curve-fitting and P_i is the input power (in milliwatts). Hence, the conversion efficiency is a function of the input power. Specifically, it often increases with the input power to reach a peak and then decreases with the input power after the peak. Table 3.1 gives the values of the fitted parameters for different harvesters, where all parameters have been normalized by q_3 (Chen et al. 2016b). In addition to (3.22), there are other non-linear models proposed for RF energy harvesters. For example, in Chen et al. (2017d), another non-linear model was proposed as

$$P_o = \frac{aP_i + b}{P_i + c} - \frac{b}{c} \tag{3.23}$$

where a, b, and c are constants that can be obtained from curve-fitting. In Boshkovska et al. (2015)

$$P_o = \frac{\frac{M}{1 + e^{-a(P_i - b)}} - M\Omega}{1 - \Omega} \tag{3.24}$$

where $\Omega = \frac{1}{1 + e^{ab}}$ and a, b, and M are constants. Figure 3.6 compares the fitted efficiencies using (3.22) and (3.23) for the experimental data in Le et al. (2008). In general, both fitted curves track the experimental data quite well, while the model in (3.22) is slightly better than that in (3.23) by using more parameters.

The above models mainly examine the relationship between the efficiency and the input power. This is useful because it allows system designers to choose the harvester for applications with different input power to achieve the maximum efficiency.

In addition to the input power, the operating frequency and the distance have large impact on the efficiency too. Sun et al. (2013) and Song et al. (2016, 2017) studied the effect of frequency on the conversion efficiency, while Le et al. (2008) studied the effect of distance on the conversion efficiency.

Table 3.1 Fitted parameters for some energy harvesters.

Reference	p_2	p_1	p_0	q_2	q_1	q_0
Stoopman et al. (2013)	1.34	5.2e−5	1.6e−6	5.5e−2	−3.18e−4	2.87e−6
Sun et al. (2013)	4.5e5	7.6e5	685	1.11e4	1.4e4	73.1
Le et al. (2008)	78.5	−2.34	1.62	2.43	−0.482	0.0658
Masuch et al. (2012)	230	−20.5	0.623	9.24	−0.77	0.0808
Scorcioni et al. (2012a)	300	−12.7	0.135	4.4	−1e−2	−1.7e−3

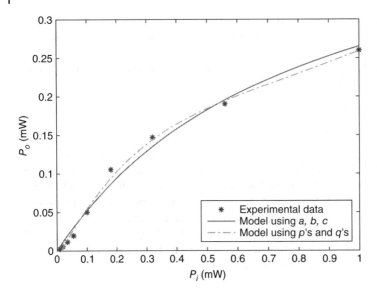

Figure 3.6 Comparison of the fitting methods in (3.22) and (3.23) with the experimental values.

For the frequency, using curve-fitting, it can be shown that the conversion efficiency is a heuristic function of the frequency following a mixture Gaussian model as

$$\eta = \sum_{i=1}^{I} a_i e^{-\frac{(f-b_i)^2}{c_i}} \tag{3.25}$$

where I is the number of Gaussian terms, f is the operating frequency, while a_i, b_i, and c_i are the parameters determined by curve-fitting. Tests show that $I = 3$ or $I = 4$ is usually enough to achieve satisfactory accuracy. The Fourier series is another good model. In this case, the conversion efficiency can be described as

$$\eta = a + \sum_{i=1}^{I} [b_i \cos(idf) + c_i \sin(idf)] \tag{3.26}$$

where a, b_i, c_i, and d are the parameters to be fitted. Figure 3.7 compares the fitted curve using the Fourier model and the experimental values in Sun et al. (2013). These models can largely track the change of the conversion efficiency caused by the frequency. They can be used to guide the choice of the operating frequency or frequencies for the RF energy harvester.

The effect of distance on the conversion efficiency can be examined in a similar way. It was found that the Gaussian mixture model and the Fourier model are applicable to the distance effect too. The polynomial model can also describe the effect of distance in some cases. In these curves, there is only one peak, where the conversion efficiency first increases and then decreases when the distance increases. Figure 3.8 compares the Fourier model with the experimental values in Le et al. (2008).

The above results have mainly focused on the PV cell and the RF energy harvester. Other energy harvesters, including piezoelectric and electrostatic harvesters that convert the mechanical energy into electricity, are also of interest. They are not discussed here.

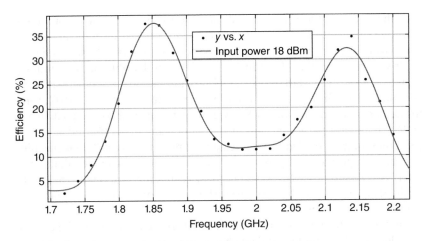

Figure 3.7 Comparison of the Fourier model and the experimental value for frequency.

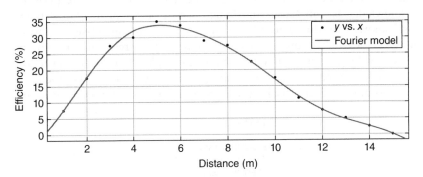

Figure 3.8 Comparison of the Fourier model and the experimental value for distance.

The models in Chapter 2 describe the change of the input power P_i. The models in the above describe the relationship between P_o and P_i. By combining them, one can model the change of the output power P_o, which is the power available for wireless communications. This can be considered as a two-step process. Alternatively, many studies have tried to model the output power P_o directly by considering the energy source and the energy harvester as a whole. In the following, we will discuss some of these modeling studies.

3.4 Overall Models

The overall model combines the energy source and the energy harvester as one whole blackbox and only models the output power of this box. There have been quite a few studies on the modeling of the harvested power.

Some of these studies used empirical methods. For example, in Lee et al. (2011), a testbed was built using the Texas Instruments eZ430-RF2500-SEH platform, which consists of a target board eZ430-RF2500T connected to a solar cell SEH-01-DK. The transmitter sends data packets in each time slot to the receiver when there is enough energy harvested from the solar cell. Otherwise, it will stay inactive. The authors measured the number of inactive time slots between two transmissions. This is an indicator of the available energy from harvesting. The measurements were then fitted using different distributions. It was reported that the transformed Poisson distribution, the uniform distribution and the two-state Markovian model provide the best fits but there is no single distribution that has the highest accuracy in all cases. In Azmat et al. (2016), combining the measurements of the ambient RF power from the cellular bands in Chen and Oh (2016) and the RF energy harvesters designed in Kotani and Ito (2007) and Scorcioni et al. (2012b), the authors modeled the harvested RF power using different machine learning algorithms. Figure 3.9 compares the actual data and the modeled data using these harvesters at 897 and 897.2 MHz, when linear regression is applied. The model can describe the data quite well. In most cases, the prediction error is less than 15%, giving an accuracy of more than 85%.

In other studies, theoretical methods have been used. These methods can be divided into two types: deterministic models; and stochastic models. The deterministic models assume that the arrival time and the amount of the harvested energy are perfectly known. This knowledge is non-causal and hence has to be obtained offline. Using this non-causal knowledge, the transmitter can optimize its transmission strategy. However, the efficiency of the optimal transmission strategy relies heavily on the accuracy of this knowledge. Hence, they are suitable for applications where the energy sources are predictable or change very slowly, such as solar/light energy. The stochastic models do not assume any non-causal knowledge and hence can be used for online

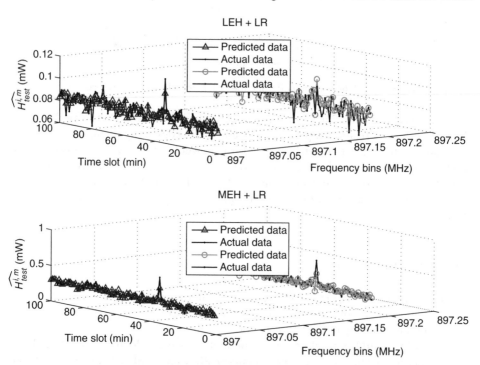

Figure 3.9 Comparison of the actual power and predicted power using linear regression (The upper part represents the results using Scorcioni's harvester and linear regression, and lower part represents the results using Kotani's harvester and linear regression.).

optimization of the transmission strategy. Instead of treating the harvested energy as a deterministic value arriving at a deterministic time, these models treat the harvested energy as a random process. They are suitable for applications where the energy source is unpredictable or random, which is the case in practice. However, there is always mismatch between the actual energy arrival and the modeled energy arrival, which may cause performance degradation.

For example, in Ozel et al. (2011), although the authors considered a Poisson counting process for the harvested energy, in the optimization, they actually assumed full knowledge of the arrival time and the amount of harvested energy at the transmitter. Hence, effectively a deterministic model was used. A similar assumption was used in Yang and Ulukus (2012) for a different optimization problem. Even if the arrival time is predictable, the amount of harvested power may still be random due to the operating environment. Hence, it would be useful to examine the effect of imperfect energy knowledge on the performances of the optimized transmission strategy in these studies.

The main focus of these modeling studies is on the stochastic model. Depending on their way of dealing with the temporal change, they can be classified as time-uncorrelated models and time-correlated models.

Examples of time-uncorrelated models include the following. In Aprem et al. (2013), a Bernoulli process is used to model the harvested energy. In this model, the time horizon is divided into equal time slots. At the beginning of the nth time slot, the harvested energy is

$$E_n = \begin{cases} E, & \text{with probability } p \\ 0, & \text{with probability } 1 - p. \end{cases} \tag{3.27}$$

In addition, $E_1, E_2, \cdots, E_n$ are independent and identically distributed. Hence, this model assumes that, during each time slot, the harvester either harvests an amount of E or does not harvest any energy. This is a simple model that has been used in a number of studies. In Mao et al. (2013), a Poisson distribution is used to model the harvested energy. Again, the harvested energy arrives at the beginning of each time slot but the amount of the harvested energy follows a Poisson distribution as

$$Pr\{E_n = k\} = \frac{e^{-\lambda}\lambda^k}{k!} \tag{3.28}$$

where $k = 0, 1, \cdots$. Similarly, $E_1, E_2, \cdots, E_n$ are independent and identically distributed. This can be considered as a generalization of the Bernoulli model from a binary case to a M-ary case. There are other similar models. A common assumption made in these models is that the harvested energies in different time slots are independent. This may not be the case in practice. For example, in the ambient 3G signal, the energy is likely to be correlated within the same phone conversation across different time slots. Thus, time-correlated models may be more useful in these applications.

The most popular time-correlated stochastic model is the Markov model. In Ho and Zhang (2012), the harvested energy was assumed to follow a discrete first-order Markov process. In this case, the probability mass function of E_n is given as

$$f(E_n|E_0) = \prod_{i=1}^{n} f(E_i|E_{i-1}) \tag{3.29}$$

One sees that the harvested energy at time i only depends on the harvested energy in the previous time slot at time $i - 1$. In Michelusi et al. (2013), a two-state Markov process

was used to model the harvested energy. In this case, the harvested energy is either 0 or
E. One has

$$Pr\{E_n = E | E_{n-1} = E\} = p \tag{3.30a}$$

$$Pr\{E_n = 0 | E_{n-1} = 0\} = q \tag{3.30b}$$

$$Pr\{E_n = 0 | E_{n-1} = E\} = 1 - p \tag{3.30c}$$

$$Pr\{E_n = E | E_{n-1} = 0\} = 1 - q \tag{3.30d}$$

where $0 < p, q < 1$ are the state transition probabilities. In this model, the harvested
energy changes between state E that harvests a fixed amount of energy and state 0
that does not harvest any energy. The transition either leads to a new state or keeps
the current state, each with certain probabilities. Figure 3.10 shows the state transition
of the two-state Markov chain model. This model has been widely used to optimize
the scheduling policies in many studies. The number of states can be more than two in
the generalized Markov process but the complexity increases with the number of states
considered.

The empirical methods can also be combined with the theoretical methods to describe
the harvested energy. For example, in Ku et al. (2015), measurements were first used to
determine the mathematical models and then used to find the parameters of the models.

There are other models for the harvested energy. For example, in Flint et al. (2014),
the stochastic geometry approach was used to model the harvested energy. In Chen
et al. (2017d), the probability density function (PDF) and the cumulative distribution
function (CDF) of the harvested RF energy from multiple antennas, multiple frequencies
or multiple time slots have been derived for Nakagami-m fading signals. For example,
if the harvested signals suffer from Rayleigh fading and I linear energy harvesters are
used to harvest these signals for energy, the PDF and CDF of the harvested energy can
be given as (Chen et al. 2017d)

$$f_{P_o}(y) = \sum_{i=1}^{I} \prod_{j=1, j \neq i}^{I} \frac{1}{2\eta_i \sigma_i^2 - 2\eta_j \sigma_j^2} e^{-\frac{y}{2\eta_i \sigma_i^2}} \tag{3.31}$$

$$F_{P_o}(y) = \sum_{i=1}^{I} \prod_{j=1, j \neq i}^{I} \frac{2\eta_i \sigma_i^2}{2\eta_i \sigma_i^2 - 2\eta_j \sigma_j^2} [1 - e^{-\frac{y}{2\eta_i \sigma_i^2}}] \tag{3.32}$$

where η_i is the conversion efficiency of the ith RF energy harvester and σ_i^2 is the average
fading power of the ith signal. Details of the derivation and more cases can be found in
Chen et al. (2017d). These models allow efficient designs of energy harvesting wireless
communications systems. Next, we will discuss the battery or supercapacitor used in
energy harvesting wireless.

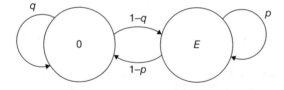

Figure 3.10 Two-state Markov chain
model.

3.5 Battery and Supercapacitor

A battery or supercapacitor is not always required in an energy harvesting communications system. It depends on the protocol of energy usage in the system. For example, in the harvest-use protocol, the energy is used immediately after being harvested so that no buffer is required (Krikidis et al. 2013). In this protocol, wireless communications is performed on a best-effort basis. When there is enough energy, it will transmit data. Otherwise, it stays inactive. The quality of service (QoS) is not guaranteed in this protocol. However, in many wireless applications, QoS is important. Thus, although the energy harvester aims to replace the battery in wireless communications, in these applications, a rechargeable battery or a supercapacitor is still required to accumulate and preserve the harvested energy so that it can be saved for later use. Also, when the harvested energy is larger than the required amount, the excess energy can also be stored to reduce the effect of variation in energy source. In this section, the battery and supercapacitor will be discussed.

3.5.1 Battery

An ideal battery contains a certain amount of energy units. When the wireless device is in operation, the operation consumes a fixed amount of energy units to reduce the battery capacity. Otherwise, the capacity stays the same for any future use. In practice, all batteries suffer from leakage. Even when the battery is not in use, its capacity decreases due to internal chemical reactions. Further, the charge and discharge of the battery will not be perfect such that their efficiency will be less than 100%, implying that some energy units will be lost during the charging and discharging processes. In addition, batteries suffer from non-linear distortions. For example, the rate-dependent capacity actually decreases in a non-linear way with the discharge rate so that a higher discharge rate leads to a much lower capacity. Also, the operating temperature can affect the discharge and the leakage considerably, as it accelerates or decelerates the chemical reactions inside the battery. Finally, the maximum capacity of the battery decreases with more charges and discharges. The voltage changes as well during discharge. These characteristics will affect the battery performance and therefore affect the performance of the energy harvesting wireless system.

To examine and describe the characteristics of rechargeable batteries, many types of battery models have been proposed (Rao et al. 2003) and this is still an on-going effort. These models can be classified as physical models, empirical models, abstract models, or their combination. The physical models simulate the actual physical and chemical processes inside the battery to understand the principles underpinning an electrochemical battery (Doyle et al. 1993). They are usually very accurate but have very high computational complexity at the same time due to the many processes that affect the performance of a battery. The empirical models use measurement data to approximate the behavior of the battery with simple equations (Syracuse and Clark 1997). They have low computational complexity but low accuracy too. Also, the models vary from battery to battery so that their generality is not high. The abstract models emulate the battery behavior by using simple equivalent representations, such as stochastic systems and circuits (Bergveld et al. 1999). Their complexity and accuracy are between the physical models and the empirical models.

3.5.2 Supercapacitor

A supercapacitor is another type of energy storage commonly used in energy harvesting communications. They are capacitors with very high capacitance in a small size and are suitable for wireless communications for several reasons. Firstly, unlike batteries that can only be charged and discharged for a finite number of times (1000 cycles in many cases), supercapacitors can be charged and discharged almost for an unlimited number of times. Thus, they last longer and are more suitable for energy harvesting wireless communications that aim to minimize maintenance cost. Secondly, unlike batteries that need some protection circuits, supercapacitors only need very simple charging circuits and can be charged more quickly with higher efficiencies. Finally, they are more environmentally friendly than batteries in terms of disposal. Similar to batteries, supercapacitors also suffer from charging and discharging losses with less than 100% efficiencies, as well as leakage and changing voltage and capacity.

In general, supercapacitors have smaller capacities than rechargeable batteries due to their size limitation. Hence, leakage becomes very important for supercapacitors, as the leaked electricity could be significant compared with the energy harvested by and stored in supercapacitors. The leakage depends on several important factors, such as the capacitance, the amount of stored energy and the temperature. It is very hard to find theoretical models to describe the leakage as a function of these factors. Hence, the leakage pattern is often examined experimentally. For example, based on measurements, constant leakage current, exponential leakage voltage, polynomial approximation or piecewise linear approximation can be used (Renner et al. 2009; Zhu et al. 2009). One major disadvantage of supercapacitors is that their voltage decreases linearly to zero as the stored energy is being used so that their residual energy cannot be used in wireless communications, as most wireless nodes have a minimum voltage requirement below which they do not work.

3.6 Summary

In this chapter, we have examined the energy harvesters used for energy harvesting wireless communications. We have discussed the principles and theories for two commonly used energy harvesters: the RF energy harvester that converts RF power into electricity; and the PV cell that converts solar/light power into electricity.

The PV cell uses the photoconductivity effect to convert the incident light power into electricity. This is a well-established technique and can provide a modest level of energy that is sufficient for most energy harvesting wireless applications. The PV cells can be modeled by equivalent circuits using one or two diodes. Their input–output relationships are quite simple, although the determination of the model parameters requires considerable effort.

The RF energy harvester has unique advantages in wireless communications as many wireless communications are implemented using RF signals. It mainly consists of an antenna that captures the radio waves and a rectifier that rectifies the captured waves into direct current for use in wireless communications. The amount of energy harvested from the RF power and its conversion efficiency depend on the techniques used. Near-field harvesting can provide a harvested power of several watts with efficiencies higher than 80%, while far-field harvesting can only provide a harvested power of several milli- or microwatts with efficiencies between 5% and 60%. The actual efficiency depends on various factors, such as the input power, the frequency and the operating distance. Their non-linear relationships have been discussed.

A supplementary component of energy harvesting wireless communications is energy storage. Both rechargeable batteries and supercapacitors have been discussed. The supercapacitors have advantages in charging efficiency, charging time and lifetime but disadvantages in leakage and residual energy, over the rechargeable batteries. Also, to understand how batteries work and how the leakage occurs, different models have been proposed.

The ultimate goal of the discussion on the energy harvester and the discussion on the energy source is to understand how the energy flows from the source to the harvester and then from the harvester to the communications device. These two processes can be combined in one study to model the energy available for communications directly. Various models have been proposed to model this energy. Among them, the two-state Markov model is widely used.

Next, we will move on from the energy harvester to the communications device to understand how the communications device can be redesigned based on the characteristics of the energy harvesting process.

4

Physical Layer Techniques

4.1 Introduction

Communication is the transmission of information from a source to one or more destinations. To enable this, communications protocols or communications models are required. The most widely used communications protocol is the transmission control protocol/internet protocol (TCP/IP) model, which has four layers stacked from the bottom to the top as the network interface layer, the internet layer, the transport layer, and the application layer. Another frequently used model is the open systems interconnection (OSI) model, which has seven layers stacked from the bottom to the top as the physical layer, the link layer, the network layer, the transport layer, the session layer, the presentation layer, and the application layer. These two models have certain equivalence in terms of functions. In wireless communications, it is often the physical layer and the link layer in the OSI model, equivalent to the network interface layer in the TCP/IP model, that are of interest. The physical layer performs the transmission and reception of the physical bits, while the link layer provides control functions to enable the efficient operations of transmission and reception. Figure 4.1 shows diagrams of these models. One sees that the physical layer lies at the bottom of the whole system. Any issue in this layer will affect the whole system. Thus, this layer is the most important layer in any communications systems. It will be studied first.

The main purpose of the physical layer is to provide reliable and efficient transmission and reception of the information. Many communications techniques, such as modulation, channel encoding, multiple-input-multiple-output (MIMO) and orthogonal-frequency-division-multiplexing (OFDM), are designed to achieve this purpose. At the core of all these techniques, no matter how the signals are modulated or encoded and no matter which antennas or subcarriers the transmitter uses, the received signals must be detected to recover the transmitted information. Therefore, signal detection is the most important task of the physical layer.

There are many different types of signal detectors, such as coherent detectors, non-coherent detectors, energy detectors, and differential detectors (Kay 1998). They have different performances and can be used for different applications. In applications where performance is of utmost importance, coherent detectors can be used. In these detectors, in order to achieve the best performance, the system state information, such as the channel parameters and the transmitter parameters, is often useful to facilitate the signal detection. Such information can be provided by performing signal estimation

Energy Harvesting Communications: Principles and Theories, First Edition. Yunfei Chen.
© 2019 John Wiley & Sons Ltd. Published 2019 by John Wiley & Sons Ltd.

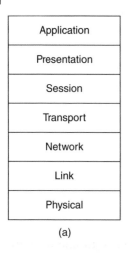

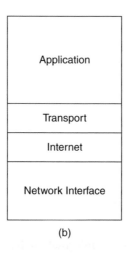

Application
Presentation
Session
Transport
Network
Link
Physical

(a)

Application
Transport
Internet
Network Interface

(b)

Figure 4.1 Diagrams of (a) OSI and (b) TCP/IP models.

or parameter estimation. Thus, signal estimation is another important part of the physical layer (Kay 1993).

In addition to signal detection and signal estimation, which are mainly conducted at the receiver or the destination, in some cases, the transmitter or the source can also perform pre-distortion or pre-coding for best performance. For example, in conventional communications systems with information only, the transmitter can perform beamforming to achieve diversity gain or Nyquist pulse shaping to avoid inter-symbol interference. Similarly, in energy harvesting communications systems with both information and power, the transmitter can also adopt waveform designs to make the power transfer most efficient.

Next, we will first evaluate the effect of energy harvesting on communications, and then we will study signal detection, channel estimation and waveform design for energy harvesting communications systems. After that, we will have a brief discussion on other important issues and techniques in the physical layer of the energy harvesting communications systems.

4.2 Effect of Energy Harvesting

The main effect of energy harvesting on wireless communications is the dynamic power supply. In this book, the radio frequency (RF) energy harvester will be used as an example to examine this effect. Unlike other energy sources, such as solar/light and mechanical energies, the RF signal can be used for both power transfer and information delivery. It is also relatively easier to control. Thus, it provides a good example of how the energy harvesting capability can be integrated with existing communications techniques to implement energy harvesting communications. To do this, first, we will derive the distribution of the random transmission power in some special cases. The randomness comes from the random channel between the power transmitter and the RF energy harvester in this case. Then, using this random power supply, we will discuss a tradeoff between the transmission delay and the transmission probability, when the harvested energy is accumulated for fixed-power transmission. After that, we will use the random power supply to derive the bit error rate (BER) and the achievable rate of the wireless system, where BER measures the reliability of the system while the achievable rate measures the capacity of the system, when the harvested energy is used immediately after being harvested for variable-power transmission. Finally, we will discuss some general information theoretic limits for all energy sources, including the RF power.

4.2.1 Distribution of Transmission Power

Chapter 3 considered the general case when different harvesters are adopted to harvest energy from signals over different antennas, frequency bands or time slots with independent but non-identically distributed channels. This generality has led to the complicated expressions for the probability density function of the harvested power. In this subsection, we will only consider the special case when the same harvester is employed to harvest energy over different time slots or when the channels are independent and identically distributed.

More specifically, consider an energy harvesting wireless system. In this system, the base station or the access point are power sources that transfer a certain amount of wireless power to the remote user devices for charging first. The remote devices then use only the harvested energy to transmit their data back to the base station. This could be a wireless sensor network where the access point collects data from sensors in the field or could be a cellular network where mobile users receive power supply from a power beacon. All devices are assumed half-duplex with a single antenna. Each power transfer or data transmission is performed in a time slot of T seconds. In order to harvest enough energy for transmission, the remote device has to harvest energy from K time slots with a total of KT seconds.

Using the above assumptions, one can divide the whole communication process into two stages. In the first stage, the RF power source (base station or access point) transfers RF power to the remote device as

$$y_k = h_k \sqrt{P_s} s_k + n_k \tag{4.1}$$

where $k = 1, 2, \cdots, K$ denote the K time slots over which the remote device harvests energy, h_k is the fading coefficient from the power source to the remote device and is a complex Gaussian random variable with mean s and variance $2\alpha^2$, P_s is the transmission power for power transfer (different from the transmission power for information delivery discussed later), s_k is the energy signal with $s_k = 1$ for simplicity, and n_k is the complex additive white Gaussian noise (AWGN) with mean zero and variance $2\sigma^2$. In this case, $s = 0$ gives the Rayleigh fading channels, and $s \neq 0$ gives the general Rician fading channels. It is easy to derive that the mean of y_k in (4.1) is

$$\mu = s \sqrt{P_s} \tag{4.2}$$

as $s_k = 1$ and the variance of y_k in (4.1) is

$$2\beta^2 = 2\alpha^2 P_s + 2\sigma^2. \tag{4.3}$$

Using (4.1), the total amount of energy harvested over K time slots can be derived as

$$E = \eta \sum_{k=1}^{K} |y_k|^2 T = \eta T \sum_{k=1}^{K} |h_k \sqrt{P_s} + n_k|^2 \tag{4.4}$$

where η is the conversion efficiency of the energy harvester and is the same for all K signals, as the same harvester is used for K times.

In the second stage, the harvested energy in (4.4) is used by the remote device to transmit its data back to the access point. One has the received signal at the access point as

$$r = g\sqrt{P_x}x + n \tag{4.5}$$

where g is the fading coefficient in this channel, P_x is the transmission power for information delivery and it may be different from P_s for power transfer, x is the transmitted symbol, and n is the complex AWGN with mean zero and variance $2\sigma_d^2$. Assume that g is complex Gaussian distributed with mean s_d and variance $2\alpha_d^2$. Using (4.4), the signal transmission power satisfies $P_x \leq P$, where P is the harvested power given by

$$P = \frac{E}{T} = \eta\beta^2 W \tag{4.6}$$

with

$$W = \sum_{k=1}^{K} \left| \frac{h_k\sqrt{P_s} + n_k}{\beta} \right|^2 . \tag{4.7}$$

If the remote device has a limited battery capacity, such as a supercapacitor, the energy needs to be used as soon as it is harvested. In this energy usage protocol, $P_x = P$. Since W is a random variable, both P and P_x will randomly change, leading to a variable-power transmission. If the remote device has a large battery capacity, such as a rechargeable battery, the harvested energy can be stored and accumulated until it is needed. In this case, P_x can be a fixed value, leading to a fixed-power transmission. In this case, for conventional communications, the system performance is only affected by the random fading g, while for energy harvesting communications, the system performance is affected by both the random fading g and the random power P caused by the random fading during power transfer.

In the general Rician fading channels, W is a sum of the squares of $2K$ (both real and imaginary parts of y_k) independent Gaussian random variables with non-zero mean and unit variance. Thus, it can be derived as a non-central chi-square random variable with $2K$ degrees of freedoms and non-centrality parameter $\lambda = KP_s\frac{|s|^2}{\beta^2}$ (Proakis 2001). Using this fact and the relationship of $P = \eta\beta^2 W$, one has the probability density function (PDF) and the cumulative distribution function (CDF) of P as

$$f_P(y) = \frac{e^{-\frac{y}{2\eta\beta^2} - \frac{KP_s|s|^2}{2\sigma^2}}}{2\eta\beta^2(\eta KP_s|s|^2)^{\frac{K-1}{2}}} y^{\frac{K-1}{2}} I_{K-1}\left(\sqrt{\frac{KP_s|s|^2 y}{\eta\beta^4}}\right) \tag{4.8}$$

and

$$F_P(y) = 1 - Q_K\left(\sqrt{\frac{KP_s|s|^2}{\beta^2}}, \sqrt{\frac{y}{\eta\beta^2}}\right) \tag{4.9}$$

respectively, where $I_{K-1}(\cdot)$ is the $(K-1)$th order modified Bessel function of the first kind Gradshteyn and Ryzhik (2000, eq. (8.406.1)) and $Q_K(\cdot,\cdot)$ is the Kth order Marcum Q function Proakis (2001, eq. (2.1–122)).

In the special case of Rayleigh fading channels, $s = 0$. Then, it can be derived that W is a central chi-square random variable with $2K$ degrees of freedom, equivalent to a Gamma random variable with shape parameter K and scale parameter 2. Again, since $P = \eta\beta^2 W$, the PDF and CDF of P are derived, respectively, as

$$f_P(y) = \frac{y^{K-1}}{(2\eta\beta^2)^K \Gamma(K)} e^{-\frac{y}{2\eta\beta^2}} \tag{4.10}$$

and

$$F_P(y) = \frac{\gamma(K, \frac{y}{2\eta\beta^2})}{\Gamma(K)} \tag{4.11}$$

where $\Gamma(\cdot)$ is the Gamma function Gradshteyn and Ryzhik (2000, eq. (8.310.1)) and $\gamma(\cdot, \cdot)$ is the incomplete Gamma function Gradshteyn and Ryzhik (2000, eq. (8.350.1)). These distributions will be used in the following subsections to examine the effect of energy harvesting on wireless performances. In contrast, for a conventional wireless communications system, P and P_x are fixed.

4.2.2 Transmission Delay and Probability

Consider the fixed-power transmission using a rechargeable battery first. In this case, the remote device harvests energy from K time slots until the harvested power P is larger than or equal to the required fixed transmission power P_x. Then, the remote device uses the harvested energy to transmit its data to the base station in the next time slot for T seconds. There is a tradeoff here. If the remote device harvests the energy for a longer time, there will be a higher probability that it has enough energy for information transmission or a larger transmission probability. On the other hand, a longer harvesting time will cause a larger transmission delay and thus reduces the system throughput, as this time could have been used for information transmission. An optimum value of K exists, and will be studied next.

To study this tradeoff, define the transmission probability as

$$P_T = Pr\{P \geq P_x\} = Pr\{\Lambda \geq \Lambda_x\} \tag{4.12}$$

where $\Lambda = \frac{P}{2\sigma_d^2}$ and $\Lambda_x = \frac{P_x}{2\sigma_d^2}$. The first equation in (4.12) makes sure that the harvested power is larger than or equal to the required power in order to transmit the information. Such a fixed value of P_x can be required to guarantee certain quality of service (QoS). The second equation translates the power limitation into the signal-to-noise ratio (SNR) limitation, where the transmission SNR must be larger than or equal to the required SNR. Since the noise power $2\sigma_d^2$ is a constant, these two limitations are equivalent.

On the other hand, the transmission delay is KT seconds, caused by energy harvesting. Since the actual time for information transmission is T seconds, the effective throughput is $\frac{1}{K+1}$. This effective throughput is only possible when the remote device actually performs transmission. Thus, the average throughput can be defined as

$$C = \frac{1}{K+1} P_T. \tag{4.13}$$

This value will be our performance measure used to consider the tradeoff between the transmission delay and the transmission probability. Its calculation boils down to the calculation of P_T, which requires the CDF of P.

Using (4.9), the average throughput in Rician fading channels can be calculated as

$$C = \frac{1}{K+1} Q_K \left(\sqrt{\frac{K P_s |s|^2}{\beta^2}}, \sqrt{\frac{2\sigma_d^2 \Lambda_x}{\eta \beta^2}} \right) \tag{4.14}$$

and using (4.11), the average throughput in Rayleigh fading channels becomes

$$C = \frac{1}{K+1} \left[1 - \frac{\gamma \left(K, \frac{2\sigma_d^2 \Lambda_x}{2\eta \beta^2} \right)}{\Gamma(K)} \right]. \tag{4.15}$$

One notes that (4.15) can also be obtained from (4.14) by setting $s = 0$ and using the relationship between the Marcum Q function and the Gamma function. One can see from these equations that C decreases when $K + 1$ increases but increases when K in the Gamma function or the Marcum Q function increases. Thus, an optimum K may exist. Note again that this is the case when the remote device can use a large rechargeable battery for storage so that fixed-power transmission is possible to guarantee the QoS. This is not possible for supercapacitors that have very limited storage.

Next, some numerical examples will be given to show the effect of energy harvesting on this tradeoff. In these examples, we examine the values of Λ_x from 0 to 20 dB with a step size of 2 dB and the values of K from 1 to 20 with a step size of 1 for the average throughput C. We set $\eta = 0.5$, which corresponds to a 50% efficiency for the RF energy harvester. We also define $\omega = \frac{\Omega P_s}{2\sigma^2}$ as the average SNR in the power transfer channel, where $\Omega = |s|^2 + 2\alpha^2$. This value indicates the quality of the random channel from the power source to the remote device. Figures 4.2–4.4 show the relationship between C and K under different conditions. For Rayleigh fading channels, $s = 0$.

From Figure 4.2, first, one can see that, for $\Lambda_x = 2$ dB, C monotonically decreases with K, while for other values of Λ_x, C first increases then decreases with K. This

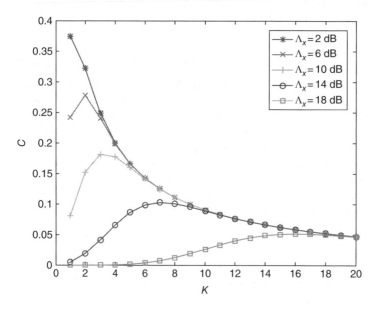

Figure 4.2 C versus K in Rayleigh fading channels when $\omega = 10$ dB.

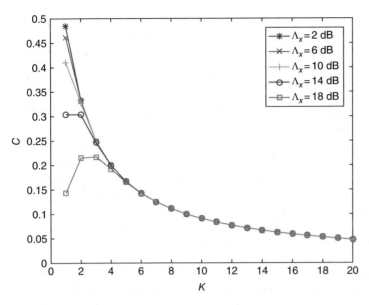

Figure 4.3 *C* versus *K* in Rayleigh fading channels when $\omega = 20$ dB.

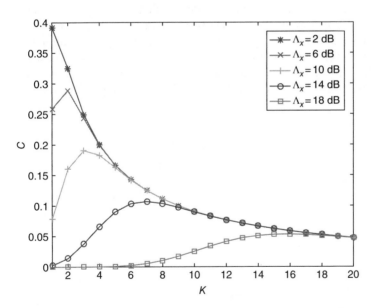

Figure 4.4 *C* versus *K* in Rician fading channels when $\omega = 10$ dB and $\frac{|s|^2}{2\alpha^2} = 1$.

implies that the optimum value of K exists, as expected. Based on this observation, in order to achieve a balance between transmission probability and transmission delay, for $\Lambda_x = 2$ dB, one should choose $K = 1$, for $\Lambda_x = 6$ dB, one should choose $K = 2$, and so on. Thus, these curves give very useful guidance on how the harvesting time should be chosen. Secondly, the optimum value of C in general decreases with Λ_x, implying that, if one requires a higher transmission power or data rate for information transmission, it will lead to an overall larger delay or smaller throughput. Thirdly, when K is large, the values of C for different values of Λ_x tend to overlap with each other. This can be explained using the results for Rayleigh fading channels. Using an equation in Gradshteyn and Ryzhik (2000, eq. (8.352.1)), $C = \frac{1}{K+1} e^{-\frac{2\sigma_d^2 \Lambda_x}{2\eta\beta^2}} \sum_{m=0}^{K-1} \frac{(\frac{2\sigma_d^2 \Lambda_x}{2\eta\beta^2})^m}{m!}$ in (4.15).

When $K \to \infty$, $\sum_{m=0}^{K-1} \frac{(\frac{2\sigma_d^2 \Lambda_x}{2\eta\beta^2})^m}{m!} \to e^{\frac{2\sigma_d^2 \Lambda_x}{2\eta\beta^2}}$ according to the Taylor series expansion of exponential functions. Thus, $C \to \frac{1}{K+1}$, independent of Λ_x.

Also, comparing Figure 4.3 with Figure 4.2, one can see that, when ω increases because of either an increase in $2\alpha^2$ and P_s or a decrease in $2\sigma^2$, C increases in general. In this case, the optimum value of K decreases, as expected, as less time slots will be required if less power is lost in a better channel from the power source to the remote device. Finally, comparing Figure 4.4 with Figure 4.2, one can see that, when the channel condition changes from Rayleigh to Rician or when $|s|^2$ increases, C increases slightly.

Figures 4.5 and 4.6 show C versus Λ_x to examine the effect of Λ_x on the system performance more clearly.

From Figure 4.5, one sees that C does not change with Λ_x and then gradually decreases to zero, when Λ_x increases, in most cases considered. This suggests that Λ_x cannot be chosen too large in the design. Particularly, in Figure 4.5, Λ_x should be smaller than 8 dB for $K = 4$, 12 dB for $K = 6$, and so on.

Beyond these threshold values, the transmission delay will be too large to be compensated by the harvested energy in the average throughput. Note that C generally increases when K decreases due to smaller delay but there is a range determined by Λ_x. For example, when Λ_x is less than 12 dB, $K = 4$ has a larger C than $K = 6$, but when Λ_x is larger than 12 dB, C for $K = 4$ decreases faster so that it is disadvantageous to use a smaller K. These threshold values change with K. Finally, comparing Figure 4.6 with

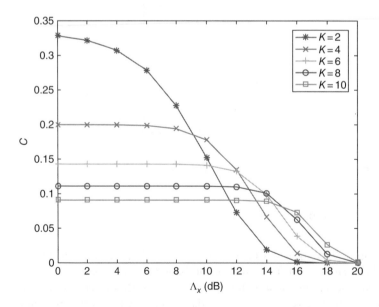

Figure 4.5 C versus Λ_x in Rayleigh fading channels when $\omega = 10$ dB.

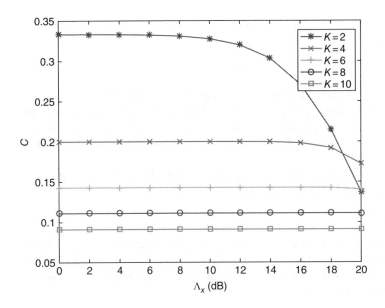

Figure 4.6 C versus Λ_x in Rayleigh fading channels when $\omega = 20$ dB.

Figure 4.5, the range where C remains constant increases when ω increases, implying that there will be more choices or more flexibility when the channel conditions are better.

From the above results, for energy harvesting communications, the choices of harvesting time and QoS are quite important. For the best tradeoff between transmission probability and transmission delay, the optimum number of harvesting time slots should be used within a certain range of QoS. For conventional communications, there is no delay caused by energy harvesting so the value of C is always 1, which is larger than that for energy harvesting communications. Thus, the randomness in power supply causes performance degradation.

4.2.3 Bit Error Rate

In the previous subsection, the fixed-power transmission scheme was studied. In this subsection and the next subsection, the variable-power transmission scheme is considered. In this scheme, the remote device uses a supercapacitor with a limited capacity such that the harvested energy is used as soon as possible. One practical motivation for this scheme is that most batteries suffer from leakage. The longer the energy is stored, the more leakage it will incur. Also, to have fixed-power transmission, as in the previous subsection, power management will be required. This may complicate the energy harvester design and thus may not be desirable or necessary in simple low-power applications.

The error rate or the probability of error measures the reliability of a wireless system. It will be studied in this subsection. The achievable rate measures the capacity of a wireless channel. It will be examined in the next subsection. Using (4.5), the overall SNR of the received data signal can be derived as

$$\epsilon = \frac{|g|^2}{2\sigma_d^2}P_x = \gamma P_x \tag{4.16}$$

where $\gamma = \frac{|g|^2}{2\sigma_d^2}$ is the SNR from fading. Since the energy is used as soon as it is harvested, one has $P_x = P$ so that

$$\epsilon = \gamma P. \tag{4.17}$$

As mentioned before, for conventional communications, the only randomness in the received SNR comes from the fading coefficient g, while for energy harvesting communications, the randomness in the received SNR comes from both the fading coefficient g and the transmission power P. Next, we use binary signaling as an example to analyze the BER.

From (4.5), it can be shown using results in Simon and Alouini (2005) that binary signals with coherent detection have BERs of

$$P_e(\epsilon) = Q(\sqrt{2u\epsilon}) \tag{4.18}$$

where $u = 1$ for binary phase shift keying (BPSK), $u = \frac{1}{2}$ for binary frequency shift keying (BFSK), $Q(\cdot)$ is the Gaussian Q function Proakis (2001, eq. (2.1–97)), and ϵ is given in (4.17). Also, from Simon and Alouini (2005), for binary signals with non-coherent detection, their BERs are given by

$$P_e(\epsilon) = \frac{1}{2}e^{-u\epsilon} \tag{4.19}$$

where $u = 1$ for differential phase shift keying (DPSK) and $u = \frac{1}{2}$ for BFSK. These expressions are conditional BERs, conditioned on the SNR ϵ. Next, we need to find the distribution of ϵ to calculate the unconditional BERs.

Consider the simpler Rayleigh fading channels first. In this case, it can be shown that the PDF of γ is given by

$$f_\gamma(x) = \frac{1}{\bar{\gamma}}e^{-\frac{x}{\bar{\gamma}}} \tag{4.20}$$

where $\bar{\gamma} = \frac{2\alpha_d^2}{2\sigma_d^2}$ is the average SNR of the channel. Using (4.20) and (4.10), the PDF of ϵ is calculated as

$$f_\epsilon(z) = \frac{2z^{\frac{K-1}{2}}}{(2\eta\beta^2\bar{\gamma})^{\frac{K+1}{2}}\Gamma(K)}K_{K-1}\left(\sqrt{\frac{2z}{\eta\beta^2\bar{\gamma}}}\right) \tag{4.21}$$

where $K_{K-1}(\cdot)$ is the $(K-1)$th order modified Bessel function of the second kind Gradshteyn and Ryzhik (2000, eq. (8.407.1)). Similarly, the CDF of ϵ can be calculated as

$$F_\epsilon(z) = 1 - \frac{z^{\frac{K}{2}}}{(2\eta\beta^2\bar{\gamma})^{\frac{K}{2}}\Gamma(K)}K_K\left(\sqrt{\frac{2z}{\eta\beta^2\bar{\gamma}}}\right). \tag{4.22}$$

Using (4.19) and (4.21), the average BER for binary signaling with non-coherent detection is obtained as

$$\bar{P}_e = \frac{e^{\frac{1}{4u\eta\beta^2\bar{\gamma}}}}{2(2u\eta\beta^2\bar{\gamma})^{\frac{K}{2}}}W_{-\frac{K}{2},\frac{K-1}{2}}\left(\frac{1}{2u\eta\beta^2\bar{\gamma}}\right) \tag{4.23}$$

where an equation in Gradshteyn and Ryzhik (2000, eq. (6.631.3)) has been used and $W(\cdot,\cdot)$ is the Whittaker function Gradshteyn and Ryzhik (2000, eq. (9.220.4)). Also, using (4.18) and (4.22), the average BER of binary signaling with coherent detection is derived as

$$\bar{P}_e = \frac{1}{2} - \frac{\sqrt{u\eta\beta^2\bar{\gamma}/2}\,\Gamma(K+0.5)}{2(2u\eta\beta^2\bar{\gamma})^{\frac{K}{2}}\Gamma(K)} e^{\frac{1}{4u\eta\beta^2\bar{\gamma}}}\, W_{-\frac{K}{2},\frac{K}{2}}\left(\frac{1}{2u\eta\beta^2\bar{\gamma}}\right).$$
(4.24)

These expressions can be used to examine the choice of the number of time slots. Next, we study the more complicated Rician fading channels.

In Rician fading channels, the PDF of γ can be derived as

$$f_\gamma(x) = \frac{1}{\bar{\gamma}} e^{-\frac{|s_d|^2}{2\alpha_d^2} - \frac{x}{\bar{\gamma}}} I_0\left(\sqrt{\frac{2x|s_d|^2}{\bar{\gamma}\alpha_d^2}}\right).$$
(4.25)

Then, using (4.25) and (4.8), the PDF of ϵ is

$$f_\epsilon(z) = \frac{e^{-\frac{KP_s|s|^2}{2\beta^2} - \frac{|s_d|^2}{2\alpha_d^2}}}{2\eta\beta^2\bar{\gamma}(\eta KP_s|s|^2)^{\frac{K-1}{2}}} \int_0^\infty x^{\frac{K-3}{2}} e^{-\frac{x^2}{2\eta\beta^2} - \frac{z}{x\bar{\gamma}}}$$
$$I_{K-1}\left(\sqrt{\frac{KP_s|s|^2 x}{\eta\beta^4}}\right) I_0\left(\sqrt{\frac{2z|s_d|^2}{\bar{\gamma}\alpha_d^2 x}}\right) dx.$$
(4.26)

The CDF of ϵ can be derived similarly as

$$F_\epsilon(z) = 1 - \frac{e^{-\frac{KP_s|s|^2}{2\beta^2}}}{2\eta\beta^2(\eta KP_s|s|^2)^{\frac{K-1}{2}}} \int_0^\infty x^{\frac{K-1}{2}} e^{-\frac{x}{2\eta\beta^2}}$$
$$I_{K-1}\left(\sqrt{\frac{KP_s|s|^2 x}{\eta\beta^4}}\right) Q_1\left(\frac{|s_d|}{\alpha_d}, \sqrt{\frac{2z}{\bar{\gamma}x}}\right) dx.$$
(4.27)

Thus, (4.26) can be used to calculate the average BER for non-coherent detection as

$$\bar{P}_e = \frac{e^{-\frac{KP_s|s|^2}{2\beta^2} - \frac{|s_d|^2}{2\alpha_d^2}}}{4\eta\beta^2(\eta KP_s|s|^2)^{\frac{K-1}{2}}} \int_0^\infty \frac{x^{\frac{K-1}{2}}}{\sqrt{\frac{|s_d|^2}{2\alpha_d^2}(1+\bar{\gamma}ux)}} e^{-\frac{x}{2\eta\beta^2} + \frac{|s_d|^2/(2\alpha_d^2)}{2(1+\bar{\gamma}ux)}}$$
$$I_{K-1}\left(\sqrt{\frac{KP_s|s|^2 x}{\eta\beta^4}}\right) M_{-\frac{1}{2},0}\left(\frac{|s_d|^2/(2\alpha_d^2)}{1+\bar{\gamma}ux}\right) dx$$
(4.28)

where an equation in Gradshteyn and Ryzhik (2000, eq. (6.614.3)) has been used and $M(\cdot,\cdot)$ is another type of Whittaker function defined by Gradshteyn and Ryzhik (2000, eq. (9.220.2)). Also, (4.27) can be used with integration by parts to give the average BER for coherent detection as

$$\bar{P}_e = \frac{e^{-\frac{KP_s|s|^2}{2\beta^2} - \frac{|s_d|^2}{2\alpha_d^2}}}{4\eta\beta^2(\eta KP_s|s|^2)^{\frac{K-1}{2}}} \int_0^\infty x^{\frac{K-1}{2}} e^{-\frac{x}{2\eta\beta^2}} \sqrt{\frac{\bar{\gamma}ux}{\bar{\gamma}ux+1}} I_{K-1}\left(\sqrt{\frac{KP_s|s|^2 x}{\eta\beta^4}}\right)$$
$$\left[\Phi_1\left(0.5, 1, 1; \frac{1}{\bar{\gamma}ux+1}; \frac{|s_d|^2/(2\alpha_d^2)}{\bar{\gamma}ux+1}\right) - {}_1F_1\left(0.5; 1; \frac{|s_d|^2/(2\alpha_d^2)}{\bar{\gamma}ux+1}\right)\right] dx. \quad (4.29)$$

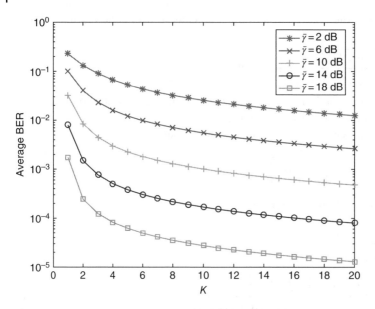

Figure 4.7 Average BER versus K in Rayleigh fading channels.

where an elegant result in Sofotasios et al. (2015) has been used, $\Phi_1(\cdot,\cdot,\cdot;\cdot;\cdot)$ and $_1F_1(\cdot;\cdot;\cdot)$ are the hypergeometric functions (Sofotasios et al. 2015).

Figures 4.7 and 4.8 give the average BER. For simplicity, we only consider coherent detection of BFSK in the Rayleigh fading channels and assume that $s_d = s = 0$, $\alpha_d^2 = \alpha^2$, and $\sigma_d^2 = \sigma^2$. Thus, $\omega = \bar{\gamma}$ in this case. From Figure 4.7, one can see that the average BER first decreases quickly with K but then is only marginally reduced when K keeps increasing. Since a larger value of K requires a larger capacity of the supercapacitor and also possibly causes more leakage, it is desirable to use a small value of K, less than $K = 3$ in most cases considered. This is also confirmed by Figure 4.8, where the average BER also decreases with K but the performance gain at small values of K is larger than that at large values of K. The BER performance reaches a lower limit as K keeps increasing.

Figure 4.9 compares the BER performances of coherent BFSK in conventional communications and energy harvesting communications when $K = 1$. For the conventional communications in the figure, the SNR is given by $\epsilon = 2K\eta\beta^2\gamma$, where $2K\eta\beta^2$ is the average of P in energy harvesting communications so that both cases have the same average overall SNR for a fair comparison. One can see that the BER performance is degraded by energy harvesting, due to the random variation in power supply, as expected.

One sees from these figures that, in general, the effect of energy harvesting on the BER diminishes quickly when the harvesting time increases. Taking this observation and the practical limitation on the supercapacitor into account, the number of time slots for energy harvesting should be chosen as small as possible. Note also that the above results effectively evaluate the average BER averaged over ϵ, which is a product of γ and P. Similar results have also been obtained in conventional communications without energy harvesting. For example, in dual-hop wireless relaying, the BER can be averaged over the cascaded channel power, which is a product of the channel power in the first hop and the channel power in the second hop (Chen et al. 2012). In back-scatter communications,

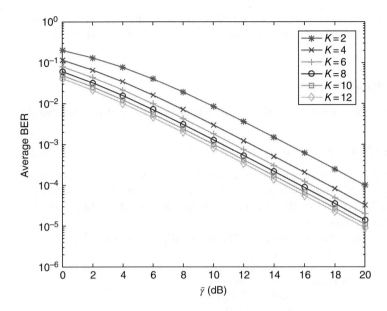

Figure 4.8 Average BER versus $\bar{\gamma}$ in Rayleigh fading channels.

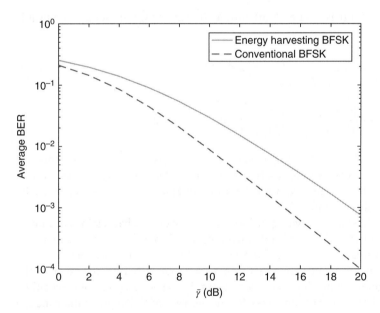

Figure 4.9 Comparison of conventional BFSK and energy harvesting BFSK in Rayleigh fading channels when $K = 1$.

the BER is averaged over the product of the forward link gain and the reverse link gain (Gao et al. 2016a). These results can be adapted to the BER analysis for energy harvesting communications.

4.2.4 Achievable Rate

For variable-power transmission, using the same SNR as in (4.17), it is easy to give the achievable rate as

$$R = \log_2(1 + \gamma P) = \log_2(1 + \epsilon). \tag{4.30}$$

Again, for conventional communications, the achievable rate only needs to be averaged over γ with P being a fixed value, while for energy harvesting communications, it needs to be averaged over both γ and P. We only need the distribution of ϵ, which has been derived in the previous subsection.

Thus, for Rayleigh fading channels, using (4.21) and (4.30), the average achievable rate is calculated as

$$\bar{R} = \int_0^\infty \log_2(1 + z) f_\epsilon(z) dz = \frac{2}{(2\eta\beta^2\bar{\gamma})^{\frac{K+1}{2}}\Gamma(K)}$$
$$\int_0^\infty z^{\frac{K-1}{2}} \log_2(1 + z) K_{K-1}\left(2\sqrt{\frac{z}{2\eta\beta^2\bar{\gamma}}}\right) dz. \tag{4.31}$$

For Rician fading channels, using (4.26) and (4.30), similarly, the average achievable rate is

$$\bar{R} = \frac{e^{-\frac{KP_s|s|^2}{2\beta^2} - \frac{|s_d|^2}{2a_d^2}}}{2\eta\beta^2\bar{\gamma}(\eta K P_s|s|^2)^{\frac{K-1}{2}}} \int_0^\infty \int_0^\infty x^{\frac{K-3}{2}} e^{-\frac{x^2}{2\eta\beta^2} - \frac{z}{x\bar{\gamma}}}$$
$$I_{K-1}\left(\sqrt{\frac{KP_s|s|^2 x}{\eta\beta^4}}\right) I_0\left(\sqrt{\frac{2z|s_d|^2}{\bar{\gamma}a_d^2 x}}\right) \log_2(1 + z) dx dz. \tag{4.32}$$

This integration could be simplified by using the series expansion of $I_0(\cdot)$ but this will not be discussed here. Next, we use (4.31) to show the effect of energy harvesting on the achievable rate. The settings are similar to those in Figures 4.7–4.9 except that the average achievable rate is examined.

Figures 4.10 and 4.11 show the average achievable rate of energy harvesting communications. Similar to BER, the achievable rate increases with K but the increase is marginal when K is large. This can also be seen from Figure 4.11, where the rate increase from $K = 10$ to $K = 12$ is much smaller than that from $K = 2$ to $K = 4$. This suggests that the harvesting time should not be chosen too long from the capacity's point of view either. The effect of energy harvesting diminishes with the increase of K. Finally, comparing the conventional and energy harvesting communications in Figure 4.12, one sees that energy harvesting reduces the average achievable rate, which agrees with what is observed in Figure 4.9.

Note that both the BER analysis and the rate analysis show that energy harvesting can cause performance degradation but its effect diminishes with the harvesting time. This conclusion is based on the assumption that the average ϵ is the same for both conventional communications and energy harvesting communications, or the average

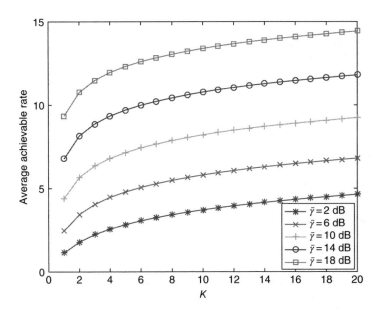

Figure 4.10 Average achievable rate (bits/s/Hz) versus K in Rayleigh fading channels.

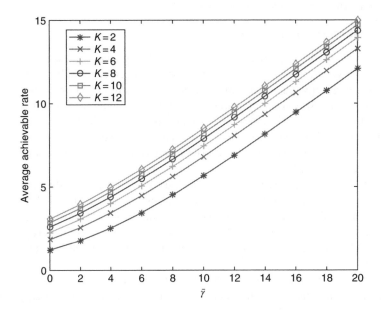

Figure 4.11 Average achievable rate (bits/s/Hz) versus $\bar{\gamma}$ in Rayleigh fading channels.

harvested power is the same as the fixed transmission power in conventional communications. In this case, the variation of power supply in energy harvesting communications is shown to degrade the system reliability or capacity. If the average harvested power is always larger than the fixed transmission power in conventional communications, these conclusions may not be valid.

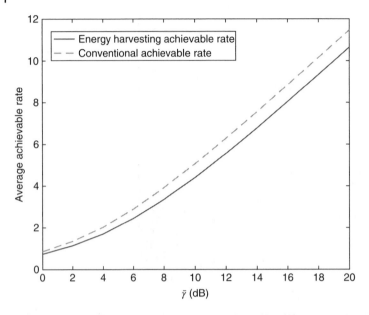

Figure 4.12 Comparison of average achievable rates (bits/s/Hz) for conventional and energy harvesting communications in Rayleigh fading channels when $K = 1$.

4.2.5 General Information Theoretic Limits

The above results have assumed energy harvesting from the RF signals. Consequently, the randomness of the power supply comes mainly from the random channel between the power source and the energy harvester. This allows us to obtain an accurate performance analysis in terms of BER and achievable rate. In a more general setting, the harvested energy can be modeled as any random process, and the randomness could come from various factors, such as the solar activity for solar power and the wind speed for wind power. In this subsection, we will give a brief discussion on how the general randomness of the power supply will affect the system capacity.

Consider a scalar AWGN channel with input X, output Y, and noise N that has zero-mean and unit-variance for simplicity. The transmission power is supplied by a battery with infinite size. The energy is harvested as a random process. In the ith channel use, an amount of E_i units will be harvested with $E\{E_i\} = \bar{E}$ as the average harvested power for all channel uses. Also, $E_i \geq 0$ and they are independent and identically distributed.

While energy is harvested, it is also used by the transmitter for data transmission. Assume that, in the ith channel use, X_i^2 units of energy are used. This gives the energy constraint as

$$\sum_{i=1}^{k} X_i^2 \leq \sum_{i=1}^{k} E_i, \ k = 1, 2, \cdots, n \tag{4.33}$$

which states that the consumed energy must be smaller than the harvested energy in any channel use. Thus, the first channel use is constrained as $X_1^2 < E_1$, the second channel use is constrained as $X_1^2 + X_2^2 < E_1 + E_2$, and so on. One sees that this energy constraint introduces memory into the channel inputs X by correlating them with each other. Thus, in energy harvesting communications, both the dynamic power supply and the random channel introduce memory and randomness into the system, unlike conventional communications where the memory and randomness mainly come from the channel.

Define the codebook as $(n, 2^{nR_n}, \epsilon_n)$, where n is the code length, R_n is the coding rate, 2^{nR_n} is the code size, and ϵ_n is the probability of error. The messages in the set $\{1, 2, \cdots, 2^{nR_n}\}$ are equally likely. Note that an error will occur when there is a decoding error at the receiver, similar to conventional systems. However, for energy harvesting systems, an extra type of error will occur when there is an energy shortage at the transmitter due to the randomness of power supply. This can be considered as an error caused by outage. Thus, ϵ_n accounts for the probability of both errors in this case. It can be shown that the capacity of this system is upper bounded as

$$C \leq \frac{1}{2}\log_2(1 + \bar{E}) \tag{4.34}$$

where $\bar{E}$ is the average of energy arrival defined before. This upper bound can be achieved via two schemes outlined below.

In the variable-power transmission scheme discussed before, or sometimes called the "harvest-use" or "best-effort-transmit" schemes in references, the transmitter can save for $h(n)$ symbols (harvests energy only or transmits zero) first, and then it can transmit the remaining $n - h(n)$ symbols by choosing them as independent samples from a Gaussian distribution with mean zero and variance P_{avg}. It can be proved that this scheme can achieve the upper bound in (4.34) while satisfying the input constraints in (4.33) with probability arbitrarily close to one, if $P_{avg} < \bar{E}$, when $n \to \infty$. When n is finite, $\epsilon > 0$ but the capacity is still close to the upper bound. The value of $h(n)$ should be from a class of functions that scale slower than n so that both $h(n) \to \infty$ and $n - h(n) \to \infty$ when $n \to \infty$, such as $h(n) = \log(n)$. In this scheme, the transmitter does not know how much energy is harvested or available. It uses a fixed codeword.

On the other hand, in the fixed-power transmission scheme discussed before, or sometimes called the "harvest-save-use" and "save-transmit" schemes in references, the transmitter transmits if the energy at the beginning of the ith channel use is larger than the fixed transmission power X_i^2, and otherwise, it transmits a zero or keeps harvesting energy. When it transmits, it transmits a random codeword as independent samples from a Gaussian distribution with mean zero and variance P_{avg}. In this scheme, the transmitter does know how much energy is available before it decides what to transmit. This is different from the other scheme where the transmitter transmits a fixed codeword.

In both schemes, an infinite battery is assumed and circuit power consumption is ignored. More details on these schemes can be found in Ozel and Ulukus (2012). In Zenaidi et al. (2017) and other similar studies, the scheduling problem considering energy arrival has been considered. Essentially, the transmitter needs to use the harvested energy wisely based on its arrival process to avoid any energy outage. In these studies, the energy causality or the energy neutrality constraint is the key.

4.3 Energy Harvesting Detection

The physical layer lies at the bottom of a wireless system and therefore it is the most important layer in the system. The main function of the physical layer is to ensure the reliable transmission and reception of data bits over the physical channel. Thus, an important task of the physical layer is to detect the transmitted data bits at the receiver for information decoding. Classical detectors for conventional communications systems are well studied. For example, for coherent detection of BPSK and BFSK signals, a correlator structure can be used. For non-coherent detection of BPSK signals, a differential structure can be used, and for non-coherent detection of BFSK signals, an energy detector can be used. All these detectors stem from the statistical theories based on different assumptions of channel knowledge and randomness. In energy harvesting communications, as discussed before, there is extra randomness from the power supply. Taking this extra randomness into account, new detectors are required for energy harvesting communications. In this section, we aim to derive new detectors by accounting for the characteristics of energy harvesting. Similar to the previous section, we will use RF energy harvesting as an example but the results can be easily extended to other energy sources, as long as the distributions of their transmission power are available.

Consider a simple point-to-point communications system. The received signal can be written as

$$r = g\sqrt{P_x}x + n \tag{4.35}$$

where g is the channel gain, P_x is the transmission power, x is the transmitted symbol, and n is the complex AWGN with mean zero and variance $2\sigma_d^2$. For fixed-power transmission, P_x is a fixed value. However, in order to accumulate a fixed amount of energy, the harvesting time becomes random, which leads to a random transmission delay and then a random transmission time. For variable-power transmission, $P_x = P$ is a random value. However, there is no transmission delay with fixed transmission time. Thus, either the transmission power or the transmission time are random in energy harvesting communications. In the following, we only consider the variable-power transmission where $P_x = P$. This gives

$$r = g\sqrt{P}x + n. \tag{4.36}$$

If both g and P are known via channel estimation at the receiver, for BPSK with $x = \pm 1$, the likelihood function can be derived as

$$f(r|x = \pm 1) = \frac{1}{2\pi\sigma_d^2}e^{-\frac{|r \mp \sqrt{P}g|^2}{2\sigma_d^2}}. \tag{4.37}$$

Using the likelihood ratio test and assuming equal probabilities for $x = +1$ and $x = -1$, one has

$$\frac{f(r|x = +1)}{f(r|x = -1)} \overset{+1}{\underset{-1}{\gtrless}} 1 \tag{4.38}$$

and the coherent detector for BPSK is derived as

$$\Re\{r \cdot g^*\} \overset{+1}{\underset{-1}{\gtrless}} 0 \tag{4.39}$$

where $\Re\{\cdot\}$ takes the real part of a complex number. This detector is the same as the coherent detector for BPSK in conventional communications (Proakis 2001). If BFSK is used, one has $x = 0$ or $x = \sqrt{2}$ with equal probabilities. The reason for using $x = \sqrt{2}$ instead of $x = 1$ is to ensure the same average power for both BFSK and BPSK in the comparison. In this case, the likelihood functions can be derived as

$$f(r|x = 0) = \frac{1}{2\pi\sigma_d^2} e^{-\frac{|r|^2}{2\sigma_d^2}} \tag{4.40}$$

$$f(r|x = \sqrt{2}) = \frac{1}{2\pi\sigma_d^2} e^{-\frac{|r - \sqrt{2P}g|^2}{2\sigma_d^2}}. \tag{4.41}$$

Using them in the likelihood ratio test, one has

$$\mathfrak{R}\{r \cdot g^*\} \underset{0}{\overset{\sqrt{2}}{\gtrless}} \sqrt{\frac{P}{2}}|g|^2. \tag{4.42}$$

The above detector is again the same as that in conventional communications without energy harvesting.

Next, we discuss several special cases when the detectors for energy harvesting are different. In the first special case, if g is known via channel estimation at the receiver but the transmission power P is unknown, the likelihood function in this case can be derived as

$$f(r|x = \pm 1) = \frac{1}{2\pi\sigma_d^2} \int_0^\infty e^{-\frac{|r \mp \sqrt{y}g|^2}{2\sigma_d^2}} f_P(y)dy.$$

$$= \frac{e^{-\frac{|r|^2}{2\sigma_d^2}}}{2\pi\sigma_d^2} \int_0^\infty e^{-\frac{|g|^2}{2\sigma_d^2}y \pm \frac{\mathfrak{R}\{rg^*\}}{\sigma_d^2}\sqrt{y}} f_P(y)dy. \tag{4.43}$$

When the channel from the power source to the energy harvester suffers from Rayleigh fading, the PDF of P is given by (4.10). Using (4.10) in (4.43), solving the integral using an equation in Gradshteyn and Ryzhik (2000, eq. (3.462.1)), and finally applying the solved integral in the likelihood ratio test (4.38), one has the new coherent detector for BPSK with unknown transmission power as

$$\frac{D_{-2K}\left(-\frac{\mathfrak{R}\{r \cdot g^*\}}{\sigma_d\sqrt{|g|^2 + \sigma_d^2/(\eta\beta^2)}}\right)}{D_{-2K}\left(\frac{\mathfrak{R}\{r \cdot g^*\}}{\sigma_d\sqrt{|g|^2 + \sigma_d^2/(\eta\beta^2)}}\right)} \underset{-1}{\overset{+1}{\gtrless}} 1 \tag{4.44}$$

where $D(\cdot)$ is the parabolic cylinder function Gradshteyn and Ryzhik (2000, eq. (9.240)) by solving the integration in (4.43) and all the other symbols are defined as before. Further, if one uses the expression of the parabolic cylinder function in Gradshteyn and Ryzhik (2000, eq. (9.240)), it can be shown that (4.44) is equivalent to (4.39). This is expected, as for phase modulation, the amplitude does not affect the performance and hence it does not matter whether P is known or unknown. One can use a similar method to derive the new detector for Rician fading channels. The derivation is omitted here. For BFSK, the likelihood function for $x = \sqrt{2}$ in a Rayleigh fading channel is derived as

$$f(r|x = \sqrt{2}) = \frac{1}{2\pi\sigma_d^2} \int_0^\infty e^{-\frac{|r-\sqrt{2}\gamma g|^2}{2\sigma_d^2}} f_P(y)dy \tag{4.45}$$

$$= \frac{e^{-\frac{|r|^2}{2\sigma_d^2}}}{2\pi\sigma_d^2} \int_0^\infty e^{-\frac{|g|^2}{\sigma_d^2}y + \frac{\sqrt{2}y\Re\{rg^*\}}{\sigma_d^2}} f_P(y)dy$$

$$= \frac{2e^{-\frac{|r|^2}{2\sigma_d^2}}\Gamma(2K)e^{\frac{\Re^2\{rg^*\}}{4|g|^2\sigma_d^2 + 2\sigma_d^4/(\eta\beta^2)}}}{2\pi\sigma_d^2(2\eta\beta^2)^K\Gamma(K)\left(\frac{2|g|^2}{\sigma_d^2} + \frac{1}{\eta\beta^2}\right)^K} D_{-2K}\left(-\frac{\Re\{rg^*\}}{\sqrt{|g|^2\sigma_d^2 + \sigma_d^4/(2\eta\beta^2)}}\right)$$

where we have used the equation in Gradshteyn and Ryzhik (2000, eq. (3.462.1)) again. Then, the new coherent detector for BFSK in energy harvesting communications with unknown transmission power can be derived as

$$\frac{2\Gamma(2K)e^{\frac{\Re^2\{rg^*\}}{4|g|^2\sigma_d^2 + 2\sigma_d^4/(\eta\beta^2)}}}{(2\eta\beta^2)^K\Gamma(K)\left(\frac{2|g|^2}{\sigma_d^2} + \frac{1}{\eta\beta^2}\right)^K} D_{-2K}\left(-\frac{\Re\{rg^*\}}{\sqrt{|g|^2\sigma_d^2 + \sigma_d^4/(2\eta\beta^2)}}\right) \underset{0}{\overset{\sqrt{2}}{\gtrless}} 1. \tag{4.46}$$

In another special case, if g is unknown but P is known, using the likelihood ratio test and assuming that the channel gain is complex Gaussian with mean zero and variance $2\alpha_d^2$, one has the non-coherent detector for BFSK as

$$|r|^2 \underset{0}{\overset{\sqrt{2}}{\gtrless}} \frac{(\sigma_d^2 + 2P\alpha_d^2)\sigma_d^2}{P\alpha_d^2} \ln\left[\frac{\sigma_d^2 + 2P\alpha_d^2}{\sigma_d^2}\right] \tag{4.47}$$

which is the traditional energy detector.

When both g and P are unknown, for energy harvesting communications, the new non-coherent detector for BFSK can be derived as

$$\frac{\sigma_d^2 e^{\frac{|r|^2}{2\sigma_d^2}}}{(2\eta\beta^2)^K\Gamma(K)} \int_0^\infty \frac{x^{K-1}}{\sigma_d^2 + 2\alpha_d^2 x} e^{-\frac{x}{2\eta\beta^2} - \frac{|r|^2}{2(\sigma_d^2 + 2\alpha_d^2 x)}} dx \underset{0}{\overset{\sqrt{2}}{\gtrless}} 1. \tag{4.48}$$

The integration does not have a closed-form expression but simplification is possible when the average SNR $\frac{\alpha_d^2}{\sigma_d^2}$ is very large. For BPSK, the non-coherent detector requires differential encoding at the transmitter. The derivation is similar to the above, albeit more complicated. It is not presented here.

The above results are only applicable to binary signaling. However, one can easily extend them to M-ary signaling by calculating the likelihood function for each transmitted symbol. Also, we have only considered RF energy harvesting, where the distribution of the harvested power or the power supply is derived in Section 4.2.1 as either a central chi-square distribution or a non-central chi-square distribution. For other harvesting techniques, as long as the distribution of the harvested power is available, similar detectors can be obtained. Next, we show the performances of the newly derived detectors for RF energy harvesting communications using BFSK and $\eta = 0.5$. Figure 4.13 compares the coherent detector for conventional communications in (4.42) with that for energy harvesting communications in (4.46), where K is the number of time slots used for harvesting. Figure 4.14 compares the non-coherent detector for conventional communications in (4.47) with that for energy harvesting communications in (4.48).

One sees from these two figures that the detector in conventional communications always outperforms that in energy harvesting communications, due to the extra variation in power supply for energy harvesting communications. This agrees with our observations before. However, the performance difference is very small for the non-coherent detectors. Also, the difference decreases when K increases. Statistically, when the shape parameter K increases, the Gamma distribution of the harvested power in Rayleigh fading channels becomes more impulsive so that less variation will occur.

The above discussion has presented some very simple results on how new detectors can be designed for energy harvesting communications. Recall that for conventional communications the randomness mainly comes from the communications channel, while for energy harvesting communications the randomness comes from both the communications channel and the power supply. Thus, in most cases, the extra randomness from the power supply needs to be considered for efficient detection of signals in energy harvesting communications.

Bearing this main difference in mind, one can easily obtain results for energy harvesting communications in other cases. For example, if imperfect channel knowledge

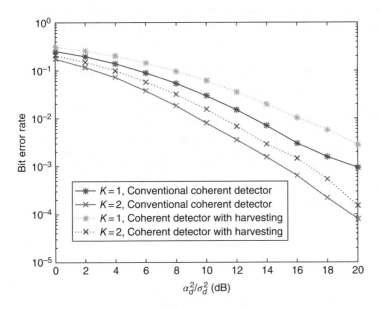

Figure 4.13 Comparison of conventional and energy harvesting communications using coherent detectors.

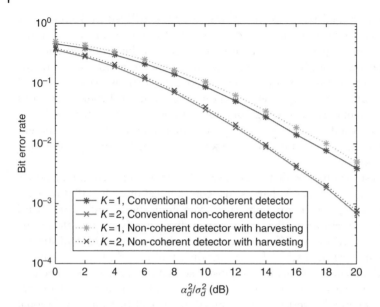

Figure 4.14 Comparison of conventional and energy harvesting communications using non-coherent detectors.

is considered due to channel estimation errors, one may introduce extra errors in the estimation of the transmission power too for a similar discussion. Also, there will be four cases, known channel and known transmission power, known channel and unknown transmission power, unknown channel and known transmission power, and unknown channel and unknown transmission power, for energy harvesting communications. Finally, detectors for the fixed-power transmission schemes can be derived too. These schemes have a fixed transmission power but a random transmission time. Both fixed-power transmission and variable-power transmission depend on the random energy arrival process.

Note that, in the general case, the random channel should be independent of the random power source. For example, if solar-powered devices are used for radio communications, the randomness of the power supply comes from the solar activities, while the randomness of the channel comes from fading and noise. However, in some cases, the random channel and the random power source may be correlated. For example, if both power transfer and information transmission share the same radio channel, the fading h_k in y_k in (4.1) may be correlated with the fading g in r in (4.35) so that P might be correlated with g in the detection. This is the case in wireless powered communications. Assume that the fading channel remains the same during power transfer and information transmission such that $h_k = g$ for $k = 1, 2, \cdots, K$. In this case,

$$y_k = g\sqrt{P_s}s_k + n_k. \tag{4.49}$$

Thus, conditioned on g, the PDF of the harvested power P will become a non-central chi-square distribution with PDF

$$f_P(y) = \frac{e^{-\frac{y}{2\eta\sigma^2} - \frac{KP_s|g|^2|s|^2}{2\sigma^2}}}{2\eta\sigma^2(\eta KP_s|g|^2|s|^2)^{\frac{K-1}{2}}} y^{\frac{K-1}{2}} I_{K-1}\left(\sqrt{\frac{KP_s|g|^2|s|^2 y}{\eta\sigma^4}}\right) \tag{4.50}$$

for any fading channels. This PDF can then be used to calculate the likelihood function when P is unknown but g is known in (4.44) and (4.46), etc. These likelihood functions will lead to different detectors. Details of these derivations are omitted here. The method is the same as before. Some similar studies have also been conducted in Liu et al. (2015a) for wireless relaying systems with two hops.

In summary, two important issues in energy harvesting detection are the extra randomness from the power supply and the relationship between this extra randomness and the usual channel randomness. They lead to new and different detector designs. Indeed, these two issues also affect the performances of other physical layer techniques, such as channel estimation, which will be discussed in the next section.

4.4 Energy Harvesting Estimation

In this section, we discuss another important physical layer technique, channel estimation. Most modern communications systems acquire channel knowledge via channel estimation in order to provide satisfactory QoS for users. There are many different types of channel estimators, depending on the channel conditions and system requirements. The most reliable channel estimators are pilot-based estimators, where the transmitter sends a number of known pilot symbols embedded as part of the header in the data frame and the receiver uses the received signals of the known pilot symbols for channel estimation to detect the received data symbols in the same or different frames (Chen and Beaulieu 2007). Several important issues arise in energy harvesting estimation.

First, channel estimator designs often depend on the statistics of the received pilot symbols. If the transmitter has used the dynamic harvested power in energy harvesting communications to transmit the pilot symbols, in the received pilot symbols, in addition to the channel randomness, one also has the extra randomness from the power supply. Consequently, the statistics of the received pilot symbols will change so that channel estimators for conventional communications cannot be used. Take r in (4.46) as an example. Assume that the received pilot symbols are given by

$$r_i = g\sqrt{P}x_i + n_i \tag{4.51}$$

where $x_i = 1$ is the pilot value and $i = 1, 2, \cdots, I$ index the pilot symbols. In the conventional communications, the only randomness comes from n_i. The moment-based (MB) estimator for g can be easily derived as

$$\hat{g} = \frac{1}{I\sqrt{P}} \sum_{i=1}^{I} r_i. \tag{4.52}$$

In energy harvesting communications, $\sqrt{P}$ is random too so that (4.52) cannot be used directly. Instead, $\sqrt{P}$ should be replaced by its average $E\{\sqrt{P}\}$ in (4.52) to give

$$\hat{g} = \frac{1}{IE\{\sqrt{P}\}} \sum_{i=1}^{I} r_i \tag{4.53}$$

where $E\{\cdot\}$ denotes the expectation operation. Similarly, for maximum likelihood (ML) estimators, minimum mean squared error (MSE) estimators and other estimation methods, the extra randomness from P needs to be considered for energy harvesting communications. This process is very similar to that in the previous section for signal detection. This will not be discussed further here.

Secondly, if the random power source and the random channel are independent but channel estimation and energy harvesting are performed by the same remote device, the addition of the energy harvesting capability may reduce the system resources for other tasks. For example, in the conventional communications, for a data frame with fixed length, one has to allocate the number of symbols between channel estimation and data transmission. In energy harvesting communications, if the data frame still has a fixed length, one has to allocate the number of symbols between channel estimation, energy harvesting and data transmission now, similar to the work in Wang et al. (2014). Also, if the random power source and the random channel are correlated, such as in wireless powered communications, energy harvesting and channel estimation might be correlated. In this case, the signals for energy harvesting contain the same channel gain to be estimated and thus this information can be shared to improve estimation accuracy.

Next, we will focus on the second issue. This will be first studied in the context of a wireless relaying system, where each communication is performed in two hops: from source to relay; and from relay to destination. In energy harvesting channel estimation, the transmission of the pilot symbols in the second hop is powered by energy harvested from the first hop. Then, a one-hop system without relaying will be discussed.

4.4.1 With Relaying

Consider a wireless relaying network with one source, one relay and one destination. The signal is transmitted from the source to the destination via the relay. There is no direct link between the source and the destination. Assume that a total of K pilots are used for both energy harvesting and channel estimation. Each pilot occupies a time duration of T_p. We consider six schemes of energy harvesting channel estimation.

4.4.1.1 Scheme 1

In Scheme 1, the relay node obtains energy from the source that requires the channel state information. The harvested energy will be used by the relay node to forward pilots from the source node and transmit extra pilots to the destination node so that channel estimation can be performed. In this case, energy harvesting estimation works in the following way.

First, one has the source node that will transmit I pilots to the relay node for energy harvesting. The received signals of these pilots are given by

$$y_{r-eh}^{(i)} = \sqrt{P_s}hs + n_{r-eh}^{(i)} \tag{4.54}$$

where $i = 1, 2, \cdots, I$, P_s is the transmission power of the source node, h is the channel coefficient of the source-to-relay link and is a complex Gaussian random variable with mean zero and variance $2\alpha^2$, s is the pilot value and it is assumed that $s = 1$ in the following, and n_{r-eh} is the AWGN with mean zero and variance $2\sigma^2$. All the noise in this section is assumed circularly symmetric. Using (4.54), the harvested energy is

$$E_h = \eta P_s |h|^2 I T_p \tag{4.55}$$

where η is the conversion efficiency of the energy harvester that has been discussed in Chapter 3 and IT_p is the total time for energy harvesting. Note that $P_s|h|^2$ is the amount of radiated power from the source node picked up by the harvester as its input. Due to path loss and fading, this amount is often small.

In the second step, the source node will transmit another J_1 pilots to the relay node. These pilots will be forwarded to the destination for channel estimation using the harvested energy. The received signals of these pilots at the destination is

$$y_{d-s}^{(j_1)} = \sqrt{P_r} g a y_{r-ce}^{(j_1)} + n_{d-s}^{(j_1)}, \tag{4.56}$$

where $y_{r-ce}^{(j_1)} = \sqrt{P_s} h + n_{r-ce}^{(j_1)}$ is the pilot that is forwarded by the relay, $j_1 = 1, 2, \cdots, J_1$, $n_{r-ce}^{(j_1)}$ is the AWGN at the relay, P_r is the transmission power of the relay node that will be calculated later, g is the channel coefficient of the relay-to-destination link and it is a complex Gaussian random variable with mean zero and variance $2\alpha^2$, a is the amplification factor, and $n_{d-s}^{(j_1)}$ is the AWGN at the destination. Both $n_{r-ce}^{(j_1)}$ and $n_{d-s}^{(j_1)}$ are complex Gaussian with mean zero and variance $2\sigma^2$. This chapter assumes identical noise variances for the relay and the destination. This is the case when the source, the relay and the destination are peer nodes in the network such that they have similar receivers. Nevertheless, it is also straightforward to extend the result to the case when the relay and the destination have different noise variances. In Scheme 1, since the relay node does not perform channel estimation, fixed-gain relaying can be used such that one can set a as a constant for simplicity.

Finally, the relay node also uses the harvested energy to transmit J_2 pilots from its own to the destination, and their received signals are given by

$$y_{d-r}^{(j_2)} = \sqrt{P_r} g + n_{d-r}^{(j_2)} \tag{4.57}$$

where $j_2 = 1, 2, \cdots, J_2$, $n_{d-r}^{(j_2)}$ is the AWGN at the destination and is again complex Gaussian with mean zero and variance $2\sigma^2$. This signal only has the pilot value and it does not have any forwarded signal. The relay uses the harvested energy in (4.55) to forward J_1 pilots from the source node and transmit J_2 pilots of its own. Thus, the transmission power of the relay can be written as

$$P_r = \frac{E_h}{J T_p} = \eta P_s |h|^2 \frac{I}{J} \tag{4.58}$$

where $J = J_1 + J_2$. Using $y_{d-s}^{(j_1)}$ in (4.56) and $y_{d-r}^{(j_2)}$ in (4.57), we can estimate the channel coefficients g and h.

From (4.57), one has

$$y_{d-r}^{(j_2)} = \sqrt{\eta \frac{I}{J} P_s} |h| g + n_{d-r}^{(j_2)} \tag{4.59}$$

and from (4.56), one has

$$y_{d-s}^{(j_1)} = \sqrt{\eta \frac{I}{J} P_s} |h| g h a + \sqrt{\eta \frac{I}{J} P_s} |h| g a n_{r-ce}^{(j_1)} + n_{d-s}^{(j_1)}. \tag{4.60}$$

It is well-known that the MB estimators are often simpler than other estimators. In some cases, they also provide efficient estimation. Thus, they are considered first. The first-order moments of (4.59) and (4.60) are

$$E\{y_{d-r}^{(j_2)}\} = \sqrt{\eta \frac{I}{J} P_s} |h| g \tag{4.61}$$

$$E\{y_{d-s}^{(j_1)}\} = \sqrt{\eta \frac{I}{j} P_s} |h| gha. \tag{4.62}$$

One can approximate $E\{y_{d-r}^{(j_2)}\}$ using $\frac{1}{J_2} \sum_{j_2=1}^{J_2} y_{d-r}^{(j_2)}$. Also, the value of $E\{y_{d-s}^{(j_1)}\}$ can be approximated as $\frac{1}{J_1} \sum_{j_1=1}^{J_1} y_{d-s}^{(j_1)}$. Solving the equations in (4.61) and (4.62) for g and h, one has the MB estimators for g and h in Scheme 1 as

$$\hat{g}_1 = \frac{\frac{1}{J_2} \sum_{j_2=1}^{J_2} y_{d-r}^{(j_2)} |\frac{1}{J_2} \sum_{j_2=1}^{J_2} y_{d-r}^{(j_2)}|}{\frac{1}{a} \sqrt{\eta \frac{I}{j}} |\frac{1}{J_1} \sum_{j_1=1}^{J_1} y_{d-s}^{(j_1)}|} \tag{4.63}$$

$$\hat{h}_1 = \frac{1}{\sqrt{P_s} a} \frac{\frac{1}{J_1} \sum_{j_1=1}^{J_1} y_{d-s}^{(j_1)}}{\frac{1}{J_2} \sum_{j_2=1}^{J_2} y_{d-r}^{(j_2)}}, \tag{4.64}$$

respectively. Note that other orders of moments can also be used but the lower the order of moment is, the better the MB estimator will be. Thus, we use the first order.

The ML estimators can be derived as in the following. Denote $G_y = \sqrt{\eta \frac{I}{j} P_s} |h| g$. Using (4.57) and the ML method, the log-likelihood function can be derived as

$$llf_1 = -J_2 \ln(2\pi\sigma_d^2) - \frac{1}{2\sigma_d^2} \sum_{j_2=1}^{J_2} |y_{d-r}^{(j_2)} - G_y|^2. \tag{4.65}$$

Thus, by differentiating (4.65) with respect to G_y, setting the derivative to zero and solving the equation for G_y, the ML estimate of G_y can be derived as

$$\hat{G}_y = \frac{1}{J_2} \sum_{j_2=1}^{J_2} y_{d-r}^{(j_2)} = \sqrt{\eta \frac{I}{j} P_s} |\hat{h}| \hat{g}. \tag{4.66}$$

Also, denote $H_y = \sqrt{P_s} h$. Using (4.60) and the ML method, another log-likelihood function can be derived as

$$llf_2 = -J_1 \ln(2\pi(1 + |G_y|^2 a^2)\sigma_d^2)$$

$$- \frac{1}{2(1 + |G_y|^2 a^2)\sigma_d^2} \sum_{j_1=1}^{J_1} |y_{d-s}^{(j_1)} - G_y H_y a|^2. \tag{4.67}$$

By differentiating (4.67) with respect to H_y, setting the derivative to zero and solving the equation for H_y, the ML estimate of H_y can be derived as

$$\hat{H}_y = \frac{1}{J_1 \hat{G}_y a} \sum_{j_1=1}^{J_1} y_{d-s}^{(j_1)} = \sqrt{P_s} \hat{h}. \tag{4.68}$$

The invariance principle of ML estimation states that a function of ML estimate is the ML estimate of that function. Using this principle, the ML estimates of g and h can be derived by solving (4.66) and (4.68) for $\hat{g}$ and $\hat{h}$, which are the same as the MB estimators using the first-order moments. Thus, results for the MB estimators are also applicable to the ML estimators. Since both $y_{d-s}^{(j_1)}$ and $y_{d-r}^{(j_2)}$ are samples received at the destination, in this scheme, the relay does not perform channel estimation. Only the destination performs channel estimation. This reduces the complexity at the relay, which is desirable to encourage idle nodes to take part in relaying. The first- and second-order moments of the estimates can also be derived.

For Scheme 1, denote $y_r = \frac{1}{J_2}\sum_{j_2=1}^{J_2} y_{d-r}^{(j_2)} = r_{y_r}e^{j\theta_{y_r}}$ and $y_s = \frac{1}{J_1}\sum_{j_1=1}^{J_1} y_{d-s}^{(j_1)} = r_{y_s}e^{j\theta_{y_s}}$. One sees that y_r and y_s are complex Gaussian random variables with means $S_{y_r} = \sqrt{\eta_j^l P_s}|h|g$ and $S_{y_s} = \sqrt{\eta_j^l P_s}|h|gha$ and variances $2\beta_{y_r}^2 = \frac{2\sigma^2}{J_2}$ and $2\beta_{y_s}^2 = \frac{2\sigma^2}{J_1}(1 + \eta_j^l P_s|h|^2|g|^2 a^2)$, respectively. Hence, r_{y_r} and r_{y_s} are Rician random variables.

From (4.63), one has

$$E\{\hat{g}_1\} = \frac{a}{\sqrt{\eta_j^l}}E\{r_{y_r}^2 e^{j\theta_{y_r}}\}E\left\{\frac{1}{r_{y_s}}\right\} \tag{4.69}$$

where

$$E\{r_{y_r}^2 e^{j\theta_{y_r}}\} = \frac{3\beta_{y_r}^2}{e^{\frac{|S_{y_r}|^2}{2\beta_{y_r}^2}}\pi}\int_0^{2\pi} e^{j\theta_{y_r} + \frac{|S_{y_r}|^2\cos^2(\theta_{y_r}+\epsilon)}{4\beta_{y_r}^2}} D_{-4}\left(-\frac{|S_{y_r}|\cos(\theta_{y_r}+\epsilon)}{\beta_{y_r}}\right)d\theta_{y_r} \tag{4.70}$$

using Gradshteyn and Ryzhik (2000, eq. (3.462.1)) and Stuber (2001, eq. (A.29))

$$E\left\{\frac{1}{r_{y_s}}\right\} = \frac{\sqrt{\pi}e^{-\frac{|S_{y_s}|^2}{4\beta_{y_s}^2}}}{\sqrt{2\beta_{y_s}^2}}I_0\left(\frac{|S_{y_s}|^2}{4\beta_{y_s}^2}\right) \tag{4.71}$$

using Gradshteyn and Ryzhik (2000, eq. (6.618.4)) and Stuber (2001, eq. (2.45)), ϵ is the negative of the phase angle of g, $D_{-4}(\cdot)$ is the parabolic cylinder function (Gradshteyn and Ryzhik 2000, eq. (9.240)) and $I_0(\cdot)$ is the zeroth-order modified Bessel function of the first kind (Gradshteyn and Ryzhik 2000 eq. (8.406.1)).

Also, from (4.64), one has

$$E\{\hat{h}_1\} = \frac{S_{y_s}}{\sqrt{P_s}a}E\left\{\frac{1}{r_{y_r}}e^{-j\theta_{y_r}}\right\} \tag{4.72}$$

where

$$E\left\{\frac{e^{-j\theta_{y_r}}}{r_{y_r}}\right\} = \frac{1}{\sqrt{2\pi\beta_{y_r}^2}}\int_0^{2\pi} e^{-j\theta_{y_r} - \frac{|S_{y_r}|^2\sin^2(\theta_{y_r}+\epsilon)}{2\beta_{y_r}^2}} Q(-|S_{y_r}|\cos(\theta_{y_r}+\epsilon)/\beta_{y_r})d\theta_{y_r} \tag{4.73}$$

and $Q(\cdot)$ is the Gaussian Q function. The second-order moments can be derived in the following.

From (4.63), one has

$$E\{|\hat{g}_1|^2\} = \frac{a^2}{\eta_j^l}E\{r_{y_r}^4\}E\left\{\frac{1}{r_{y_s}^2}\right\} \tag{4.74}$$

where

$$E\{r_{y_r}^4\} = 2(2\beta_{y_r}^2)^2 + 4(2\beta_{y_r}^2)|S_{y_r}|^2 + |S_{y_r}|^4 \tag{4.75}$$

using moments of a Rician random variable and

$$E\left\{\frac{1}{r_{y_s}^2}\right\} = \int_0^\infty \frac{1}{\beta_{y_s}^2 x} e^{-\frac{x^2+|S_{y_s}|^2}{2\beta_{y_s}^2}} I_0\left(\frac{x|S_{y_s}|}{\beta_{y_s}^2}\right) dx \qquad (4.76)$$

using Stuber (2001, eq. (2.45)). Also, from (4.64), one has

$$E\{|\hat{h}_1|^2\} = \frac{1}{P_s a^2} E\{r_{y_s}^2\} E\left\{\frac{1}{r_{y_r}^2}\right\} \qquad (4.77)$$

where

$$E\{r_{y_s}^2\} = 2\beta_{y_s}^2 + |S_{y_s}|^2 \qquad (4.78)$$

and

$$E\left\{\frac{1}{r_{y_r}^2}\right\} = \int_0^\infty \frac{1}{\beta_{y_r}^2 x} e^{-\frac{x^2+|S_{y_r}|^2}{2\beta_{y_r}^2}} I_0\left(\frac{x|S_{y_r}|}{\beta_{y_r}^2}\right) dx. \qquad (4.79)$$

4.4.1.2 Scheme 2

Scheme 2 works in a similar way to Scheme 1, but Scheme 2 harvests the energy by splitting the signal in the power domain. Details of these energy harvesting strategies can also be found in Chapter 6. First, the source transmits K_1 pilots to the relay. The received signal at the relay is divided into two parts. One part is used for channel estimation as $z_{r-ce}^{(k_1)} = \sqrt{(1-\rho)P_s}h + n_{r-ce}^{(k_1)}$, which is forwarded to the destination to give

$$z_{d-s}^{(k_1)} = \sqrt{P_r}gaz_{r-ce}^{(k_1)} + n_{d-s}^{(k_1)} \qquad (4.80)$$

where $k_1 = 1, 2, \cdots, K_1$ index the pilots from the source, ρ is the important power splitting factor, $n_{r-ce}^{(k_1)}$ and $n_{d-s}^{(k_1)}$ are the AWGN with means zero and variances $2\sigma_r^2$ and $2\sigma_d^2$, respectively. Another part of the received signal at the relay is harvested as $E_h = \eta\rho P_s|h|^2 K_1 T_p$.

Secondly, the relay also transmits K_2 own pilots to the destination such that the received signal at the destination is

$$z_{d-r}^{(k_2)} = \sqrt{P_r}g + n_{d-r}^{(k_2)} \qquad (4.81)$$

where $k_2 = 1, 2, \cdots, K_2$ and $n_{d-r}^{(k_2)}$ is the AWGN with mean zero and variance $2\sigma_d^2$.

Since the relay forwards K_1 pilots from the source and also transmits K_2 pilots from itself, a total of $K = K_1 + K_2$ pilots will be received at the destination as

$$P_r = \frac{E_h}{K T_p} = \eta\rho P_s|h|^2 \frac{K_1}{K}. \qquad (4.82)$$

Thus, one can obtain

$$z_{d-r}^{(k_2)} = \sqrt{\eta\rho P_s \frac{K_1}{K}}|h|g + n_{d-r}^{(k_2)} \qquad (4.83)$$

and

$$z_{d-s}^{(k_1)} = \sqrt{\eta\rho(1-\rho)\frac{K_1}{K}P_s}|h|gha + \sqrt{\eta\rho P_s \frac{K_1}{K}}|h|gan_{r-ce}^{(k_1)} + n_{d-s}^{(k_1)}. \qquad (4.84)$$

The MB estimators are derived first. One has the first-order moments of $z_{d-r}^{(k_2)}$ and $z_{d-s}^{(k_1)}$ as

$$E\{z_{d-r}^{(k_2)}\} = \sqrt{\eta \rho P_s \frac{K_1}{K}} |h| g \qquad (4.85)$$

$$E\{z_{d-s}^{(k_1)}\} = \sqrt{\eta \rho (1-\rho) \frac{K_1}{K} P_s} |h| gha. \qquad (4.86)$$

Thus, it is quite straightforward to derive the MB estimators for g and h from (4.85) and (4.86) as

$$\hat{g}_2 = \frac{a\sqrt{1-\rho}}{\sqrt{\eta \rho \frac{K_1}{K}}} \frac{\frac{1}{K_2}\sum_{k_2=1}^{K_2} z_{d-r}^{(k_2)} |\frac{1}{K_2}\sum_{k_2=1}^{K_2} z_{d-r}^{(k_2)}|}{|\frac{1}{K_1}\sum_{k_1=1}^{K_1} z_{d-s}^{(k_1)}|} \qquad (4.87)$$

and

$$\hat{h}_2 = \frac{1}{\sqrt{(1-\rho)P_s} a} \frac{\frac{1}{K_1}\sum_{k_1=1}^{K_1} z_{d-s}^{(k_1)}}{\frac{1}{K_2}\sum_{k_2=1}^{K_2} z_{d-r}^{(k_2)}}, \qquad (4.88)$$

respectively.

The ML estimators are derived in the following. Denote $G_z = \sqrt{\eta \rho P_s \frac{K_1}{K}} |h| g$ and $H_z = \sqrt{(1-\rho)P_s} h$. Similar to before, using the ML method, the ML estimate of G_z can be derived as

$$\hat{G}_z = \frac{1}{K_2}\sum_{k_2=1}^{K_2} z_{d-r}^{(k_2)} = \sqrt{\eta \rho P_s \frac{K_1}{K}} |\hat{h}| \hat{g} \qquad (4.89)$$

and using the ML method, the ML estimate of H_z can be derived as

$$\hat{H}_z = \frac{1}{K_1 \hat{G}_z a}\sum_{k_1=1}^{K_1} z_{d-s}^{(k_1)} = \sqrt{(1-\rho)P_s} \hat{h}. \qquad (4.90)$$

Using the invariance principle, the ML estimators for g and h can be obtained by solving (4.89) and (4.90), which are the same as the MB estimators using the first-order moments. Again, only the destination needs to perform channel estimation to reduce complexity at the relay. The ML estimation and the MB estimator are the same for both Scheme 1 and Scheme 2. From here on, we will not derive them separately and will only focus on the simpler MB estimators.

Next, we derive the first- and second-order moments of $\hat{g}_2$ and $\hat{h}_2$. Denote $z_r = \frac{1}{K_2}\sum_{k_2=1}^{K_2} z_{d-r}^{(k_2)} = r_{z_r} e^{j\theta_{z_r}}$ and $z_s = \frac{1}{K_1}\sum_{k_1=1}^{K_1} z_{d-s}^{(k_1)} = r_{z_s} e^{j\theta_{z_s}}$, which are complex Gaussian random variables with means $S_{z_r} = \sqrt{\eta \rho P_s \frac{K_1}{K}} |h| g$ and $S_{z_s} = \sqrt{\eta \rho (1-\rho) \frac{K_1}{K} P_s} |h| gha$ and variances $2\beta_{z_r}^2 = \frac{2\sigma^2}{K_2}$ and $2\beta_{z_s}^2 = \frac{2\sigma^2}{K_1}(1 + \eta \rho P_s |h|^2 |g|^2 a^2 \frac{K_1}{K})$, respectively.

From (4.87), the first-order moment of $\hat{g}_2$ can be derived as

$$E\{\hat{g}_2\} = \frac{a}{\sqrt{\eta \frac{K_1}{K}}} \sqrt{\frac{1-\rho}{\rho}} E\{r_{z_r}^2 e^{j\theta_{z_r}}\} E\{\frac{1}{r_{z_s}}\} \qquad (4.91)$$

where $E\{r_{z_r}^2 e^{j\theta_{z_r}}\}$ and $E\{\frac{1}{r_{z_s}}\}$ are similar to those in Scheme 1, except that S_{y_r}, β_{y_r}, S_{y_s}, and β_{y_s} are replaced by S_{z_r}, β_{z_r}, S_{z_s}, and β_{z_s}, respectively. The first-order moment of $\hat{h}_2$ is derived from (4.88) as

$$E\{\hat{h}_2\} = \sqrt{\eta \rho P_s \frac{K_1}{K}}|h|ghE\left\{\frac{1}{r_{z_r}}e^{-j\theta_{z_r}}\right\} \tag{4.92}$$

where $E\{\frac{1}{r_{z_r}}e^{-j\theta_{z_r}}\}$ is similar to before except that S_{y_r} and β_{y_r} are replaced by S_{z_r} and β_{z_r}, respectively.

For the second-order moments, one has

$$E\{|\hat{g}_2|^2\} = \frac{a^2}{\eta \frac{K_1}{K}} \frac{1-\rho}{\rho} E\{r_{z_r}^4\} E\left\{\frac{1}{r_{z_s}^2}\right\} \tag{4.93}$$

and

$$E\{|\hat{h}_2|^2\} = \frac{1}{(1-\rho)P_s a^2} E\{r_{z_s}^2\} E\left\{\frac{1}{r_{z_r}^2}\right\} \tag{4.94}$$

where $E\{r_{z_r}^4\}$, $E\{\frac{1}{r_{z_s}^2}\}$, $E\{r_{z_s}^2\}$, and $E\{\frac{1}{r_{z_r}^2}\}$ are derived by replacing S_{y_r}, β_{y_r}, S_{y_s}, and β_{y_s} with S_{z_r}, β_{z_r}, S_{z_s}, and β_{z_s} in the results for Scheme 1.

4.4.1.3 Scheme 3

In Scheme 1 and Scheme 2, channel estimation is only performed at the destination node by using pilots forwarded from the source node that contain the cascaded channel coefficient $|h|gh$ as well as pilots of the relay node's own that contain $|h|g$. This reduces the complexity at the relay node. The accuracy of channel estimation can be further improved if the relay node performs channel estimation. This is the case, for example, in variable-gain relaying.

In Scheme 3, the source node sends J_1 pilots to the relay node and the relay node uses these pilots to estimate the source-to-relay link h. Then, the source node sends I pilots to the relay node for energy harvesting. Using the harvested energy, the relay node sends J_2 pilots of its own to the destination node to estimate the relay-to-destination link g.

First, the source node sends J_1 pilots to the relay node such that the received signal at the relay node is

$$u_{r-ce}^{(j_1)} = \sqrt{P_s}h + n_{r-ce}^{(j_1)} \tag{4.95}$$

where $j_1 = 1, 2, \cdots, J_1$ and $n_{r-ce}^{(j_1)}$ is the AWGN with mean zero and variance $2\sigma^2$. Using (4.95), the relay node can estimate h as

$$E\{u_{r-ce}^{(j_1)}\} = \sqrt{P_s}h. \tag{4.96}$$

Secondly, the source node sends I pilots to the relay node for energy harvesting. The harvested energy is given by

$$E_h = \eta P_s |h|^2 I T_p. \tag{4.97}$$

Finally, the relay node uses the harvested energy to transmit J_2 pilots of its own to the destination node. The received signal at the destination is

$$u_{d-r}^{(j_2)} = \sqrt{P_r}g + n_{d-r}^{(j_2)} \tag{4.98}$$

where $j_2 = 1, 2, \cdots, J_2$. Since the harvested energy is used to transmit J_2 pilots, the transmission power of the relay is

$$P_r = \frac{E_h}{J_2 T_p} = \eta P_s |h|^2 \frac{I}{J_2}. \tag{4.99}$$

Using (4.99) in (4.98), one has

$$u_{d-r}^{(j_2)} = \sqrt{\eta P_s \frac{I}{J_2}} |h|g + n_{d-r}^{(j_2)}. \tag{4.100}$$

Thus,

$$E\{u_{d-r}^{(j_2)}\} = \sqrt{\eta P_s \frac{I}{J_2}} |h|g. \tag{4.101}$$

Using (4.96) and (4.101), the pilot-based MB estimators for g and h can be readily derived as

$$\hat{g}_3 = \frac{\frac{1}{J_2} \sum_{j_2=1}^{J_2} u_{d-r}^{(j_2)}}{\sqrt{\eta \frac{I}{J_2}} |\frac{1}{J_1} \sum_{j_1=1}^{J_1} u_{r-ce}^{(j_1)}|} \tag{4.102}$$

and

$$\hat{h}_3 = \frac{1}{\sqrt{P_s}} \frac{1}{J_1} \sum_{j_1=1}^{J_1} u_{r-ce}^{(j_1)}. \tag{4.103}$$

In Scheme 3, the relay node performs channel estimation of h and the estimate of h will be sent to the destination node via control channels for the estimation of g at the destination node. Thus, this scheme is more complicated than Scheme 1 and Scheme 2.

The first- and second-order moments can also be derived. We denote $u_r = \frac{1}{J_2} \sum_{j_2=1}^{J_2} u_{d-r}^{(j_2)} = r_{u_r} e^{j\theta_{u_r}}$ and $u_s = \frac{1}{J_1} \sum_{j_1=1}^{J_1} u_{r-ce}^{(j_1)} = r_{u_s} e^{j\theta_{u_s}}$. Then, u_r and u_s are complex Gaussian random variables with means $S_{u_r} = \sqrt{\eta P_s \frac{I}{J_2}} |h|g$ and $S_{u_s} = \sqrt{P_s}h$ and variances $2\beta_{u_r}^2 = \frac{2\sigma^2}{J_2}$ and $2\beta_{u_s}^2 = \frac{2\sigma^2}{J_1}$, respectively.

In this case, the first-order moment of $\hat{g}_3$ is given by

$$E\{\hat{g}_3\} = \sqrt{P_s} |h|gE\left\{\frac{1}{r_{u_s}}\right\} \tag{4.104}$$

where $E\{\frac{1}{r_{u_s}}\}$ can be obtained by replacing S_{y_s} and β_{y_s} with S_{u_s} and β_{u_s}. It is also quite obvious that the first-order moment of $\hat{h}_3$ is

$$E\{\hat{h}_3\} = h \tag{4.105}$$

so that $\hat{h}_3$ is an unbiased estimator.

For the second-order moments, one has

$$E\{|\hat{g}_3|^2\} = \frac{J_2}{\eta I}E\{r_{u_r}^2\}E\left\{\frac{1}{r_{u_s}^2}\right\}$$

(4.106)

with $E\{r_{u_r}^2\} = 2\beta_{u_r}^2 + |S_{u_r}|^2$ and $E\{\frac{1}{r_{u_s}^2}\}$ are obtained by replacing S_{y_s} and β_{y_s} with S_{u_s} and β_{u_s}, respectively. Similarly, one has

$$E\{|\hat{h}_3|^2\} = \frac{2\sigma^2}{P_s J_1} + |h|^2.$$

(4.107)

4.4.1.4 Scheme 4

Scheme 4 is similar to Scheme 3, where the relay node estimates the source-to-relay link and the destination node estimates the relay-to-destination link, except that the relay node uses power splitting to harvest energy from the pilots sent by the source node that are also used for channel estimation at the relay node.

In this scheme, the source node sends K_1 pilots to the relay node, part of which is received for channel estimation as

$$v_{r-ce}^{(k_1)} = \sqrt{(1-\rho)P_s}h + n_{r-ce}^{(k_1)}$$

(4.108)

for $k_1 = 1, 2, \cdots, K_1$ and part of which is harvested with the harvested energy given by

$$E_h = \eta\rho P_s|h|^2K_1 T_p.$$

(4.109)

The relay node uses (4.108) to estimate the source-to-relay link h as

$$E\{v_{r-ce}^{(k_1)}\} = \sqrt{(1-\rho)P_s}h.$$

(4.110)

Next, the relay node uses the harvested energy in (4.109) to transmit K_2 pilots of its own such that the received signal at the destination is

$$v_{d-r}^{(k_2)} = \sqrt{P_r}g + n_{d-r}^{(k_2)}$$

(4.111)

for $k_2 = 1, 2, \cdots, K_2$, where the transmission power is given by

$$P_r = \frac{E_h}{K_2 T_p} = \eta\rho P_s|h|^2\frac{K_1}{K_2}.$$

(4.112)

Thus,

$$E\{v_{d-r}^{(k_2)}\} = \sqrt{\eta\rho P_s\frac{K_1}{K_2}}|h|g.$$

(4.113)

Using (4.110) and (4.113), the MB estimators for g and h can be derived, respectively, as

$$\hat{g}_4 = \frac{\frac{1}{K_2}\sum_{k_2=1}^{K_2} v_{d-r}^{(k_2)}}{\sqrt{\eta\frac{K_1}{K_2}\frac{\rho}{1-\rho}}|\frac{1}{K_1}\sum_{k_1=1}^{K_1} v_{r-ce}^{(k_1)}|}$$

(4.114)

and

$$\hat{h}_4 = \frac{1}{\sqrt{(1-\rho)P_s}}\frac{1}{K_1}\sum_{k_1=1}^{K_1} v_{r-ce}^{(k_1)}.$$

(4.115)

This scheme also requires channel estimation at both the relay and the destination.

In Scheme 4, let $v_r = \frac{1}{K_2} \sum_{k_2=1}^{K_2} v_{d-r}^{(k_2)} = r_{v_r} e^{j\theta_{v_r}}$ and $v_s = \frac{1}{K_1} \sum_{k_1=1}^{K_1} v_{r-ce}^{(k_1)} = r_{v_s} e^{j\theta_{v_s}}$ so that v_r is a complex Gaussian random variable with mean $S_{v_r} = \sqrt{\eta \rho P_s \frac{K_1}{K_2}} |h| g$ and variance $2\beta_{v_r}^2 = \frac{2\sigma^2}{K_2}$, and v_s is a complex Gaussian random variable with mean $S_{v_s} = \sqrt{(1-\rho)P_s} h$ and variance $2\beta_{v_s}^2 = \frac{2\sigma^2}{K_1}$.

Then, following similar procedures, one has

$$E\{\hat{g}_4\} = \sqrt{(1-\rho)P_s} |h| g E\left\{\frac{1}{r_{v_s}}\right\}, \tag{4.116}$$

$$E\{\hat{h}_4\} = h, \tag{4.117}$$

$$E\{|\hat{g}_4|^2\} = \frac{2\sigma^2 + \eta K_1 \rho P_s |h|^2 |g|^2}{\eta K_1 \rho/(1-\rho)} E\left\{\frac{1}{r_{v_s}^2}\right\}, \tag{4.118}$$

$$E\{|\hat{h}_4|^2\} = \frac{2\sigma^2}{(1-\rho)P_s K_1} + |h|^2, \tag{4.119}$$

where $E\{\frac{1}{r_{v_s}}\}$ and $E\{\frac{1}{r_{v_s}^2}\}$ are obtained by replacing S_{y_s} and β_{y_s} with S_{v_s} and β_{v_s} in $E\{\frac{1}{r_{y_s}}\}$ and $E\{\frac{1}{r_{y_s}^2}\}$, respectively.

4.4.1.5 Scheme 5

In Scheme 5, the relay node does not send any pilots of its own. Instead, the relay node uses the pilots from the source node to estimate the source-to-relay link and the destination node uses the same pilots forwarded by the relay to estimate the relay-to-destination link. In contrast, in Scheme 1 and Scheme 2, the destination node uses the pilots from the relay node to estimate the relay-to-destination link and uses the pilots forwarded from the source node to estimate the source-to-relay link, while in Scheme 3 and Scheme 4 the relay node uses pilots from the source node to estimate the source-to-relay link and the destination node uses pilots from the relay node to estimate the relay-to-destination link.

First, I pilots are sent from the source node to the relay node for energy harvesting. Using time switching, the received signal at the relay node is given by (4.54). Thus, the harvested energy is the same as before.

Secondly, J pilots are sent from the source to the relay node for channel estimation. The received signal at the relay node is

$$w_{r-ce}^{(j)} = \sqrt{P_s} h + n_{r-ce}^{(j)} \tag{4.120}$$

where $j = 1, 2, \cdots, J$ and $n_{r-ce}^{(j)}$ is the complex AWGN with mean zero and variance $2\sigma^2$. This received signal is first used for channel estimation at the relay to give

$$E\{w_{r-ce}^{(j)}\} = \sqrt{P_s} h. \tag{4.121}$$

Then, the signal in (4.120) is forwarded by the relay node using the harvested energy as

$$w_{d-r}^{(j)} = \sqrt{P_r} g a w_{r-ce}^{(j)} + n_{d-r}^{(j)} \tag{4.122}$$

where the transmission power can be derived as $P_r = \frac{E_h}{JT_p} = \eta P_s |h|^2 \frac{I}{J}$. Thus, one has from (4.122)

$$E\{w_{d-r}^{(j)}\} = \sqrt{\eta \frac{I}{J} P_s} |h| gha.$$ (4.123)

The MB estimators for g and h can be derived as

$$\hat{g}_5 = \frac{\frac{1}{J}\sum_{j=1}^{J} w_{d-r}^{(j)} \frac{1}{J}\sum_{j=1}^{J} w_{r-ce}^{(j)} *}{\sqrt{\eta \frac{I}{J}} a |\frac{1}{J}\sum_{j=1}^{J} w_{r-ce}^{(j)}|^3}$$ (4.124)

and

$$\hat{h}_5 = \frac{1}{\sqrt{P_s}} \frac{1}{J} \sum_{j=1}^{J} w_{r-ce}^{(j)},$$ (4.125)

respectively. Note that $I + J = K$ for this scheme.

In this case, denote $w_r = \frac{1}{J}\sum_{j=1}^{J} w_{r-ce}^{(j)} = r_{w_r} e^{j\theta_{w_r}}$, which is a complex Gaussian random variable with mean $S_{w_r} = \sqrt{P_s} h$ and variance $2\beta_{w_r}^2 = \frac{2\sigma^2}{J}$.

For $\hat{g}_5$, one can first find the conditional moment, conditioned on w_r, and then take the expectation over w_r to have

$$E\{\hat{g}_5\} = \sqrt{P_s} |h| g E\left\{\frac{1}{r_{w_r}}\right\}$$ (4.126)

and

$$E\{|\hat{g}_5|^2\} = P_s |h|^2 |g|^2 E\left\{\frac{1}{r_{w_r}^2}\right\} + \frac{2\sigma^2}{\eta I a^2} E\left\{\frac{1}{r_{w_r}^4}\right\}$$ (4.127)

where $E\{\frac{1}{r_{w_r}}\}$ and $E\{\frac{1}{r_{w_r}^2}\}$ are obtained by replacing S_{y_s} and β_{y_s} with S_{w_r} and β_{w_r} in $E\{\frac{1}{r_{y_s}}\}$ and $E\{\frac{1}{r_{y_s}^2}\}$, respectively, and

$$E\{\frac{1}{r_{w_r}^4}\} = \int_0^\infty \frac{1}{\beta_{w_r}^2 x^3} e^{-\frac{x^2+|S_{w_r}|^2}{2\beta_{w_r}^2}} I_0\left(\frac{x|S_{w_r}|}{\beta_{w_r}^2}\right) dx.$$ (4.128)

For $\hat{h}_5$, one also has

$$E\{\hat{h}_5\} = h$$ (4.129)

and

$$E\{|\hat{h}_5|^2\} = \frac{2\sigma^2}{P_s J} + |h|^2.$$ (4.130)

4.4.1.6 Scheme 6

In Scheme 6, power splitting is used to harvest the energy. The rest is similar to Scheme 5. In this case, the source node sends K pilots to the relay node, part of which is used for channel estimation at the relay node to estimate the source-to-relay link, giving

$$x_{r-ce}^{(k)} = \sqrt{(1-\rho)P_s} h + n_{r-ce}^{(k)}$$ (4.131)

where $k = 1, 2, \cdots, K$ and $n_{r-ce}^{(k)}$ is the complex AWGN with mean zero and variance $2\sigma^2$. The other part of the received power is used for energy harvesting with the harvested energy being

$$E_h = \eta \rho P_s |h|^2 K T_p. \tag{4.132}$$

Using (4.131), the source-to-relay link can be estimated as

$$E\{x_{r-ce}^{(k)}\} = \sqrt{(1-\rho)P_s} h. \tag{4.133}$$

The received signal in (4.131) is then forwarded to the destination using the harvested energy, giving

$$x_{d-r}^{(k)} = \sqrt{\eta \rho (1-\rho) P_s} |h| gha + \sqrt{\eta \rho P_s} |h| gan_{r-ce}^{(k)} + n_{d-r}^{(k)} \tag{4.134}$$

where $n_{d-r}^{(k)}$ is also AWGN with mean zero and variance $2\sigma^2$. Thus,

$$E\{x_{d-r}^{(k)}\} = \sqrt{\eta \rho (1-\rho) P_s} |h| gha. \tag{4.135}$$

Finally, one has

$$\hat{g}_6 = \frac{\frac{1}{K} \sum_{k=1}^{K} x_{d-r}^{(k)} \frac{1}{K} \sum_{k=1}^{K} x_{r-ce}^{(k)*}}{a \sqrt{\eta \frac{\rho}{1-\rho}} |\frac{1}{K} \sum_{k=1}^{K} x_{r-ce}^{(k)}|^3} \tag{4.136}$$

$$\hat{h}_6 = \frac{1}{\sqrt{(1-\rho)P_s}} \frac{1}{K} \sum_{k=1}^{K} x_{r-ce}^{(k)}. \tag{4.137}$$

The derivation in Scheme 6 is very similar to that in Scheme 5. Let $x_r = \frac{1}{K} \sum_{k=1}^{K} x_{r-ce}^{(k)} = r_{x_r} e^{j\theta_{x_r}}$, which is a complex Gaussian random variable with mean $S_{x_r} = \sqrt{(1-\rho)P_s} h$ and variance $2\beta_{x_r}^2 = \frac{2\sigma^2}{K}$.

For $\hat{g}_6$, one can also first find the conditional moment, conditioned on x_r, and then take the expectation over x_r to have

$$E\{\hat{g}_6\} = \sqrt{(1-\rho)P_s} |h| g E\left\{\frac{1}{r_{x_r}}\right\} \tag{4.138}$$

and

$$E\{|\hat{g}_6|^2\} = (1-\rho)P_s |h|^2 |g|^2 E\left\{\frac{1}{r_{x_r}^2}\right\} + \frac{2\sigma^2}{\eta K a^2 \rho/(1-\rho)} E\left\{\frac{1}{r_{x_r}^4}\right\} \tag{4.139}$$

where $E\{\frac{1}{r_{x_r}}\}$ and $E\{\frac{1}{r_{x_r}^2}\}$ are obtained by replacing S_{y} and β_{y} with S_{x_r} and β_{x_r} in $E\{\frac{1}{r_{y_s}}\}$ and $E\{\frac{1}{r_{y_s}^2}\}$, respectively, and

$$E\left\{\frac{1}{r_{x_r}^4}\right\} = \int_0^{\infty} \frac{1}{\beta_{x_r}^2 x^3} e^{-\frac{x^2 + |S_{x_r}|^2}{2\beta_{x_r}^2}} I_0\left(\frac{x|S_{x_r}|}{\beta_{x_r}^2}\right) dx. \tag{4.140}$$

For $\hat{h}_6$, its first-order moment and second-order moment are given by

$$E\{\hat{h}_6\} = h \tag{4.141}$$

and

$$E\{|\hat{h}_6|^2\} = \frac{2\sigma^2}{(1-\rho)P_sK} + |h|^2, \tag{4.142}$$

respectively.

Next, we use some graphs to show the performances of the obtained estimators for h and g in terms of the MSE. From the first- and second-order moments of the estimates, one can see that the estimator performance depends on various system parameters, including η, P_s, $2\sigma^2$, g, h, K, I and J. For η, P_s, $2\sigma^2$, and K, their effects on the estimator performance are quite straightforward. In particular, it is expected that the estimator performance improves when η, P_s, and K increase or when $2\sigma^2$ decreases, as this reduces the noise which is the main source of estimation error. Thus, in the examination, we set $\eta = 0.5$, $P_s = 1$, $K = 100$, and $2\sigma^2 = 2$ to focus on the effects of g, h, I, J_2, ρ, K_1, and K_2. Define $\gamma_g = \frac{|g|^2}{2\sigma^2}$ as the instantaneous SNR of the relay-to-destination link and $\gamma_h = \frac{|h|^2}{2\sigma^2}$ as the instantaneous SNR of the source-to-relay link. The values of g and h will change with γ_g and γ_h and their real and imaginary parts are equal to each other. The MSE is defined as $\frac{1}{R}\sum_{r=1}^{R}|g - \hat{g}_r|^2$, $\frac{1}{R}\sum_{r=1}^{R}|h - \hat{h}_r|^2$, and $\frac{1}{R}\sum_{r=1}^{R}|gh - \hat{g}_r\hat{h}_r|^2$ for $\hat{g}$, $\hat{h}$, and $\hat{g}\hat{h}$, respectively, where $R = 10\ 000$ is the total number of simulation runs and $\hat{g}_r$ and $\hat{h}_r$ are the channel estimates in the rth run.

Figures 4.15 and 4.16 show the MSE of the estimators $\hat{g}_1$ and $\hat{h}_1$ in Scheme 1 versus the values of I and J_2, respectively, when different values of SNR or different values of g and h are considered. In Figure 4.15, the values of J_1 and J_2 are set as $\frac{I}{2}$ to focus on the effect of I, and the value of I is examined from 4 to 96 with a step size of 4. In Figure 4.16, the values of I are fixed to focus on the effect of J_2, and the value of J_2 is examined from 2 to $J - 2$ with a step size of 2. Several observations can be made. First, from Figure 4.15, the

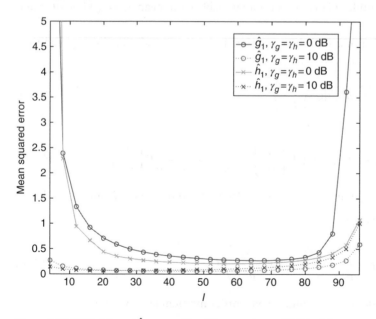

Figure 4.15 MSE of $\hat{g}_1$ and $\hat{h}_1$ versus I for different values of SNR in Scheme 1.

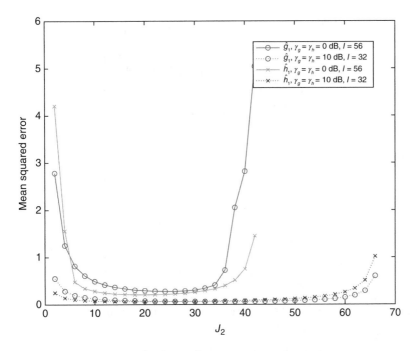

Figure 4.16 MSE of $\hat{g}_1$ and $\hat{h}_1$ versus J_2 for different values of SNR in Scheme 1.

MSE first decreases and then increases with the value of I, as expected, as a larger value of I leads to more harvested energy such that the estimation at the destination node will be more accurate but it also leads to a smaller value of J due to fixed K such that the sample size in the estimation reduces. The optimum value of I exists, and there is also a wide range of choices for I that give close-to-optimum performances. This provides flexibility in system design. Secondly, under the same conditions, the estimator generally performs better with a larger value of SNR, as expected, as a larger SNR gives relatively smaller noise, which is the main source of estimation error. Also, the performance of $\hat{g}_1$ is close to that of $\hat{h}_1$, especially near the optimum values of I. Finally, from Figure 4.16, the MSE first decreases and then increases when the value of J_2 increases. This is also expected. A larger value of J_2 leads to a better approximation of $E\{y_{d-r}^{(j_2)}\}$ but due to fixed J it also leads to a worse approximation of $E\{y_{d-s}^{(j_1)}\}$. Since both of them are needed in $\hat{g}_1$ and $\hat{h}_1$, their interaction yields an optimum value of J_2, as seen from Figure 4.16.

Figures 4.17 and 4.18 show the MSE versus ρ and K_2, respectively, for $\hat{g}_2$ and $\hat{h}_2$ in Scheme 2. In Figure 4.17, the value of ρ is examined from 0.1 to 0.9 with a step size of 0.1, when $K_2 = \frac{K}{2}$ is fixed to focus on the effect of ρ. In Figure 4.18, the value of K_2 is examined from 4 to 96 with a step size of 4, while the value of ρ is fixed. Again, optimum values of ρ and K_2 exist. For ρ, when it is large, more energy is harvested for relay transmission but the signal component in the samples will be weaker, leading to more estimation errors. Thus, a balanced choice of ρ needs to be made and it plays a similar role to $\frac{I}{K}$ in Scheme 1. From Figure 4.17, the MSE curve is relatively flat between 0.2 and 0.8, indicating that there is a wide range of choices for ρ that can achieve close-to-optimum performance, which is desirable for system design. For K_2, since K is fixed, a larger value

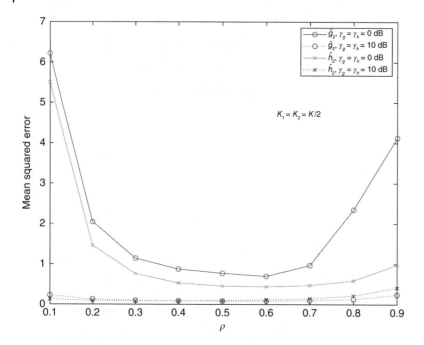

Figure 4.17 MSE of $\hat{g}_2$ and $\hat{h}_2$ versus ρ for different values of SNR in Scheme 2.

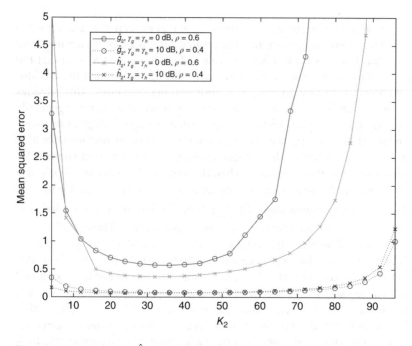

Figure 4.18 MSE of $\hat{g}_2$ and $\hat{h}_2$ versus K_2 for different values of SNR in Scheme 2.

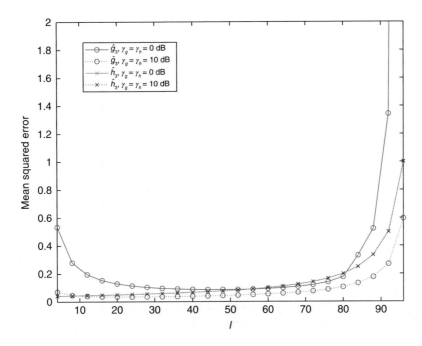

Figure 4.19 MSE of $\hat{g}_3$ and $\hat{h}_3$ versus I for different values of SNR in Scheme 3.

of K_2 gives a better approximation using the pilots from the relay node but also gives a worse approximation using the pilots from the source node, as K_1 will be smaller. In all cases, increasing the SNR reduces the MSE.

Figures 4.19 and 4.20 show the MSE versus I and J_1, respectively, for $\hat{g}_3$ and $\hat{h}_3$ in Scheme 3. In Figure 4.19, the value of I is examined from 4 to 96 with a step size of 4, when $J_1 = \frac{I}{2}$ is fixed to focus on the effect of I. Also, in Figure 4.20, the value of J_1 is examined from 2 to $J - 2$ with a step size of 2, while the value of ρ in this figure is fixed. In Figure 4.19, the optimum value of I still exists for $\hat{g}_3$, as the value of I affects both energy harvesting and the sample size for $\hat{g}_3$. However, for $\hat{h}_3$, its MSE monotonically increases with I, as the estimation of h does not rely on energy harvesting and it only relies on J_1. When I increases, for fixed K, the value of J_1 decreases such that more estimation errors occur. Also in this case, the SNR has very minor effect on $\hat{h}_3$. In Figure 4.20, it can also be seen that the optimum value of J_1 does not exist.

Figures 4.21 and 4.22 show the MSE versus ρ and K_1, respectively, for $\hat{g}_4$ and $\hat{h}_4$ in Scheme 4. In Figure 4.21, the value of ρ is studied from 0.1 to 0.9 with a step size of 0.1, while we fix $K_1 = \frac{K}{2}$ to focus on the effect of ρ. As well, in Figure 4.22, the value of K_1 is examined from 4 to 96 with a step size of 4, while the value of ρ in this figure is fixed. As can be seen from Figure 4.21, the value of ρ affects the performance of $\hat{g}_4$ both ways, making its MSE larger due to weaker signal component in the samples and less accurate $\hat{h}_4$ but also making its MSE smaller due to more harvested energy, when ρ increases. For $\hat{h}_4$, increasing ρ will increase the MSE of $\hat{h}_4$ as the signal component in $v_{r-ce}^{(k_1)}$ will be weaker. Also, similar to Figure 4.20, increasing K_1 reduces the MSE of the estimator.

For Scheme 5 and Scheme 6, the effects of the system parameters are straightforward. From their expressions of the first- and second-order moments, it can be expected that

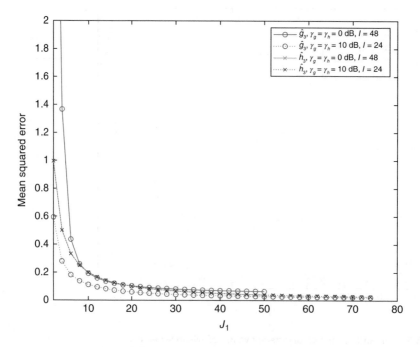

Figure 4.20 MSE of $\hat{g}_3$ and $\hat{h}_3$ versus J_1 for different values of SNR in Scheme 3.

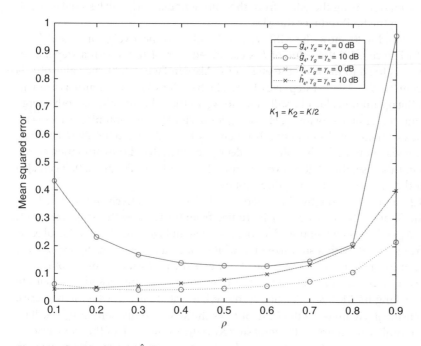

Figure 4.21 MSE of $\hat{g}_4$ and $\hat{h}_4$ versus ρ for different values of SNR in Scheme 4.

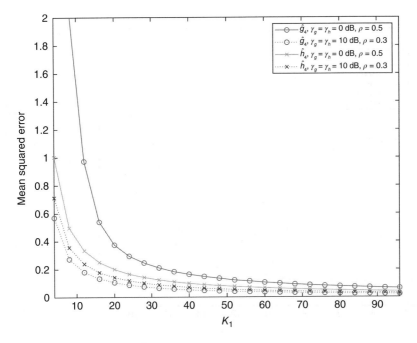

Figure 4.22 MSE of $\hat{g}_4$ and $\hat{h}_4$ versus K_1 for different values of SNR in Scheme 4.

the MSE of $\hat{h}_5$ increases when I increases and for the MSE of $\hat{g}_5$ there is an optimum I due to the interaction between energy harvesting and sample size. Also, the MSE of $\hat{h}_6$ monotonically increases when ρ increases due to weaker signal component in the samples, while the MSE of $\hat{g}_6$ first decreases then increases with ρ. They are not shown here.

Figures 4.23 and 4.24 compare the estimators in different schemes in terms of their minimum MSE of $\hat{g}\hat{h}$ achieved by performing exhaustive searches over the relevant parameters. One sees from these two figures that Scheme 5 and Scheme 6 have the best performances, followed by Scheme 3 and Scheme 4, and Scheme 1 and Scheme 2 have the worst performance. Also, Scheme 5 outperforms Scheme 6, Scheme 3 outperforms Scheme 4, and Scheme 1 outperforms Scheme 2, indicating that time switching leads to better MSE performance than power splitting. Finally, the estimator performance is more sensitive to γ_g than to γ_h. More details of derivations can be found in Chen et al. (2017b).

4.4.2 Without Relaying

In this subsection, we consider the case without relaying. In this case, the transmission contains three parts: channel estimation; energy harvesting; and data transmission. Figure 4.25 shows a diagram of the considered data frame. In this system, the data frame is first divided into two parts in time for channel estimation and data transmission. Then, each part in time is divided into two parts in amplitude so that energy harvesting is performed by taking parts of the received energy for channel estimation as well as parts of the received energy for data transmission. This is equivalent to dividing the whole data

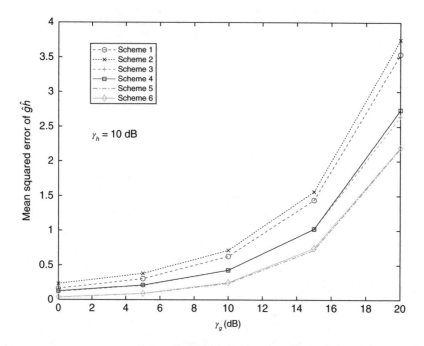

Figure 4.23 MSE of $\hat{g}\hat{h}$ versus γ_g for different schemes.

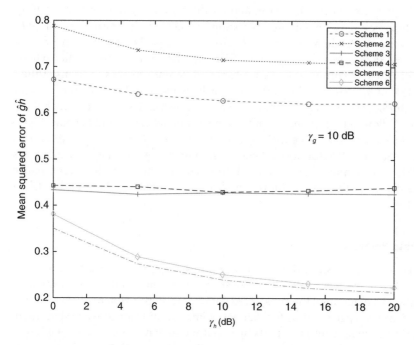

Figure 4.24 MSE of $\hat{g}\hat{h}$ versus γ_h for different schemes.

Figure 4.25 The considered data frame with three parts.

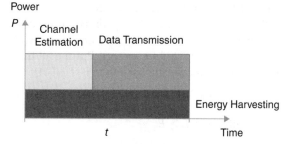

frame into three parts in time only. It is the area of the block in the figure that matters, as it represents the energy allocation.

From this diagram, the received signals for the pilots used in channel estimation can be expressed as

$$p_i = \sqrt{\rho_p P_s} h + n_i \tag{4.143}$$

where $i = 1, 2, \cdots, L$ index the pilots, P_s is the transmission power, ρ_p is the portion of power split from the received signal for channel estimation, h is the complex fading gain to be estimated, and n_i is the complex AWGN with mean zero and variance $2\sigma^2$. These pilots can be used to estimate the channel gain using the maximum likelihood method as

$$\hat{h} = \frac{1}{\sqrt{\rho_p P_s} L} \sum_{i=1}^{L} p_i = h + e \tag{4.144}$$

where the estimation error e is Gaussian with mean zero and variance $\frac{2\sigma^2}{\rho_p P_s L}$.

The received signals of the symbols used for information delivery can be expressed as

$$y_i = \sqrt{\rho_d P_s} h + n_i \tag{4.145}$$

where $i = L + 1, L + 2, \cdots, L + M$ index the data symbols to be decoded and ρ_d is the portion of power split from the received signal for data decoding. Using the estimate in (4.144), one has

$$y_i = \sqrt{\rho_d P_s} \hat{h} - \sqrt{\rho_d P_s} e + n_i. \tag{4.146}$$

Thus, the achievable rate can be derived as

$$R = \frac{M}{L + M} \log_2 \left(1 + \frac{\rho_d P_s |\hat{h}|^2}{2\sigma^2 + \rho_d \frac{2\sigma^2}{\rho_p L}} \right). \tag{4.147}$$

Finally, the total energy harvested from pilots for channel estimation and symbols for information delivery can be derived as

$$E = \eta(1 - \rho_p) P_s |h|^2 L + \eta(1 - \rho_d) P_s |h|^2 M \tag{4.148}$$

where η is the conversion efficiency defined as before. The average energy harvested over each symbol is thus

$$\bar{E} = \frac{E}{L + M} = 2\alpha^2 \eta P_s \frac{(1 - \rho_p)L + (1 - \rho_d)M}{L + M} \tag{4.149}$$

where $E\{|h|^2\} = 2\alpha^2$ has been used.

From the above, the pilot symbols are used for channel estimation. The estimated channel gain is then used for data decoding. At the same time, both the pilot and data symbols are harvested for energy. Putting all these requirements together, we can formulate an optimization problem as

$$\max_{\rho_p, \rho_d, L} R \tag{4.150}$$

$$\text{subject to } \bar{E} \geq E_0 \tag{4.151}$$

$$1 \geq \rho_p \geq 0 \tag{4.152}$$

$$1 \geq \rho_d \geq 0 \tag{4.153}$$

$$M + L \geq L \geq 0. \tag{4.154}$$

In this problem, we try to maximize the average achievable rate for information delivery with respect to the number of pilots for channel estimation as well as the two power splitting parameters ρ_p and ρ_d, while satisfying a minimum requirement on the harvested energy per symbol. The values of L, ρ_p, and ρ_d actually determine the energy allocation between channel estimation, data transmission, and energy harvesting within the transmitted data frame. Thus, this is essentially a resource allocation problem. It is related to the second issue discussed at the beginning of this section, that is, the addition of energy harvesting capability consumes resources provided for other functions and therefore, it needs to be carefully chosen. A similar problem was studied in Zhou (2015), where a minimum mean squared error channel estimator was considered, but the rest of the settings are similar to those in this subsection. Also, without energy harvesting, the problem in (4.150) becomes the classical pilot allocation problem in conventional communications, where a larger number of pilots gives better estimation accuracy but reduces the throughput of the channel (Wang et al. 2014).

In the following, we consider a simpler case when $\rho_d = \rho_p = \rho$. Also, fix $L + M = K$ so that $M = K - L$. Using them in (4.148), the optimization problem becomes

$$\max_{\rho, L} \frac{K - L}{K} \log_2 \left(1 + \frac{\rho P_s |\hat{h}|^2}{2\sigma^2 + \frac{2\sigma^2}{L}}\right) \tag{4.155}$$

$$\text{subject to } 2\alpha^2 \eta (1 - \rho) P_s K \geq E_0 \tag{4.156}$$

$$1 \geq \rho \geq 0 \tag{4.157}$$

$$K \geq L \geq 0. \tag{4.158}$$

One sees that the rate monotonically increases with ρ so that the optimum ρ is determined by the constraint on the harvested energy as $\rho_{opt} = \max\{0, 1 - \frac{E_0}{2\alpha^2 \eta P_s K}\}$. By replacing ρ with the optimum value in the rate expression, one has the rate as the function of a single variable L. By taking the first-order derivative of this function with respect to L and solving the equation for L, the optimum L can be obtained. This involves a non-linear function due to the logarithm function. We show some numerical results instead. Let $P_s = 1$, $|\hat{h}|^2 = 1$, $K = 100$, $2\sigma^2 = 1$, $\eta = 0.5$, $E_0 = 50$, and $2\alpha^2 = 1$. Figure 4.26 shows R versus L under these conditions using the optimum ρ. From this figure, the optimum L is around 8.

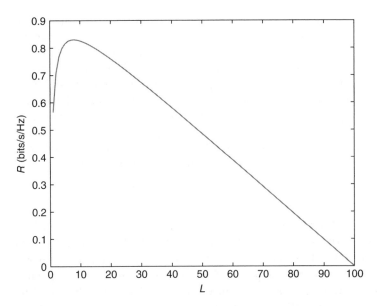

Figure 4.26 *R* versus *L*.

This problem has many variants that one can work on. For example, instead of using the instantaneous achievable rate, one may use the average achievable rate as the objective function. Thus, R will be replaced by $E_{\hat{h}}\{R\}$ in the above problems, where $E\{\cdot\}$ denotes the expectation operator. Also, one can consider the sum rate in the downlink and uplink. In this case, the uplink rate decreases with the parameter ρ due to a reduced harvested energy. Thus, the constraint on the harvested energy is not needed any more but the tradeoff will be reflected by the fact that the downlink rate increases with ρ while the uplink rate decreases with ρ.

The key point in these optimizations and in energy harvesting estimation generally is that both energy harvesting and channel estimation are part of the overhead incurred for data transmission. Hence, an increase in one part will lead to a decrease in the other part, assuming a fixed amount of resource. In Yang et al. (2014) and Zeng and Zhang (2015b), only channel estimation and energy harvesting were considered in the tradeoff, without taking into account the achievable rate for data transmission. One can either maximize the amount of harvested energy with respect to a constraint on the estimation accuracy or vice versa. Again, the fundamental problem in all these studies is the tradeoff between energy harvesting and channel estimation subject to limited resource. This is unique for energy harvesting estimation, as conventional communications do not have this allocation issue.

4.5 Energy Transmission Waveform

For information transmission, it is well known that the transmitted pulse can be optimized to achieve the largest signal-to-interference-plus-noise ratio for best performance. Similarly, for power transfer, the transmitted waveform can also be optimized to achieve the maximum transferred power. In an energy harvesting communications system, this is part of the signal design problem at the transmitter.

The transferred power needs to be harvested and maximized at the receiver. Thus, the amount of transferred power is a function of the parameters of the transmitted waveform at the transmitter, the parameters of the communications channel, and the parameters of the energy harvester at the receiver. To optimize the transmitted waveform for maximum transferred power, one must know the channel parameters and the energy harvester parameters, as the optimal waveform will be adapted to the channel conditions and the harvester designs. Next, energy transfer waveform design will be discussed.

4.5.1 Scenario

Consider a wireless energy transfer system where there are M energy sources with $M > 1$. Each source transmits an energy waveform with a complex amplitude of $w_m = \omega_m e^{j\psi_m}$, where $m = 1, 2, \cdots, M$ index the sources, $j = \sqrt{-1}$, ω_m is the magnitude of the energy waveform, and ψ_m is the phase of the energy waveform. These ω_m and ψ_m are the parameters to be optimized. The waveforms are transmitted over Rayleigh fading channels. The faded signals are then harvested by the energy harvesters at the receiver. These M energy sources could be M simultaneously transmitting users. In this case, only one harvester is needed to harvest all energies. These M energy sources could also be M antennas, frequency bands or time slots. In this case, either M harvesters will be needed or the same harvester will be used for M times.

In the multi-user case, if all users transmit their waveforms simultaneously in the same frequency band using the same antenna, the received signal at the single harvester is

$$z = \sum_{m=1}^{M} w_m h_m + n \tag{4.159}$$

where h_m is the fading gain from the mth user to the energy harvester and n is the AWGN with mean zero and variance $2\sigma^2$. The channel gain h_m is a circularly symmetric complex Gaussian random variable with mean zero and variance $2\alpha_m^2$. For example, in a two-way relaying system or a non-orthogonal multiple access (NOMA) system, the received signals come from all users. Also, if co-channel interference is harvested, the received energy is made of signals from all interfering users too. Using (4.159), if a linear energy harvester is used, the harvested power (the term power is used interchangeably with the term energy here due to the fixed time slot) is

$$P_o = \eta |z|^2 \tag{4.160}$$

where η is the conversion efficiency of the harvester. Thus, the harvested power is a function of the waveform parameters ω_m and ψ_m, the channel parameters h_m, and the harvester parameter η, as mentioned before. If a non-linear energy harvester is used, as discussed in Chen et al. (2017d), the harvested power is

$$P_o = \frac{a|z|^2 + b}{|z|^2 + c} - \frac{b}{c} \tag{4.161}$$

where a, b, and c are the parameters of the harvester that can be determined by standard curve-fitting methods.

On the other hand, if the sources operate over different antennas, frequency bands, or time slots, one needs M energy harvesters, or one harvester used for M times. For the mth harvester, the received signal to be harvested is given by

$$z_m = w_m h_m + n_m \tag{4.162}$$

where n_m is the AWGN with mean zero and variance $2\sigma^2$, and all the other symbols are defined as before. This is, for example, the case when harvesters are tuned at different frequencies (Sun et al. 2013) or antennas (He et al. 2015) to increase the amount of harvested energy, or when the energy is accumulated from transmissions in different time slots. In this case, using the linear harvester, the total harvested energy is given by

$$P_o = \sum_{m=1}^{M} \eta_m |z_m|^2 \tag{4.163}$$

and using the non-linear harvester, the total harvested energy is

$$P_o = \sum_{m=1}^{M} \left(\frac{a_m |z_m|^2 + b_m}{|z_m|^2 + c_m} - \frac{b_m}{c_m} \right), \tag{4.164}$$

where η_m is the conversion efficiency of the mth harvester. In this case, the individual signals are harvested and combined.

4.5.2 Energy Waveform Optimization

Here we consider the optimization of the average total harvested power, subject to the constraint that the total transmitted power is fixed for all M sources. Thus, the optimization problem is written as

$$\{\hat{w}_1, \cdots, \hat{w}_M\} = \max_{w_1, \cdots, w_M} \{E\{P_o\}\}, \text{ with } \sum_{m=1}^{M} |w_m|^2 = P \tag{4.165}$$

where $E\{\cdot\}$ is the expectation operation and P is the total power limit. This optimization depends on the harvester model and the channel condition.

4.5.2.1 Linear Harvester
When a single linear energy harvester is used, the average harvested power can be calculated from (4.160) as

$$E\{P_o\} = \eta \left[\sum_{m=1}^{M} (2\alpha_m^2) \omega_m^2 + 2\eta\sigma^2 \right]. \tag{4.166}$$

This value can be easily maximized using linear programming to give the maximum average harvested power as

$$E\{P_o\}_{max} = 2\eta P \alpha_{\hat{m}}^2 + 2\sigma^2 \tag{4.167}$$

where $\hat{m} = \max_{m=1,2,\cdots,M} \{2\alpha_m^2\}$. Thus, the optimum waveforms are given by $\omega_m^2 = P$ when $m = \hat{m}$ and $\omega_m^2 = 0$ when $m \neq \hat{m}$. In other words, the maximum harvested power can be achieved by transmitting all the power P at the user with the highest average channel fading power to the energy harvester and switch off all other users. In doing so, the phase of the optimum waveform can be an arbitrary value.

In the second case when multiple linear harvesters are used over different frequencies, antennas or time slots, similarly, the average total harvested power is calculated from (4.163) as

$$E\{P_o\} = 2 \sum_{m=1}^{M} \eta_m \alpha_m^2 \omega_m^2 + 2 \sum_{m=1}^{M} \eta_m \sigma^2. \tag{4.168}$$

Again, using linear programming, this value can be maximized as

$$E\{P_o\}_{max} = 2\eta_{\hat{m}} P \alpha_{\hat{m}}^2 + 2 \sum_{m=1}^{M} \eta_m \sigma^2 \tag{4.169}$$

where $\hat{m} = \max_{m=1,2,\cdots,M} \{2\eta_m \alpha_m^2\}$ in this case and the optimum waveform is achieved by setting $\omega_m^2 = P$ when $m = \hat{m}$ and $\omega_m^2 = 0$ when $m \neq \hat{m}$.

4.5.2.2 Non-Linear Harvester

In many applications, the energy harvester is not linear. Consider the single harvester case first. In this case, the average harvested power can be derived from (4.161) by taking the expectation to give

$$E\{P_o\} = a - \frac{b}{c} + \frac{b - ac}{2\beta^2} e^{\frac{c}{2\beta^2}} \left[-Ei\left(-\frac{c}{2\beta^2} \right) \right] \tag{4.170}$$

where $2\beta^2 = 2 \sum_{m=1}^{M} \omega_m^2 \alpha_m^2 + 2\sigma^2$ and $Ei(\cdot)$ represents the exponential integral Gradshteyn and Ryzhik (2000, eq. (8.211.1)). If one denotes $g(x) = xe^x[-Ei(-x)]$, one can show that $g(x)$ is a monotonically increasing function of x. Thus, the maximization of the average harvested power in (4.170) is equivalent to the maximization of $2\beta^2$ in that function. However, $2\beta^2$ is a quadratic form of ω_m. Thus, the maximum harvested power is

$$E\{P_o\}_{max} = a - \frac{b}{c} + \frac{b - ac}{2\hat{\beta}^2} e^{\frac{c}{2\hat{\beta}^2}} \left[-Ei\left(-\frac{c}{2\hat{\beta}^2} \right) \right] \tag{4.171}$$

where $2\hat{\beta}^2 = 2P\alpha_{\hat{m}}^2 + 2\sigma^2$ and $\hat{m} = \max_{m=1,2,\cdots,M} \{2\alpha_m^2\}$. The optimum waveforms are $\omega_m^2 = P$ when $m = \hat{m}$ and $\omega_m^2 = 0$ when $m \neq \hat{m}$. Again, in this case, one allocates the full power to the source with the best channel condition and switches off all other sources.

In the case when multiple non-linear harvesters are used, one can use the Lagrange multiplier to find the optimum solution. In this case, the function $g(x)$ defined before can be approximated as $g(x) \approx \frac{0.98x+0.12}{x+0.86}$ by curve-fitting for $0 < x < 30$. Then, one has the target function for optimization as

$$\sum_{m=1}^{M} \left(a_m - \frac{b_m}{c_m} \right) + \sum_{m=1}^{M} (b_m - a_m c_m) \frac{0.98c_m + 0.12(\omega_m^2 2\alpha_m^2 + 2\sigma^2)}{c_m + 0.86(\omega_m^2 2\alpha_m^2 + 2\sigma^2)}$$

$$+ \lambda \left(P - \sum_{m=1}^{M} \omega_m^2 \right) \tag{4.172}$$

where λ is the Lagrange multiplier. Using (4.172), one has the optimum waveforms as

$$\omega_m^2 = \frac{\sqrt{1.2\alpha_m^2 c_m (a_m c_m - b_m)/\lambda_0} - c_m - 1.72\sigma^2}{1.72\alpha_m^2} \tag{4.173}$$

where $\lambda_0 = \left(\dfrac{\sum_{m=1}^{M} \frac{\sqrt{1.2a_m^2 c_m (a_m c_m - b_m)}}{1.72 a_m^2}}{P + \sum_{m=1}^{M} \frac{c_m + 1.72 \sigma^2}{1.72 a_m^2}} \right)^2$. One sees that, in this case, the optimum waveforms

are not to transmit the full power over the best channel any more. Instead, the power needs to be distributed among different frequencies, antennas, or time slots. In all the above cases, the phase of the optimum waveform can be arbitrary. This gives flexibility in the system design. More details on this work can be found in Chen et al. (2017d), where in addition to Rayleigh fading, the general Rician fading channel and the Gamma-shadowed Rician fading channel have also been discussed.

The above discussion considered the use of a single harvester for signals from multiple users or multiple time slots, or the use of multiple harvesters for signals from multiple antennas and frequencies. In Clerckx and Bayguzina (2016), the authors considered the other case when a single harvester is used for a sum of signals from multiple antennas and frequency bands and the noise can be ignored. Specifically, the received signal at the single harvester is given by Clerckx and Bayguzina (2016, eq. (4))

$$y = \Re \left\{ \sum_{n=1}^{N} \mathbf{h}_n \mathbf{w}_n e^{j2\pi f_n t} \right\} \tag{4.174}$$

where $\Re\{\cdot\}$ takes the real part of a complex value, $n = 1, 2, \cdots, N$ index N frequency bands or subcarriers within the signal bandwidth, f_n is the nth frequency band, $\mathbf{h}_n = [h_{n,1} h_{n,2} \cdots h_{n,M}]$ is the channel gain in the nth frequency for M transmitting antennas, and $\mathbf{w}_n = [w_{n,1} w_{n,2} \cdots w_{n,M}]^T$ is the waveform amplitude to be determined in the nth frequency band for M transmitting antennas, $(\cdot)^T$ is the transpose operation, N is the total number of frequencies, and M is the total number of antennas. Compared with (4.159), one can notice several differences. First, the noise has been ignored in (4.174). Secondly, only the real part of the received signal is harvested for energy. The imaginary part has been discarded. Thirdly, instead of having a sum of signals from multiple users, (4.174) considers a sum of signals from multiple frequency bands and multiple antennas. Finally, instead of harvesting each signal and adding the harvested energy up, as in (4.163) and (4.164), (4.174) adds up the signals before harvesting them.

Using the received signal in (4.174), they aimed to maximize the output current at the energy harvester to give the following optimization problem as

$$\sum_{i \geq 2, i \text{ even}}^{n_0} k_i E\{y^i\}, \text{ with } \sum_{n=0}^{N-1} \sum_{m=1}^{M} |w_{n,m}|^2 < P \tag{4.175}$$

where n_0 is a constant that determines how much non-linearity of the energy harvester one can account for in the optimization, k_i is the parameter of the harvester, and other symbols are defined as before. When $n_0 = 2$, this model gives the linear harvester. In this case, it was shown in Clerckx and Bayguzina (2016) that the optimum waveforms are to transmit the full power P over the best channel among all frequencies and antennas. This agrees with our discussion before. If $n_0 > 2$, higher-order terms are included in the optimization and thus, the harvester is non-linear. In this case, it was reported in Clerckx and Bayguzina (2016) that one has to distribute the power over all different frequencies or antennas in order to maximize the output of the energy harvest. This again agrees with the observation from Chen et al. (2017d).

Both Chen et al. (2017d) and Clerckx and Bayguzina (2016) conclude that, if a linear harvester is used or the harvester operates in its linear region, the optimum strategy is to put the full power at the source with the best channel condition, and if a non-linear harvester is used or the harvester operates in its non-linear region, the optimum strategy is to distribute the power across all sources.

In another related work (Collado and Georgiadis 2014), the authors also investigated the effect of waveform on the conversion efficiency of the energy harvester. This is different from the results in Chen et al. (2017d) and Clerckx and Bayguzina (2016) in that here the conversion efficiency η is a function of the waveform, while it is a constant in most studies. The motivation of this work was that waveforms with higher peak-to-average-power ratio (PAPR) can activate the energy harvester with a lower threshold. The study in Collado and Georgiadis (2014) showed that the chaotic waveform has the highest conversion efficiency and thus, the largest harvested power, as it has the largest PAPR.

4.6 Other Issues and Techniques

In the previous sections, we have discussed energy harvesting detection, energy harvesting estimation, and energy harvesting waveform. These are the main design problems in the physical layer to solve the fundamental issues of any communications systems. In addition to these techniques, other advanced techniques may also be employed in the physical layer to solve one or two specific issues, such as security, spectral efficiency, and energy efficiency, to improve energy harvesting communications. Next, we will discuss some of these issues and techniques, including circuit power consumption, physical layer security, NOMA, and joint detection and estimation.

4.6.1 Circuit Power Consumption

Circuit power consumption is an important issue for energy harvesting communications. Many current energy harvesting communications systems are designed for low-power applications, such as wireless sensing and radio frequency identification (Sudevalayam and Kulkarni 2011). Consequently, the circuit power may not be negligible compared with the transmission power. In this subsection, we consider the case when the harvested power needs to cover both the circuit power and the transmission power for the remote device. Figure 4.27 shows a diagram of the circuit of the remote

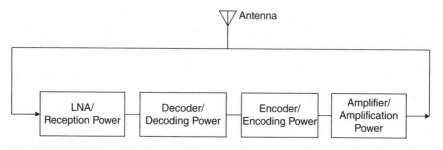

Figure 4.27 A diagram of the circuit of the remote device.

device considered in this book. The remote device needs to receive information from the access point and then responds to the information by sending its own data. Thus, four main parts of circuit power are considered: reception power; decoding power; encoding power; and amplification power.

Assume that the remote device uses a transmission power of P_x plus a circuit power of $P_c = P_{re} + P_{dc} + P_{ec} + P_{am}$, where P_{re} is the reception power, P_{dc} is the decoding power, P_{ec} is the encoding power, and P_{am} is the amplification power. It was reported in Bjornson et al. (2014) that, based on the central limit theorem, one has the amplification power and the reception power following Gaussian distributions with $\sqrt{P_{am}} \sim C\mathcal{N}(0, \delta_t^2 P_x)$ and $\sqrt{P_{re}} \sim C\mathcal{N}(0, \delta_r^2 P_r)$, respectively, where δ_r^2 and δ_t^2 are the parameters related to the receiver and transmitter structures, respectively, and P_r is the received signal power. Furthermore, one has $P_{dc} = F_{de} \cdot 2^{2B}$, $P_{ec} = F_{co} \cdot 2^{2B}$, where B is the bit resolution of the signal, and F_{de} and F_{co} are constants depending on the decoder and encoder structures, respectively (Chen et al. 2013).

Using all the above assumptions, the total power consumption including circuit power and transmission power can be written as

$$P_{tot} = P_x + \delta_t^2 P_x + F_{de} \cdot 2^{2B} + F_{co} \cdot 2^{2B} + \delta_r^2 P_r \tag{4.176}$$

where $P_{am} \approx E\{P_{am}\} = \delta_t^2 P_x$ and $P_{re} \approx E\{P_{re}\} = \delta_r^2 P_r$ have been used. Thus, if the power harvested from the access point is $P = \eta \beta^2 W$ as in (4.6), one has

$$\eta \beta^2 W = P_x + \delta_t^2 P_x + F_{de} \cdot 2^{2B} + F_{co} \cdot 2^{2B} + \delta_r^2 P_r. \tag{4.177}$$

This gives the new transmission power of the energy harvesting device as

$$P_x = \frac{1}{1 + \delta_t^2} [\eta \beta^2 W - F_{de} \cdot 2^{2B} - F_{co} \cdot 2^{2B} - \delta_r^2 P_r]. \tag{4.178}$$

This new transmission power can then be used in energy harvesting detection and energy harvesting estimation to design new detectors and estimators that consider the circuit power consumption. These designs will not be presented here as they are quite straightforward.

4.6.2 Physical Layer Security

Physical layer security has become increasingly important in recent communications systems. In fact, security is a fundamental problem in wireless communications due to the broadcast nature of the wireless medium. Traditional security measures often employ encryption algorithms at the upper layers. The most recent advance is to use the physical (PHY) layer security technique by exploiting the characteristics of the wireless fading channels for perfect secrecy (Liu and Trappe 2009). There are two unique aspects of the physical layer security technique in energy harvesting communications.

First, since the transmission power of an energy harvesting device is dynamic, the secrecy of the communications system becomes dynamic too. The secrecy rate is defined as the difference between the rate over the desired channel and the rate over the eavesdropping channel. Hence, the secrecy rate changes with the energy arrival process, and the average secrecy rate is often lower in such cases, similar to the BER analyzed before.

Secondly, for certain energy harvesting communications systems, such as the wireless powered systems to be discussed in Chapter 6, RF signals are employed to carry energy as well as information, and the receiver receives the RF signals for both information decoding and energy harvesting. To do this, in practice, the transmitter has to increase its transmission power in order to facilitate energy harvesting at the receiver as extra energy is required at the receiver for harvesting. However, this also increases information leakage, as higher information energy is more susceptible to eavesdropping when there is an eavesdropper trying to intercept the information. Moreover, at the receiver, only a portion of the energy is used for information delivery, leading to a reduced rate. Thus, the security issue in the hybrid information and energy transmission will loom larger as a portion of the energy has to be used for power transfer.

Assume that Alice is an energy harvesting device and she sends information to Bob using the harvested energy. This communication is intercepted by an eavesdropper Eve. The received signal at Bob is given by

$$y_b = \sqrt{P_x} h_b s + n_b \tag{4.179}$$

where P_x is the transmitted power of Alice, h_b is the channel gain from Alice to Bob, s is the normalized transmitted symbol, and n_b is the AWGN with mean zero and variance $2\sigma^2$ at Bob. Similarly, the transmitted symbol will also be intercepted by Eve to have

$$y_e = \sqrt{P_x} h_e s + n_e \tag{4.180}$$

with an independent channel gain of h_e and noise of n_e. The secrecy rate is defined as

$$C_s = \max \left\{ \log_2 \left(1 + \frac{P_x |h_b|^2}{2\sigma^2} \right) - \log_2 \left(1 + \frac{P_x |h_e|^2}{2\sigma^2} \right), 0 \right\}. \tag{4.181}$$

In conventional communications, only the channel gains are random so that C_s needs to be averaged over the fading distributions to obtain the ergodic secrecy rate. In energy harvesting communications, both the channel gains and the transmission power can be random so that the averaging is performed over the channel gains as well as P_x. This is similar to the error rate analysis in Section 4.2.3. Also, the secrecy outage probability is defined as the probability that the rate in the desired channel is larger than that in the eavesdropping channel. Unlike the secrecy rate, energy harvesting does not change this secrecy outage probability.

On the other hand, if Alice is not an energy harvesting device but Bob is so that Alice needs to transmit her information as well as transfer a certain amount of energy to Bob so that Bob can harvest the energy for a response, the secrecy rate becomes

$$C_s = \max \left\{ \log_2 \left[1 + \frac{(P_x + P_0)|h_b|^2}{2\sigma^2} \right] - \log_2 \left[1 + \frac{(P_x + P_0)|h_e|^2}{2\sigma^2} \right], 0 \right\} \tag{4.182}$$

where P_x is the power used to guarantee the quality of information decoding at Bob, while P_0 is the extra power required by Bob for harvesting. Since the logarithm operation is non-linear, in general, the value of P_0 will lead to a different secrecy rate. This is the second effect of energy harvesting on the physical layer security.

There are many studies on other security issues in the literature. For example, in energy harvesting communications systems, some nodes are dedicated to energy harvesting while other nodes are dedicated to information decoding. Hence, the energy receivers could be potential eavesdroppers, or internal eavesdroppers (Pan et al. 2016). One may use interference alignment to nullify the signals for enhanced security (Zhao et al. 2017a). They are not discussed here.

4.6.3 Non-orthogonal Multiple Access

NOMA is one of the most recent advances in wireless communications. It aims to solve the spectral efficiency problem by using non-orthogonal channels in the power or code domain, compared with the conventional orthogonal multiple access (OMA) systems. In the following, we focus on the power domain NOMA.

In the power domain NOMA, signals for multiple users are superposed and transmitted over the same time slot, the same frequency, or the same code (Ding et al. 2017a). Multiple access is achieved by allocating different power coefficients to different users. Consider a two-user system as an example. The superposed signal is given by $a_1 s_1 + a_2 s_2$, where a_1 and a_2 are the power coefficients for user 1 and user 2 with $a_1^2 + a_2^2 = 1$, respectively, s_1 and s_2 are the transmitted signals for user 1 and user 2, respectively, with unit power. This signal is transmitted by the access point or base station. The received signal at user 1 is

$$y_1 = h_1 \sqrt{P_x}(a_1 s_1 + a_2 s_2) + n_1 \tag{4.183}$$

where h_1 is the channel gain from the base station to user 1 and n_1 is the AWGN with mean zero and variance $2\sigma^2$. The received signal at user 2 is

$$y_2 = h_2 \sqrt{P_x}(a_1 s_1 + a_2 s_2) + n_2 \tag{4.184}$$

where h_2 is the channel gain from the base station to user 2 and n_2 is the AWGN with mean zero and variance $2\sigma^2$.

The key idea of power domain NOMA is to allocate more power to the user with poorer channel condition. Thus, channel state information must be available. Assume that $|h_1| \leq |h_2|$. Then, $a_1 \geq a_2$. In this case, user 1 will decode its received signal y_1 directly so that its rate is

$$R_1 = \log_2\left(1 + \frac{|h_1|^2 a_1^2}{|h_1|^2 a_2^2 + \frac{2\sigma^2}{P_x}}\right) \tag{4.185}$$

and user 2 will decode user 1's signal first and then use successive interference cancellation to decode its own signal so that its rate is

$$R_2 = \log_2\left(1 + \frac{P_x |h_2|^2 a_2^2}{2\sigma^2}\right) \tag{4.186}$$

assuming perfect cancellation. The sum rate is therefore

$$R = \log_2\left(1 + \frac{|h_1|^2 a_1^2}{|h_1|^2 a_2^2 + \frac{2\sigma^2}{P_x}}\right) + \log_2\left(1 + \frac{P_x |h_2|^2 a_2^2}{2\sigma^2}\right). \tag{4.187}$$

It can be shown that this sum rate is larger than the sum rate in OMA in many cases. Thus, NOMA can achieve higher spectral efficiency. For more than two users, similar procedures can be applied, where the channel gains are first ordered and the power coefficients are assigned based on the order. Based on its power coefficient, each user either decodes its own signal directly by treating other users as interference or adopts successive interference cancellation before decoding.

For energy harvesting communications, if the access point or the base station harvest energy, the first effect of energy harvest is that the transmission power P_x may become

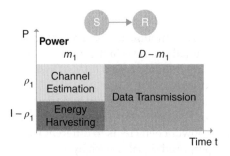

Figure 4.28 The data packet structure that splits pilots.

a random variable. Hence, the sum rate R is randomly varying, even if the channel gains are static. This is applicable to all energy harvesting communications systems where the access point harvests energy from the sun, wind or RF signals, etc. The energy arrival process will affect the power allocation within users and therefore the efficiency of NOMA. If the users harvest the RF signal from the base station, the second effect of energy harvesting is that this may change their order of decoding, depending on the portion of energy they harvest. Both effects can be examined by using a modified version of the sum rate. For example, the average sum rate can be obtained over P_x before the optimization of a_1 and a_2. Power splitting can be applied to y_1 and y_2 before deriving the rates.

Also, from the energy harvesting point of view, the non-orthogonal channels in NOMA actually provide more energy than the orthogonal channels in OMA, due to the extra multi-user interference. Interference degrades data performance but provides an extra source of energy.

4.6.4 Joint Detection and Estimation

The purpose of energy harvesting estimation is to provide the channel estimates for signal detection. Thus, signal detection and channel estimation need to be jointly considered with energy harvesting, similar to Section 4.4.2. In this case, there are three parts in the data packet that need to be balanced, as illustrated in Figure 4.25: energy harvesting; channel estimation; and data transmission. When the total resource is limited, there exists an optimum tradeoff between them. In this subsection, using a relaying system, we will discuss more cases of the tradeoff between data transmission, channel estimation and energy harvesting in a fixed data packet. The data packet structures studied in this subsection are shown in Figures 4.29, and 4.30. These data packets are for the source-to-relay link consisting of three parts: pilots for channel estimation; pilots for energy harvesting; and data for signal detection. For the relay-to-destination link, the data packet only has two parts, pilots for channel estimation and data for signal detection, as the destination does not need any energy harvesting.

Figure 4.28 shows data packet structure where the pilots for channel estimation have been split in power. In this structure, the source node only sends one group of pilots but each pilot is split in power for both energy harvesting and channel estimation. The data transmission happens in a time division manner. Figure 4.29 shows the data packet structure where the data symbols have been split in power. In this structure, the source sends one group of pilots for channel estimation, but the data symbols are split in power for both energy harvesting and data transmission. Figure 4.30 shows the data packet

Figure 4.29 The data packet structure that splits data.

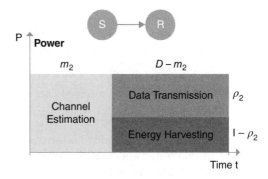

Figure 4.30 The data packet structure that splits both pilots and data.

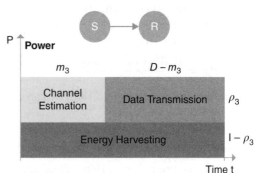

structure where both pilots and data symbols have been split in power. In this structure, the source sends one group of pilots for channel estimation and one group of data symbols for information delivery. The energy is harvested by splitting all symbols in the data packet. This structure is similar to Figure 4.25.

The considered system is an amplify-and-forward relaying system with one source, one relay, and one destination, which means the signal is only amplified at the relay before being forwarded. There are two hops: the hop from source to relay (SR); and the hop from relay to destination (RD). The first hop assumes the structures in Figures 4.28–4.30, while the second hop only has one part for channel estimation and one part for data transmission without any energy harvesting. There is no direct link between source and destination due to obstacles. Time division is used in all the structures to achieve orthogonal channels. Therefore, the source first sends the data packet to the relay, and then the relay sends the data packet to the destination. Also, a total of D symbols are used in each structure for channel estimation, data transmission, and energy harvesting. Each symbol occupies a time duration of T seconds. All block fading channels are Rayleigh. All the values of m_1, m_2, and m_3 are integers and smaller than D. Also, $0 \leq \rho_1, \rho_2, \rho_3 \leq 1$.

In the first structure, there are three parts in the first hop: pilots for channel estimation and energy harvesting, and data symbols for information transmission. First, one has the received pilots at the relay for channel estimation as

$$y_r^{(i_1)} = \sqrt{\rho_1 P_{s1}} h_1 x^{(i_1)} + n_1^{(i_1)} \tag{4.188}$$

where $i_1 = 1, 2, \ldots, m_1$ represent the pilots in the data packet, ρ_1 is the power splitting factor, $0 < m_1 < D$ is an integer, P_{s1} is the transmitted power of the source, h_1 is the fading gain in the SR channel and is complex Gaussian with mean zero and variance $2\alpha^2$, $x^{(i_1)} = 1$ is the transmitted pilot, and $n_1^{(i_1)}$ is the complex AWGN with mean zero and noise power $2\sigma^2$. The received pilots for energy harvesting can be expressed as

$$y_r^{(i_1)} = \sqrt{(1-\rho_1)P_{s1}}h_1 x^{(i_1)} + n_1^{(i_1)}. \tag{4.189}$$

In the same structure, the received data symbols are expressed as

$$y_r^{(j_1)} = \sqrt{P_{s1}}h_1 x^{(j_1)} + n_1^{(j_1)} \tag{4.190}$$

where $j_1 = m_1 + 1, \ldots, D$, and $x^{(j_1)}$ is the transmitted data symbol with unit power. Using (4.189), the harvested energy at the relay is derived as $E_{r1} = \eta P_{s1}|h_1|^2(1-\rho_1)m_1$, where η is the conversion efficiency of the energy harvester and we have assumed $T = 1$ for simplicity. The harvested energy will be used to transmit m_1 pilots to the destination for the channel estimation and $D - m_1$ data symbols from the source in the second hop in order to maintain the data rate. Thus, the transmission power of the relay is

$$P_{r1} = \frac{\eta P_{s1}|h_1|^2(1-\rho_1)m_1}{D}. \tag{4.191}$$

Also, using the signals in (4.188), we can get an estimate of h_1 as

$$\hat{h}_1 = h_1 + \varepsilon_1 \tag{4.192}$$

where $\varepsilon_1 = \frac{\sum_{i_1=1}^{m_1} n_1^{(i_1)}}{m_1 \sqrt{\rho_1 P_{s1}}}$ is the estimation error. Thus, one has $h_1 = \hat{h}_1 - \varepsilon_1$.

The received signal in (4.190) will be amplified and forwarded to the destination by using the harvested energy. The amplification factor can be written as

$$\hat{a}_{1var}^{\,2} = \frac{1}{P_{s1}|\hat{h}_1|^2 + 2\sigma^2} \tag{4.193}$$

where $\hat{h}_1$ is the estimated channel gain for the first hop between source node and relay node in (4.192).

In the second hop, the signal is transmitted from the relay to the destination. In addition to sending m_1 pilots to the destination for channel estimation, the relay also forwards the $D - m_1$ data symbols from the source to the destination. In this hop, the received pilots for channel estimation at the destination node can be written as

$$y_d^{(i_2)} = \sqrt{P_{r1}}g_1 \hat{a}_{1var} x^{(i_2)} + n_2^{(i_2)} \tag{4.194}$$

where $i_2 = 1, 2, \ldots, m_1$, $x^{(i_2)} = 1$ is the pilot value, $n_2^{(i_2)}$ is the AWGN with zero-mean and variance $2\sigma^2$, $\hat{a}_{1var}$ is the amplification factor, P_{r1} is the relay transmission power given in (4.191), and g_1 is the fading channel coefficient of the RD link and is a complex Gaussian random variable with mean zero and variance $2\alpha^2$.

Also, the received signals of the data symbols at the destination are given by

$$y_d^{(j_2)} = \sqrt{P_{r1}}g_1 \hat{a}_{1var}(\sqrt{P_{s1}}h_1 x^{(j_1)} + n_1^{(j_1)}) + n_2^{(j_2)} \tag{4.195}$$

where $n_2^{(j_2)}$ is the AWGN at the destination node with zero mean and variance $2\sigma^2$.

By using the received signals in (4.194), the channel gain of the RD link can be estimated as

$$\hat{g}_1 = \frac{\sqrt{P_{r1}}}{\sqrt{\hat{P}_{r1}}}g_1 + \varepsilon_2 \tag{4.196}$$

where $\varepsilon_2 = \dfrac{\sum_{i_2=1}^{m_1} n_2^{(i_2)}}{m_1 \hat{a}_{1var} \sqrt{\hat{P}_{r1}}}$ and $\hat{P}_{r1} = \dfrac{\eta P_{s1} |\hat{h}_1|^2 (1-\rho_1) m_1}{D}$.

By using the channel estimates in (4.195), the received signal at the destination can be rewritten as

$$
\begin{aligned}
y_d^{(j_2)} = & \sqrt{\hat{P}_{r1}} \hat{g}_1 \hat{a}_{1var} \sqrt{P_{s1}} \hat{h}_1 x^{(j_1)} \\
& - \sqrt{\hat{P}_{r1}} \hat{g}_1 \hat{a}_{1var} \sqrt{P_{s1}} \varepsilon_1 x^{(j_1)} \\
& + \sqrt{\hat{P}_{r1}} \hat{g}_1 \hat{a}_{1var} n_1^{(j_1)} \\
& - \sqrt{\hat{P}_{r1}} \sqrt{P_{s1}} \hat{h}_1 \hat{a}_{1var} \varepsilon_2 x^{(j_1)} \\
& + \sqrt{\hat{P}_{r1}} \sqrt{P_{s1}} \hat{a}_{1var} \varepsilon_1 \varepsilon_2 x^{(j_1)} \\
& - \sqrt{\hat{P}_{r1}} \varepsilon_2 \hat{a}_{1var} n_1^{(j_1)} + n_2^{(j_2)}.
\end{aligned}
\tag{4.197}
$$

From the above, the end-to-end SNR expression can be derived as

$$
\gamma_{1end} = \frac{P_{s1} |\hat{g}_1|^2 |\hat{h}_1|^2}{\upsilon_1}
\tag{4.198}
$$

where $\upsilon_1 = P_{s1} |\hat{g}_1|^2 \varepsilon_{1var} + 2\sigma^2 |\hat{g}_1|^2 + P_{s1} \varepsilon_{2var} |\hat{h}_1|^2 + P_{s1} \varepsilon_{1var} \varepsilon_{2var} + 2\sigma^2 \varepsilon_{2var} + \dfrac{2\sigma^2}{\hat{P}_{r1} \hat{a}_{1var}^{\ 2}}$
and $E[|\varepsilon_1|^2] = \varepsilon_{1var}$, $E[|\varepsilon_2|^2] = \varepsilon_{2var}$.

To derive the cumulative distribution function of the end-to-end SNR, we first calculate $Var(\varepsilon_1)$ and $Var(\varepsilon_2)$. One has already had

$$
\varepsilon_1 = \frac{\sum_{i_1=1}^{m_1} n_1^{i_1}}{m_1 \sqrt{\rho_1 P_{s1}}}.
\tag{4.199}
$$

It has a mean of zero. Also, its variance is

$$
Var(\varepsilon_1) = E\{|\varepsilon_1|^2\} = \frac{2\sigma^2}{m_1 \rho_1 P_{s1}}.
\tag{4.200}
$$

Similarly, the variance of ε_2 can be calculated as

$$
Var(\varepsilon_2) = \frac{2\sigma^2 D (P_{s1} |\hat{h}_1|^2 + 2\sigma^2)}{[\eta P_{s1} |\hat{h}_1|^2 (1 - \rho_1) m_1] m_1}.
\tag{4.201}
$$

These two equations can be used in the end-to-end SNR.

Further, $|\hat{h}_1|^2$ is an exponential random variable with scale parameter

$$
\lambda_1 = \frac{1}{2\alpha^2 + \frac{2\sigma^2}{m_1 \rho_1 P_{s1}}}.
\tag{4.202}
$$

The probability density function (PDF) of $|\hat{h}_1|^2$ can be written as

$$
f_{|\hat{h}_1|^2}(x) = \lambda_1 e^{-\lambda_1 x}.
\tag{4.203}
$$

Its CDF is

$$
F_{|\hat{h}_1|^2}(x) = 1 - e^{-\lambda_1 x}.
\tag{4.204}
$$

Similarly, let $\lambda_2 = \dfrac{1}{2a^2 + \frac{|2\sigma^2|}{m_1\rho_1 P_{s1}} + \frac{|2\sigma^2||m_2+D-m_1|\rho_1|}{|m_2|\eta|1-\rho_1|m_1}}$. Its PDF can be approximated as

$$f_{|\tilde{g}_1|^2}(x) = \lambda_2 e^{-\lambda_2 x} \tag{4.205}$$

and its CDF can be approximated as

$$F_{|\tilde{g}_1|^2}(x) = 1 - e^{-\lambda_2 x}. \tag{4.206}$$

By using these expressions, the CDF of γ_{1end} can be derived as

$$F_{\gamma_{1end}}(x) = P_{1r}\{\gamma_{1end} < x\} = I_{11} + I_{12} \tag{4.207}$$

with

$$I_{11} = 1 - e^{-\frac{\frac{2\sigma^2 x}{m_1\rho_1 P_{s1}} + \frac{2\sigma^2 x}{P_{s1}}}{2a^2 + |\frac{2\sigma^2}{m_1\rho_1 P_{s1}}|}} \tag{4.208}$$

and

$$I_{12} = \frac{1}{P_{s1}\left(2a^2 + \frac{2\sigma^2}{m_1\rho_1 P_{s1}}\right)} e^{-\frac{2\sigma^2 x + 2\sigma^2 xm_1\rho_1}{2a^2 m_1\rho_1 P_{s1} + 2\sigma^2}} \frac{2a^2 m_1\rho_1 P_{s1} + 2\sigma^2}{m_1\rho_1}$$

$$- \frac{1}{\left(2a^2 + \frac{2\sigma^2}{m_1\rho_1 P_{s1}}\right)} \tag{4.209}$$

$$\cdot e^{-\frac{\lambda_2 m_1 x 2\sigma^2 \rho_1(D)}{\eta\rho_1 m_1(1-\rho_1)m_1^2 P_{s1}y} - \frac{2\sigma^2 x + 2\sigma^2 xm_2}{2a^2 m_1\rho_1 P_{s1} + 2\sigma^2}}$$

$$\cdot \frac{2}{P_{s1}} \left(\frac{z_1(x)}{(\eta m_1(1-\rho_p)m_1^2)m_1}\right)^{\frac{1}{2}} K_1\left(2\sqrt{\frac{z_2(x)}{w_4}}\right)$$

where $w_4 = (\eta m_1(1-\rho_1)m_1^2)(2a^2 m_1\rho_1 P_{s1} + 2\sigma^2)$, $z_1(x) = (\dfrac{1}{2a^2 + \frac{2\sigma^2}{m_1\rho_1 P_{s1}} + \frac{2\sigma^2 D\rho_1}{m_1\eta|1-\rho_1|m_1}})(4\sigma^4 xD + $

$4\sigma^4 m_1\rho_1 xD + 2\sigma^2 xm_1 m_1\rho_1 D + 2\sigma^2 m_1 xD * \frac{2\sigma^2 x\rho_1}{m_1\rho_1} + 2\sigma^2 x\rho_1)$ $(2a^2 m_1\rho_1 P_{s1} + 2\sigma^2)$ and

$z_2(x) = (\dfrac{1}{2a^2 + \frac{2\sigma^2}{m_1\rho_1 P_{s1}} + \frac{2\sigma^2 D\rho_1}{m_1\eta|1-\rho_1|m_1}})$ $(4\sigma^4 xD + 4\sigma^4 m_1\rho_1 xD$ $+2\sigma^2 xm_1 m_1\rho_1 D + 2\sigma^2 m_1 xD * $

$\frac{2\sigma^2 x\rho_1}{m_1\rho_1} + 2\sigma^2 x\rho_1)m_1$.

The outage probability can then be derived as

$$P_0(\gamma_{01}) = F_{\gamma_{1end}}(\gamma_{01}). \tag{4.210}$$

Moreover, the BER can be calculated as

$$BER_1 = \frac{1}{2}\int_0^\infty \frac{e^{-x}}{\sqrt{x * \pi}} F_{\gamma_{1end}}(x)dx. \tag{4.211}$$

where erfc(x) is the complementary error function. The structures in Figures 4.29 and 4.30 can be analyzed in a similar way. The analysis is ignored here for brevity.

Figures 4.31–4.34 show the performances of the first structure in Figure 4.28 for different parameters. One can see that there are optimum values of m and optimum values of ρ_1 in all the cases considered. The SNRs will change the locations of the optimum

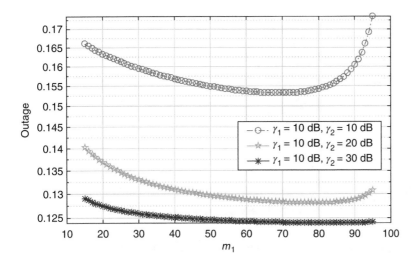

Figure 4.31 Outage versus m_1.

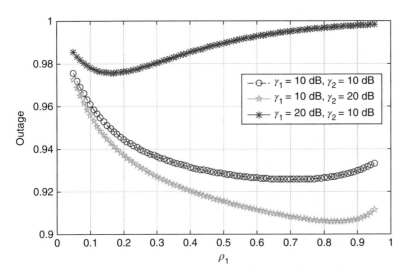

Figure 4.32 Outage versus ρ_1.

values. This is expected, as a larger value of m_1 leads to more pilots for a more accurate channel estimate but this reduces the number of data symbols that can be transmitted. Similarly, a larger value of ρ_1 allows more energy for channel estimation for a more accurate channel estimate but it also reduces the amount of energy harvested and hence a smaller transmission power for the relay, leading to poorer performances. Similar trade-offs exist for the other two data packet structures. Thus, for energy harvesting detection and estimation, it is important to allocate the appropriate amount of energy for channel estimation, energy harvesting and data transmission to achieve the optimum performances.

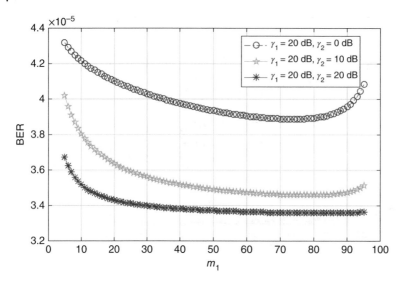

Figure 4.33 BER versus m_1.

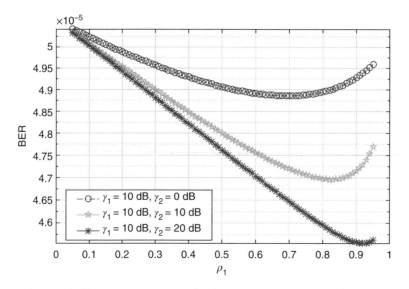

Figure 4.34 BER versus ρ_1.

4.7 Summary

In this chapter, we have discussed several main physical layer techniques in energy harvesting communications systems.

First, the effect of energy harvesting on the physical layer techniques in conventional communications systems has been examined. The examination has shown that, if the fixed-power transmission strategy is employed so that the energy harvesting device accumulates energy to transmit signals at a fixed power, there is an optimum number of time slots or harvesting time the device can choose to balance the requirement for transmission power and transmission delay. If the variable-power transmission strategy is used so that the energy harvesting device transmits as soon as the energy is harvested, the BER and rate performances will be degraded due to the random variation in the transmission power.

Secondly, new signal detectors, new channel estimators and new transmission waveforms for energy harvesting have been designed. For signal detection, the random transmission power must be considered in the detector design when it is unknown. For channel estimation, energy harvesting reduces the resources used for channel estimation and hence, an optimum tradeoff between energy harvesting, channel estimation and data transmission exists, when the total amount of resources is fixed. For waveform designs, if a linear energy harvester is applied or a single sum signal is used, the optimum strategy is always to put all the transmission power at the source with the best channel condition, while if a non-linear energy harvester is applied for a sum of harvested energy, the best strategy is to allocate different amplitudes to different sources, while the phase of the waveform can be arbitrary.

Finally, several other important physical layer issues have also been discussed briefly. For the circuit power, if it is considered in the design, it normally degrades the rate performance, as part of the harvested energy needs to be used for circuit power consumption. For the physical layer security, the random variation in the transmission power can affect the secrecy rate, while for NOMA, the decoding order and the power allocation of different users can be changed due to energy harvesting. For joint detection and estimation, the balance between channel estimation, energy harvesting and data transmission within a fixed data packet is required to achieve optimum performances.

To summarize, the physical layer techniques in energy harvesting communications have two main characteristics. The first characteristic is that the transmission power or the transmission time of an energy harvesting device can be random, due to the random energy arrival process. The second characteristic is that the resources for other functions in the physical layer can be reduced, due to the addition of the energy harvesting function, if the total resource is limited. In the next chapter, we will discuss the upper layer issues in energy harvesting wireless communications systems.

5

Upper Layer Techniques

5.1   Introduction

The upper layers refer to the layers in the communications model, as shown in Figure 4.1, that are immediately above the physical layer. In the open systems interconnection (OSI) model, these are the link layer that provides link control and the network layer that provides message routing. In the transmission control protocol/internet protocol (TCP/IP) model, they are the upper part of the network interface layer that provides link control and the internet layer that provides message routing. In both models, these layers are the layers that are the closest to the physical layer and therefore they have significant impact on the performances of transmission, detection and estimation in the physical layer. Thus, they will be considered in this chapter for energy harvesting wireless communications.

In the upper layers, the link layer provides control functions for the successful operation of the physical layer and sometimes, it is called the media access control (MAC) layer. These control functions or the MAC protocols coordinate the transmission and reception in the physical layer for each node by performing resource allocation and scheduling. For example, which user should be activated in multi-user scheduling, which channel should be assigned in access control, and how long and how much power each user should use in duty cycle and power management. These functions are vital for the smooth operation of the physical layer. In energy harvesting wireless communications, such resource allocation and scheduling have additional constraints from the energy availability. For example, some user may be activated and assigned a channel but may not have enough energy to perform data transmission. Thus, the MAC protocol designs should take the energy availability into account to maximize the overall efficiency in energy harvesting wireless communications.

The network layer in the OSI model or the internet layer in the TCP/IP model provides routing functions. In many communications systems, the source and the destination may be too far away from each other or may belong to different networks so that a proper route between them needs to be set up to allow their information exchange. The choice of the route will certainly affect the performance of the physical layer, as different routes may have different channel qualities as well as different end-to-end delay or latency due to different traffic. For energy harvesting wireless communications, new routing protocols should be designed, as the nodes on the route chosen may not have enough energy to forward the message so that the energy availability should be accounted for in the routing metric.

Energy Harvesting Communications: Principles and Theories, First Edition. Yunfei Chen.
© 2019 John Wiley & Sons Ltd. Published 2019 by John Wiley & Sons Ltd.

We will focus on the MAC protocols and the routing protocols for energy harvesting wireless communications in this chapter. There are two main changes brought up by energy harvesting. Firstly, unlike battery power or mains connection, the energy harvester could provide unlimited power. Secondly, unlike battery power or mains connection, the harvested power is not stable or has great uncertainty. The new MAC protocols and routing protocols for energy harvesting wireless communications need to take these two changes into account in their designs. We will discuss the MAC protocols for energy harvesting communications first. Then, we will discuss the routing protocols for energy harvesting communications. Finally, we will discuss other important issues for energy harvesting communications, such as scheduling and effective capacity.

5.2 Media Access Control Protocols

Since the MAC layer is the layer closest to the physical layer, it has the largest impact on the physical layer transmission and reception. Also, a wireless sensor network (WSN) is an important application of energy harvesting wireless communications. Thus, we will focus on the MAC layers for a WSN in this section. For a WSN, due to its low-power operation and large-scale deployment, it is very difficult to recharge or replace the batteries in most cases. Hence, one primary objective of a WSN is to extend its network lifetime. Many methods have been proposed to extend the lifetime of a WSN. For example, power management at the nodes can be introduced to use the energy more efficiently, but this only delays the drainage of the battery. Incremental deployment can be used to replace the old nodes with drained battery by new nodes but this is not environmentally sustainable. Energy harvesting is a promising solution to this issue.

There are many MAC protocols proposed for conventional WSNs without energy harvesting, and they can be categorized into synchronous and asynchronous protocols. The synchronous protocols include S-MAC (Ye et al. 2002), T-MAC (Dam and Langendoen 2003) and the beacon mode of IEEE 802.15.4. They synchronize the transmitting node and the receiving node so that the nodes have their active or sleep states aligned in the time domain. This reduces the listening time but incurs extra overhead on maintaining the synchronization. The asynchronous protocols include B-MAC (Polastre et al. 2004), X-MAC (Buettner et al. 2006) and RI-MAC (Sun et al. 2008), where different nodes sleep and activate at different times independently. This reduces the hardware and overhead requirements, but such an approach requires a long listening time at the transmitter to wait for the receiver to wake up in some cases. These are all based on battery power.

Energy harvesting can extend the network lifetime but also bring great uncertainty. In energy harvesting WSNs, the amount of energy harvested at different nodes may vary. Hence, the synchronous protocols may not work, as some nodes may become unusable due to insufficient energy during the communications. On the other hand, the asynchronous protocols allow nodes to operate independently and hence can be adapted to energy harvesting WSNs. The main concern in energy harvesting WSNs is the uncertainty of the energy supply, similar to the physical layer. Thus, most MAC protocol designs focus on the energy availability.

5.2.1 Duty Cycling

As mentioned before, the network lifetime is the primary objective in WSNs. Thus, many methods have been proposed to save energy so that the network lifetime can be prolonged. For example, transmission power control can be implemented at the sensing node so that a lower transmission power can be used whenever possible (Ramanathan and Hain 2000). Also, dynamic voltage scaling can be used to improve the sensing circuit for higher energy efficiency. These methods require hardware modifications to the commonly used nodes. Alternatively, one can keep the power constant and use the common circuit but control the transmission time to save energy. This leads to the duty cycling method. In the duty cycling method, there is at least one operational mode when the node consumes no energy or negligible energy, such as a sleep mode. The node goes to sleep immediately after it finishes the communications task to save energy. The smaller the duty cycle is or the longer the sleep time is, the more energy the node can save but the less tasks the node can perform. One needs to strike a balance between the sleep mode and the operation mode. This boils down to the design of the duty cycle in the MAC protocols so that the sensing node can accomplish the task while saving as much energy as possible.

In energy harvesting wireless communications, the uncertainty in energy supply brings extra challenges to the design. Specifically, the sensing node may want to transmit data but due to insufficient energy harvested, the transmission is either not possible or given up in the middle, causing an energy outage. Thus, the duty cycle should be adapted to the residual energy or the energy arrival process at the node. This is the main difference between duty cycle designs for energy harvesting systems and those for conventional systems. To this end, several duty cycle designs have been studied. They are slightly different depending on the source of energy harvested. For wireless power transfer, since this transfer is intentional, there is less uncertainty in the energy supply but the charging sequence for a mobile charger (Peng et al. 2010) or the fairness for a fixed charger (Kim et al. 2011) should be accounted for. For ambient energy harvesting, these energy sources are less controllable and hence there is more uncertainty in the energy supply so that duty cycle adjustment is more important (Hsu et al. 2006; Yoo et al. 2012). We start with the MAC protocol using wireless power transfer (Kim et al. 2011).

5.2.1.1 Wireless Power Transfer

In Kim et al. (2011), an experimental system was built and simulated. Later in Kim and Lee (2011), the performance of this system was analyzed. This protocol was called energy adaptive MAC (EA-MAC). Specifically, consider a WSN using radio frequency (RF) power transfer, which assumes a star topology with one master node and I slave nodes. At each node, there is a set of RF front dedicated to data transmission and a separate set of RF front dedicated to power transfer so that energy harvesting and data transmission can be performed at the same time without interfering with each other. The master node operates with fixed power connection and broadcasts RF power to the slave nodes. It is always active and ready to receive data from the slave nodes. The slave nodes keep harvesting the RF energy. For data transmission, they operate in two states: the sleep state when they switch off; and the active state when they transmit data. A contention-based carrier sensing multiple access with collision avoidance (CSMA/CA) is used so that in the active state the slave node only transmits the data when it acquires the channel. Otherwise, it goes back to sleep. The whole process is described by Figure 5.1.

The master node has a fixed location. Hence, the energy harvested at each slave node is different, as according to the Friis formula, the received power is proportional to the inverse of the squared distance so that slave nodes closer to the master node can

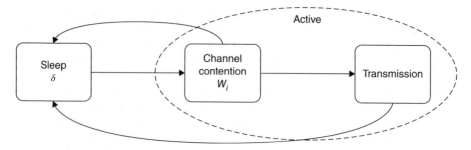

Figure 5.1 The state transition process of each node.

harvest more energy and vice versa. Consequently, the amount of energy available at different slave nodes will be different. This causes unfairness, as for the same amount of data, some slave nodes may not be able to complete the transmission due to insufficient energy. Also, due to insufficient energy, some slave nodes may have less chance to access the channel in CSMA/CA.

The EA-MAC protocol aims to tackle this fairness issue by introducing two additional changes to the MAC protocol. First, the duty cycle is adaptive to the harvested energy level at each slave node. More specifically, when the slave node wakes up from the sleep mode, it will check if its harvested energy reaches a threshold δ. This threshold is the amount of energy required to transmit one data packet. If it does, it moves to the active mode and starts to contend for the channel. If it acquires the channel, it will finish the transmission and then go back to the sleep mode. If the channel is busy or its energy is below δ, it will go back to the sleep mode and will keep checking and harvesting energy until it reaches δ before contending for the channel again. Secondly, the CSMA/CA algorithm is adaptive to the harvested energy level too. The maximum number of backoff slots will be set to $w_i * 2^{B_i} - 1$, where B_i is the backoff exponent in the usual CSMA/CA but w_i is a weighting factor as the ratio of the harvested energy level at the ith node to the average harvested energy among all slave nodes. This is compared with the conventional CSMA/CA algorithm with a maximum number of backoff slots of $2^{B_i} - 1$ at the ith slave node.

The use of δ ensures that all slave nodes transmit at the same energy level to avoid interrupted or impossible transmission due to insufficient energy. It changes the duty cycle to level up the harvested energy at different slave nodes, or it compensates the spatial difference among nodes with different duty cycles. The use of w_i ensures that nodes with less harvested energy can have a smaller backoff time so that they will not spend too much energy on contending the channel. It protects slave nodes with less harvested energy from being disadvantaged in channel access, as nodes with less energy can check the channel status less frequently than nodes with more energy so that eventually their energy levels will reach an equilibrium.

Adding a few more assumptions, the performance of EA-MAC can be analyzed. In this case, assume that all nodes can hear from each other well so that there is no hidden terminal problem. The data packet at each node has the same size and each slave node has only one data packet to transmit at a time. Each slave node has a deterministic energy harvesting rate determined by the distance only, due to the constant power broadcast from the master node. Define the time interval between two contentions as a round.

During the jth round, the ith slave node spends a time period of $T_{s,i,j}$ in the sleep mode, $T_{c,i,j}$ on the channel contention, and $T_{t,i,j}$ on packet transmission. Thus, the throughput of the ith slave node is given by (Kim and Lee 2011)

$$S_i = \frac{d \sum_{j=1}^{J} T_{p,i,j}}{\sum_{j=1}^{J} T_{c,i,j} + \sum_{j=1}^{J} T_{t,i,j} + \sum_{j=1}^{J} T_{s,i,j}} \tag{5.1}$$

where J is the total number of rounds, $T_{p,i,j}$ is the time interval of successful data transmission during the jth round, and d is the data rate.

Further, in the EA-MAC protocol, the slave node only wakes up to check the channel status and performs data transmission when its energy reaches the threshold δ determined by the amount of energy used to transmit one data packet. Thus, one has at the ith slave node

$$(P_i - P_s)T_{s,i,j} = (P_c - P_i)T_{c,i,j} + (P_t - P_i)T_{t,i,j} = \delta \tag{5.2}$$

where $P_i = \eta P_{tx} G_t G_r \left(\frac{\lambda}{4\pi d_i}\right)^2$ is the received energy at the ith slave node, η is the conversion efficiency of the energy harvester, P_{tx} is the master node transmission power, G_t and G_r are the antenna gains at the master and slave nodes, respectively, λ is the wavelength for power transfer, d_i is the distance between the master node and the ith slave node, P_s is the power consumption during the sleep mode, P_c is the power consumption for channel contention, P_t is the power consumption for data transmission, and other symbols are defined as before. Using (5.2) in (5.1), one has

$$S_i = \frac{d \sum_{j=1}^{J} T_{p,i,j}}{(1 + \alpha_i) \sum_{j=1}^{J} T_{c,i,j} + (1 + \beta_i) \sum_{j=1}^{J} T_{t,i,j}} \tag{5.3}$$

where $\alpha_i = \frac{P_c - P_i}{P_i - P_s}$ and $\beta_i = \frac{P_t - P_i}{P_i - P_s}$. Assuming that $\bar{T}_{p,i}$, $\bar{T}_{c,i}$, and $\bar{T}_{t,i}$ are the averages of $T_{p,i,j}$, $T_{c,i,j}$, $T_{t,i,j}$ across J rounds so that $\bar{T}_{p,i}J = \sum_{j=1}^{J} T_{p,i,j}$, $\bar{T}_{c,i}J = \sum_{j=1}^{J} T_{c,i,j}$, and $\bar{T}_{t,i}J = \sum_{j=1}^{J} T_{t,i,j}$. One further has

$$S_i = \frac{d \bar{T}_{p,i}}{(1 + \alpha_i)\bar{T}_{c,i} + (1 + \beta_i)\bar{T}_{t,i}}. \tag{5.4}$$

Thus, one needs the time averages to calculate the throughput.

Using a 2-D discrete-time Markov chain to describe the backoff process, the average contention time can be calculated as (Kim and Lee 2011)

$$\bar{T}_{c,i} = \frac{W}{2} \sum_{v=0}^{m-1} q_i^v (1 - q_i)[w_i 2^{B_{min}}(2^{v+1} - 1) - v - 1]$$

$$+ \frac{W}{2} q_i^m [w_i 2^{B_{min}}(2^{m+1} - 1) - m - 1] \tag{5.5}$$

where W is the time duration of each backoff slot, w_i is the weighting factor, m is the maximum number of backoff slots minus 1, q_i is the probability of a busy channel during contention at the ith slave node, and $B_{min} = \min\{B_i\}$. The details of the calculation can be found in Kim and Lee (2011). Similarly, one has

$$\bar{T}_{t,i} = T(1 - q_i^{m+1}) \tag{5.6}$$

and

$$\bar{T}_{p,i} = q_{s,i} T (1 - q_i^{m+1}) \tag{5.7}$$

where T is the time duration of one data packet and is a constant, q_i is still the probability of a busy channel during contention at the ith slave node, and $q_{s,i}$ is the probability of successful transmission without collision at the ith slave node. These two probabilities can be calculated as

$$q_i = 1 - \prod_{j \in \Phi, j \neq i} (1 - q_{t,j}) \tag{5.8}$$

with

$$q_{t,i} = \frac{\bar{T}_{t,i}}{(1 + \alpha_i) \bar{T}_{c,i} + (1 + \beta_i) \bar{T}_{t,i}} \tag{5.9}$$

and

$$q_{s,i} = \prod_{j \in \Phi, j \neq i} (1 - \tau_j q_{c,j}) \tag{5.10}$$

with

$$q_{c,i} = \frac{\bar{T}_{c,i}}{(1 + \alpha_i) \bar{T}_{c,i} + (1 + \beta_i) \bar{T}_{t,i}} \tag{5.11}$$

and

$$\tau_i = \frac{2(1 - 2q_i)(1 - q_i^{m+1})}{w_i 2^{B_{min}} (1 - (2q_i)^{m+1})(1 - q_i) + (1 - q_i^{m+1})(1 - 2q_i)} \tag{5.12}$$

where Φ represents the set of slave nodes that contend for the channel with the ith node.

From the above, the throughput of the ith slave node can be calculated by following an iterative procedure. First, initial values for q_i and $q_{s,i}$ are chosen. Using these initial values, the time averages of $\bar{T}_{c,i}$, $\bar{T}_{p,i}$, and $\bar{T}_{t,i}$ can be calculated using (5.5)–(5.7). Then, using the time averages of $\bar{T}_{c,i}$, $\bar{T}_{p,i}$, and $\bar{T}_{t,i}$, one can calculate S_i, $q_{t,i}$, $q_{c,i}$, and τ_i using (5.4), (5.9), (5.11), and (5.12), respectively. Using $q_{t,i}$, $q_{c,i}$, and τ_i, the values of q_i and $q_{s,i}$ can be updated by (5.8) and (5.10), respectively. The whole process iterates until S_i converges. It was shown in Kim and Lee (2011) that this value converges to a unique solution after around 5 iterations. It was also shown there that the analytical calculation above matches well with the simulation result and that the throughput increases with the transmission power and decreases with the distance from the master node, as expected.

Further, the Jain's fairness index of the protocol is defined as $F = \frac{(\sum_{i=1}^{I} S_i)^2}{I \sum_{i=1}^{I} S_i^2}$ and can be examined, which is a value between 0 and 1 (Jain et al. 1999). It was reported in Kim et al. (2011) that the EA-MAC protocol with adaptive contention using w_i can achieve a fairness index of around 0.9, while the EA-MAC protocol without adaptive contention when $w_i = 1$ can achieve a fairness index of around 0.4. Thus, the EA-MAC protocol improves fairness by adapting the protocol to the harvested energy.

5.2.1.2 Ambient Energy Harvesting

The work in the above subsection uses the RF power transfer. This reduces the uncertainty in power supply but leads to unfairness due to different user locations. On the other hand, if ambient energy harvesting is used, where the sensing nodes are powered by the Sun or wind, etc., there is more uncertainty in power supply but less unfairness among users. For example, the sensors in the same area may receive the same amount of solar power (except in some extreme cases where some sensors are located in shaded areas) so that the geographical locations will not affect the harvested energy much, but the Sun comes and goes causing uncertainty. In this subsection, we study the duty cycle problem in WSNs where the sensing nodes harvest the ambient energy. The fundamental problem here is very similar to what we studied in Section 4.2.5, where we have to ensure that the total energy consumed is smaller than the total energy harvested. The objective is to maximize the duty cycle, which determines the performance of the WSN, by taking the random energy arrival into account. We will consider the more complicated protocol in Hsu et al. (2006) first, followed by the less complicated protocol in Yoo et al. (2012).

To enable adaptive duty cycle using ambient energy harvesting, one needs the energy arrival model, the energy consumption model and the energy storage model when power management is used, so that the duty cycle of the sensing node can be adapted. In this case, denote $P_s(t)$ as the harvested energy at time t and $P_c(t)$ as the consumed energy at time t. The harvester has a conversion efficiency of η. Also, assume that the energy storage has an initial energy of G_0. The duty cycle is assumed to have a linear relationship with the utility or the performance of the WSN. For example, the amount of transmitted data is proportional to the transmission time. For sensing nodes that detect intrusion, the detection probability increases linearly with the duty cycle too. Also, the linear relationship is a good approximation to the non-linear relationships in some cases. Thus, to maximize the utility, we need to maximize the duty cycle whenever possible.

The duty cycle needs to be adjusted at different times. Assume that each sensor operates with a time slot of T seconds and that the adaptation of the duty cycle is performed for I time slots. Assume that P_i is the average amount of energy harvested during the ith time slot, $i = 1, 2, \cdots, I$. Also, P_c is the power consumption assumed constant during each time slot, D_i is the duty cycle in the ith time slot to be adapted, and G_i is the amount of energy in the energy storage at the beginning of the ith time slot.

There are two possible cases on the energy use in each time slot. If the power consumption P_c is larger than P_i in the ith time slot, the consumed energy will be taken from both the energy harvester and the energy storage. On the other hand, if the power consumption P_c is smaller than P_i, the consumed energy will only be taken from the energy harvester.

Thus, the energy change in the storage during the ith time slot follows

$$G_i - G_{i+1} = TD_i[P_c - P_i]^+ - \eta T(1 - D_i)P_i - \eta TD_i[P_i - P_c]^+ \tag{5.13}$$

where $[x]^+ = x$ when $x > 0$ and $[x]^+ = 0$ when $x < 0$. In (5.13), TD_i is the active time of the node and $T(1 - D_i)$ is the sleep time of the node. Also, the first term represents the energy taken from the storage when $P_c > P_i$, the second term represents the energy added to the storage when the node sleeps, and the third term represents the energy added to the storage when the node is active with $P_c < P_i$.

Using (5.13), the duty cycle adaptation problem can be formulated as (Hsu et al. 2006)

$$\max_{D_1, D_2, \cdots, D_I} \left\{ \sum_{i=1}^{I} D_i \right\} \tag{5.14}$$

$$G_i - G_{i+1} = TD_i[P_c - P_i]^+ - \eta T(1 - D_i)P_i - \eta TD_i[P_i - P_c]^+ \tag{5.15}$$

$$G_1 = G_0 \tag{5.16}$$

$$G_{I+1} \leq G_0 \tag{5.17}$$

$$D_{max} \leq D_i \leq D_{min}, i = 1, 2, \cdots, I \tag{5.18}$$

where the total duty cycle is maximized over all D_i, with a constraint on the energy change in (5.15), an initial energy of G_0 in (5.16), a constraint on the final energy that must be larger than the initial energy in (5.17), and practical limits on the duty cycle in (5.18). The constraint in (5.17) is also called the energy neutrality condition. Several comments can be made. First, the optimization above does not consider the storage capacity. In reality, the battery capacity is limited with a finite size B. This can be applied in (5.15) and (5.17). Secondly, although the optimization contains the non-linear functions $[x]^+$, these functions depend on constants only and do not depend on the variables to be optimized. Thus, standard linear programming methods can be used to solve the optimization problem. The optimum duty cycles will depend on P_i. Thus, one must have full knowledge of all P_i for $i = 1, 2, \cdots, I$. This is the deterministic model of energy arrival process discussed in Section 3.4.

Hence, the optimization algorithm can be implemented in three steps: the first step acquires knowledge of the past and future energy availability for P_i; the second step solves the optimization problem using linear programming; and the last step dynamically adapts the duty cycle based on the optimum solutions. Next, we simplify the algorithm in (5.14) further. First, we define two sets as

$$S = \{i | P_i - P_c \geq 0\} \tag{5.19a}$$

$$W = \{i | P_c - P_i > 0\}. \tag{5.19b}$$

Using (5.19a,b), (5.15) can be summed up over all I time slots to give

$$\sum_{i=1}^{I}(G_i - G_{i+1}) = \sum_{i \in W} TD_i[P_c - P_i] - \sum_{i=1}^{I} \eta TP_i + \sum_{i=1}^{I} \eta TP_iD_i - \sum_{i \in S} \eta TD_i[P_i - P_c] \tag{5.20}$$

where the term on the left-hand side of the equation is the overall energy change in the storage and the term on the right-hand side of the equation is the total energy used over I time slots. For energy neutrality operation, we would like to set $\sum_{i=1}^{I}(G_i - G_{i+1}) = 0$ so that all harvested energy is used over the I time slots to maintain the energy level in the storage. This gives

$$\sum_{i=1}^{I} P_i = \sum_{i \in W} D_i \left[\frac{P_c}{\eta} + P_i \left(1 - \frac{1}{\eta} \right) \right] + \sum_{i \in S} P_c D_i \tag{5.21}$$

from (5.20). The left-hand side of the equation represents the total energy harvested, while the right-hand side of the equation represents the energy used during W and S time slots, respectively. Using (5.21), the optimization problem is simplified as (Hsu et al. 2006)

$$\max_{D_1, D_2, \cdots, D_I} \left\{ \sum_{i=1}^{I} D_i \right\} \tag{5.22}$$

$$\sum_{i=1}^{I} P_i = \sum_{i \in W} D_i \left[\frac{P_c}{\eta} + P_i \left(1 - \frac{1}{\eta} \right) \right] + \sum_{i \in S} P_c D_i$$

$$D_{max} \le D_i \le D_{min}, i = 1, 2, \cdots, I \tag{5.23}$$

where the constraints on the energy change in each time slot are replaced by a constraint on the overall energy change over all I time slots. Hsu et al. (2006) proposed a simple iterative solution to (5.22) by using only simple arithmetic operations and sorting, which are suitable for embedded computation. Further, the error between the predicted harvested energy and the actual harvested energy was accounted for to improve the performance of the algorithm. The performances of the optimal algorithm in (5.14), the adaptive algorithm in (5.22), and the simple algorithm without duty cycle adaptation were compared using solar energy harvesting. It was shown that the performances of the optimal algorithm and the adaptive algorithm are graphically indistinguishable from each other, both of which are better than the simple algorithm without duty cycle adaptation. The performance gain varies from around 50 to 0%, depending on the conversion efficiency η. Some other variants of this problem can also be studied. For example, an energy storage with finite size can be considered. In this case, the energy change is limited by the storage size. Also, $\sum_{i=1}^{I} (G_i - G_{i+1})$ does not have to be zero and can be relaxed.

In Yoo et al. (2012), the authors reported two new MAC protocols called duty-cycle scheduling based on residual energy, DSR-MAC, and duty-cycle scheduling based on prospective increase in residual energy, DSP-MAC, to reduce the end-to-end delay and also to increase fairness among sensing nodes.

Specifically, in the DSR-MAC protocol, the duty cycle is set as (Yoo et al. 2012)

$$D_i = D_{max} - D_{max} \frac{G_i - G_t}{G_{max} - G_t}, \text{when } G_i > G_t$$

$$D_i = D_{max}, \qquad\qquad \text{when } G_i \le G_t \tag{5.24}$$

where G_i is the residual energy at the ith node, G_t is an adjustable threshold to meet the minimum requirement of the concerned application, G_{max} is the maximum possible E_i (not necessarily the battery capacity), and D_{max} is the maximum duty cycle depending on the application. Thus, in this protocol, the authors proposed to reduce the duty cycle when the residual energy increases and vice versa.

In the DSP-MAC protocol, the prediction of future energy increase is used to adjust the duty cycle more aggressively. Assume that the average harvested power over a time duration of T is P_i and that the average power consumption over this duration is P_c. Thus, the residual energy level will be increased from G_i to $G_i + (P_i - P_c)T$ at the end of time duration T, when $P_i > P_c$ and $G_i > B_t$. Thus, if the values of P_i, P_c, and T are available and when $P_i > P_c$ and $G_i > B_t$, the duty cycle of the ith node will be set as (Yoo et al. 2012)

$$D_i = D_{max} - D_{max} \frac{G_i + (P_i - P_c)T - G_t}{G_{max} - G_t} \qquad (5.25)$$

which effectively replaces G_i with $G_i + (P_i - P_c)T$ in (5.24). When $P_i < P_c$ or $G_i < B_t$, the DSP-MAC switches to the DSR-MAC mode. It is clear that the effectiveness of this protocol depends heavily on the accuracy of the estimated values of P_i, P_c, and T. A large estimation error will lead to a significant performance degradation. In Yoo et al. (2012), the authors used

$$T = \frac{G_{max} - G_i}{P_i - P_c} \qquad (5.26)$$

to calculate the value of T when $P_i > P_c$. Using the RI-MAC in Sun et al. (2008) as a benchmark, the authors showed in Yoo et al. (2012) that both the end-to-end delays and the packet delivery rates of DSR-MAC and DSP-MAC are better than the RI-MAC, and the performance gains increase when the number of nodes increases. They also showed that the DSR-MAC and DSP-MAC have similar Jain's fairness index, both of which are higher than that of the RI-MAC.

5.2.2 Other Issues in MAC Protocols

In addition to the study of duty cycle in the MAC protocols, other issues in the MAC protocols have also been investigated in the literature.

For example, in Iannello et al. (2012), the performances of two existing MAC protocols have been evaluated in the case when ambient energy harvesting is used. Two classical protocols were considered: time division multiple access (TDMA) where each node is allocated a fixed time slot and no other nodes can use it even when the concerned node does not have any data or enough energy to transmit, and the ALOHA structure where different nodes contend to access the channel. The main effect of ambient energy harvesting is the availability of energy when it needs to transmit data. The energy uncertainty was considered to study the delivery probability and the time efficiency of the two MAC protocols. Numerical results showed that TDMA always has a larger delivery probability than ALOHA, as TDMA has fixed channel access. For the time efficiency, ALOHA is not necessarily better than TDMA either. Unfortunately, the authors did not provide any comparison between energy harvesting MAC protocols and traditional MAC protocols to examine the effect of energy harvesting, which is perhaps more interesting than the comparison between TDMA and ALOHA.

In Ha et al. (2018), the authors proposed a harvest-then-transmit MAC protocol (HE-MAC) as an extension of the enhanced distributed coordination function used in current IEEE 802.11 standards considering wireless power. In this case, the sensor nodes receive data as well as wireless power from the hybrid access point. To coordinate the transmission of data and the transfer of power from the access point, the distributed coordination function needs to be re-designed. They used the Markov chain model to analyze the steady-state rate and based on this analysis, the energy harvesting rate was maximized subject to constraints on data performances. In Naderi et al. (2014), the authors optimized the wireless charging method for wireless powered WSNs.

Specifically, the position, the frequency and the number of wireless power transmitters have been investigated in terms of the sensor charging time. Based on this investigation, the wireless power transfer efficiency has been optimized, with minimum disruption to the data communication at the sensor nodes. The authors have also given some guidelines on how to choose the maximum energy harvesting threshold, the power transmitters, how to request charging, and different priorities of charging and data transmission. They showed a network throughput improvement of 300% compared with classical methods.

In Peng et al. (2010), Kim and Lee (2011), Kim et al. (2011), Naderi et al. (2014), and Ha et al. (2018), wireless power transfer is used. In this case, there is less uncertainty in the energy availability due to the intentional power transfer, and the design objective is more on the coordination between the data transmission time and the power transfer time or the adaptation to the transferred power. On the other hand, in Hsu et al. (2006), Iannello et al. (2012), and Yoo et al. (2012), ambient energy harvesting is used. In this case, there is more uncertainty in the energy availability due to the unreliable energy source so that the design objective is more on the scheduling of transmission time with energy constraints. Since the information processor and the energy harvester are normally separated in this case, there is no need for the coordination between data transmission time and power transfer time. Thus, the energy source renders a fundamental difference in the design objective.

There are other studies on MAC protocols for energy harvesting communications. For example, in Eu et al. (2011), the performances of existing CSMA and polling-based MAC protocols using ambient energy harvesting were evaluated and compared in terms of network throughput, fairness and end-to-end delay, similar to Iannello et al. (2012). The energy arrival models and the energy harvesters were studied in detail to be used for performance evaluation. In Fafoutis and Dragoni (2011), a new energy on-demand MAC protocol was proposed for WSNs using ambient energy harvesting. The idea is similar to Hsu et al. (2006) by maximizing the sensor performance using a consumed energy as close to the harvested energy as possible. This is again a design of duty cycle or transmission time based on the energy availability. A similar problem was also studied in Nguyen et al. (2014). Finally, in Yang et al. (2012) and Liu et al. (2015b), the charging time was considered with contention time and transmission time for wireless power transfer, which was also considered in Ha et al. (2018).

In summary, the energy source determines the design objective in the MAC protocols. If the energy source is the ambient environment, the data transceiver and the energy harvester operate independently at the sensor node such that the energy uncertainty is more important than the coordination between data transmission and energy harvesting. If the energy source is an intentional power transmitter that operates at the same band as data, the energy supply is more controllable with less uncertainty but the coordination between data transmission and energy harvesting due to limited channel access time becomes more important.

5.3   Routing Protocols

Routing is another main function of the upper layers. For WSNs, this function is particularly important, as most WSNs are low-power and low-data-rate applications such that the access point is normally out of the direct transmission ranges of the sensor nodes. Hence, multi-hop communications have to be used to forward the data packet from the sensing node to the access point. The physical layer performs the actual transmission and reception of the data packet, the MAC protocol coordinates the transmission and reception of the data packet at each node, while the routing protocol coordinates the transmission and reception of the data packet across the network. Specifically, the routing protocol needs to identify the best route from the sensor to the access point for multi-hop communications.

The criterion for the best route depends on the application requirements. For some applications, the network throughput is the most important metric. Thus, the best route can be chosen to maximize the network throughput. For other applications, the end-to-end delay or latency is more important. In this case, the best route can be chosen to minimize the delivery time with the quickest route. For WSNs, especially for low-power and lower-data-rate WSNs, throughput and latency are often not the main concern. The primary objective for these networks is to design a network that can transmit as many data packets as possible with a lifetime as long as possible.

The new energy harvesting feature at the sensors provides a promising solution to the network lifetime issue but also creates new problems for routing in energy harvesting systems. The main problem is that the energy supply at the sensor becomes random so that there might be energy outage when the sensor has an empty queue to forward data but it does not have enough energy to do so. Thus, the energy at the sensors must be utilized efficiently. This has been studied in the previous section on the duty cycling of the MAC protocol. For routing, in order to forward the data packet efficiently from the sensor to the access point via several hops, the helpers or the relays must be chosen carefully to avoid any energy outage.

To this end, for efficient routing, the dynamic nature of the energy availability at each sensor must be taken into account in the choice of the best route. Traditional fixed routing algorithms use predefined and fixed paths for packet forwarding. This requires the topology of the whole network as well as fixed nodes. However, the fixed routing algorithm cannot adapt to any changes in the operating environment, such as energy. Thus, they are not suitable for energy harvesting communications systems. On the other hand, opportunistic routing does not require the network topology and explores the broadcast nature of wireless to find helpers or relays en route to the access point in real-time. Thus, opportunistic routing is more suitable for energy harvesting communications systems.

To account for the dynamic nature of the energy availability in routing, some studies calculate the routing metric based on the residual energy at the nodes as current energy availability, some calculate the routing metric based on the harvesting rate as future energy availability, and some use both. In the following, we will discuss the use of the ambient energy first, followed by the use of wireless power for the routing protocols in energy harvesting communications systems.

5.3.1 Ambient Energy Harvesting

Similar to the MAC protocols, when the energy is harvested from the ambient sources, such as wind and vibration, there is great uncertainty in the energy availability. This uncertainty requires the redesign of routing protocols for energy harvesting networks.

In Shafieirad et al. (2018), the authors proposed a new energy-aware opportunistic routing protocol for large-scale WSNs. In this protocol, the sensor node sends its data packet to the access point or the fusion center via multiple hops. The candidate relays used to forward the data packet are selected to maximize the amount of delivered data packets, instead of maximizing the network throughput or minimizing the latency. This was called the "Max-SNR" routing protocol in Shafieirad et al. (2018).

Specifically, consider a network of N sensor nodes uniformly distributed within a circle of radius R that has one access point located at the center of the circle. Each node harvests energy from the environment and stores the harvested energy in a battery with infinite capacity. The whole process is divided into multiple time slots in a finite time horizon. Each node also has a data buffer with a finite size to store the data packets to be transmitted. The data packets come from the sensor node's own sensed data as well as from its neighbors that require forwarding. However, it only accepts data packets from the neighbors when its data buffer has room and when the neighbor is within its reception range of R_r. Assume that R_r is much smaller than R so that most sensor nodes need multi-hop communications to send the data packet to the access point, except those nodes within R_r of the access point. The design problem of the routing protocol is to choose the best neighbor in each hop to forward the data packet to the access point that maximizes the number of delivered packets.

The key point here is to choose the best neighbor. This choice must take the amount of available energy at each potential relay into account to avoid energy outage and it also needs to minimize the energy consumption or the number of hops so that the data packet can arrive at the access point with minimum energy. Assume that the ith node has an amount of energy E_i available at the time of selection and that the distance between this node and the access point is d_i. In Shafieirad et al. (2018), it proposed to use the selection criterion of

$$C_i = \frac{E_i}{d_i^m} \tag{5.27}$$

where m is the path loss exponent. Thus, the more energy available at the ith node, the more likely that the ith node will be chosen. Also, the shorter the distance between the ith node and the access point, the more likely that it will be chosen, as the energy consumption is inversely proportional to d_i^m. This is effectively the signal-to-noise ratio (SNR) that will be received at the access point if the ith node is chosen. Then, a timer is set at each node for a waiting time of

$$T_i = \frac{1}{C_i} \tag{5.28}$$

before the sensor node in the previous hop is allowed to forward the data packet. One can see that the larger the received SNR at the access point is, the smaller the waiting time will be and hence the more likely that it will be chosen. This is why it is called the "Max-SNR" protocol.

Using the maximum SNR criterion, the number of delivered packets has been analyzed. Denote $P_{i \to k}$ as the probability that the ith node's data packets will be forwarded by the kth node, $i, k = 1, 2, \cdots, N$. Denote $N(i)$ as the set of the ith node's neighbors that are within the transmission range R_r of the ith node with enough data buffer to store the forwarded data packets. Denote k as the node that has the maximum SNR. It was derived in Shafieirad et al. (2018) that the probability $P_{i \to k}$ is given by

$$P_{i \to k} = \int_0^\infty [1 - F_{E_k(d_k^m t)}] \sum_{n \in N(i), n \neq k} d_n^m f_{E_n}(d_n^m t) \left[\prod_{l \in N(i), l \neq n, k} F_{E_l}(d_l^m t) \right] dt \tag{5.29}$$

where $f_{E_i}(\cdot)$ and $F_{E_i}(\cdot)$ are the probability density function and the cumulative distribution function of the available energy E_i, respectively. Since the energy is harvested, these functions are determined by the energy arrival process as well as the energy conversion process at the harvester, as discussed in Section 3.4.

Using $P_{i \to k}$ in (5.29), the average number of data packets transmitted from the ith node to the kth node can be derived as

$$M_{ik} = \left(\sum_{l \in N(i)} M_{li} + M_i \right) P_{i \to k} \tag{5.30}$$

where M_i is the average number of data packets from the ith node's own sensed data, and M_{li} is the average number of data packets received from the lth neighbor that need to be forwarded to the access point. Thus, one has

$$M_{ik} - \sum_{l \in N(i)} M_{li} P_{i \to k} = M_i P_{i \to k} \tag{5.31}$$

where $M_i P_{i \to k}$ is the average number of delivered packets for the ith node's own data by the kth node, and $\sum_{l \in N(i)} M_{li} P_{i \to k}$ is the average number of data packets for the ith node's received data (to be forwarded by the ith node for its neighbors) by the kth node. Rewriting (5.31) in matrix form for all nodes, one has

$$\mathbf{AM} = \mathbf{P} \tag{5.32}$$

where $\mathbf{A}$ and $\mathbf{P}$ are the matrices determined by $P_{i \to k}$ and $\mathbf{M}$ is the vector containing all M_{ik}. The matrix equation in (5.32) can be solved for $\mathbf{M}$. Further, if the energy constraint is considered, it becomes an optimization problem of

$$\min_{\mathbf{M}} ||\mathbf{AM} - \mathbf{P}||^2 \tag{5.33}$$

$$M_{ik} \geq 0 \tag{5.34}$$

$$\sum_{l \in N(i)} M_{il} E \leq E_i \tag{5.35}$$

where the last constraint is the energy causality that the consumed energy must be smaller than the available energy, and E is the amount of energy required to transmit each data packet. The total average number of data packets transmitted by the ith node is then given by

$$M_i^t = \sum_{l \in N(i)} M_{il} \tag{5.36}$$

using the solutions from (5.33). More details of the derivation can be found in Shafieirad et al. (2018). It was shown that the "Max-SNR" protocol can achieve a delivery ratio of almost 100% for an exponential energy with parameter 1 and $N = 200$. Its performance is significantly better than the previous protocols that do not account for the energy availability, under the same conditions.

Note that, although the above results are derived for energy harvesting systems, it can be used for conventional systems too, as the derivation does not specify the randomness of the available energy E_i. Hence, the randomness could be caused by energy harvesting, or by energy consumption. Note also that the above derivation assumes independent energy availability at each node. In some energy harvesting systems, the nodes may harvest the same source and hence have correlated energies.

In Cao et al. (2016), both the energy currently available at the node and the energy to be harvested for future use at the node were used in the routing metric calculation. Specifically, in Cao et al. (2016), each node is assumed to have a table about its own energy status determined by three parameters: residual energy; energy harvesting rate; and energy harvesting density. These three parameters also determine the energy statuses for all its neighbors.

Denote E_i^r as the residual energy or the energy currently available at the ith node. For the use in this protocol, it was discretized as

$$L_i = k + 1 \tag{5.37}$$

where $k = 1, 2, \cdots, K - 1$, K is the maximum energy level allowed at the node and is predetermined, k satisfies $\frac{k}{K} < \frac{E_i^r}{E_m} \le \frac{k+1}{K}$ with E_m being the largest amount of energy or the battery capacity.

The energy consumption at the ith node is calculated as (Cao et al. 2016)

$$E_i^c = LE_0 + L\alpha d_i^m \tag{5.38}$$

where L is the number of bits in each data packet, E_0 is the energy consumption used to transmit each data bit, α is a constant representing energy loss, d is the distance between two nodes, and m is the path loss exponent.

The energy density at the ith node is calculated as (Cao et al. 2016)

$$D_i = \frac{\sum_{j \in N(i), j \neq i} (R_j T - E_j^c)}{|N(i)| + 1} \tag{5.39}$$

where $N(i)$ is the set of all neighbors of the ith node within its transmission range, $|N(i)|$ is the number of these neighbors, R_j is the energy harvesting rate, and T is the harvesting time. From (5.39), this metric takes both the harvested energy and the consumed energy into account. It will be the net energy added to the storage after the transmission. Comparing (5.27) with (5.39), one sees that (5.27) uses the ratio of current energy to the energy consumption, while (5.39) uses the difference between future energy and the energy consumption.

These values are initialized at the beginning of transmission using (5.37) and (5.39). During each transmission, the node with the largest energy density of D_i determined by (5.39) is chosen for the next hop. After each transmission, the three parameters in the tables at the jth node are updated to prepare for the next transmission. This process repeats until the data packet arrives at the access point. Simulation results have shown that this routing protocol always has a higher average residual energy compared with the conventional protocol. A similar protocol was also proposed in Kawashima and Sato (2013) by incorporating the power generation pattern, which is related to the energy harvesting rate, in the routing metric calculation.

In Martinez et al. (2014), in addition to the residual energy and the harvested energy, the energy wastage was also used in the routing metric calculation. Kawashima and Sato (2013) and Shafieirad et al. (2018) have assumed an infinite capacity for the energy storage at the sensor so that energy can be harvested as much as possible. In practice, the storage capacity is finite and hence too much energy harvesting may cause energy wastage in the network. This is really the other end of the problem, where instead of worrying about insufficient energy for data transmission, Martinez et al. (2014) worried about too much energy for data transmission.

In this case, assume a routing path of σ_n. If the ith node is on this path, it will cause an energy wastage of

$$E_i^{WN} = \max\{0, E_i + E_i^h - E_i^c - B\} \tag{5.40}$$

where E_i is the residual energy or the current energy in the battery as defined before, E_i^h is the energy to be harvested in the next time slot T, E_i^c is the energy to be consumed in the next time slot T, and B is the storage capacity. The equation in (5.40) essentially calculates the amount of energy that would exceed the storage capacity in the next time slot and hence would be wasted, if any. If the ith node is not on this path, it will cause an energy wastage of

$$E_i^{WF} = \max\{0, E_i + E_i^h - B\} \tag{5.41}$$

as it will not consume any energy in the next time slot. Hence, if the path σ_n is chosen for routing, the total energy consumption including that used for data transmission and wasted is

$$C(\sigma_n) = \sum_{i \in \sigma_n}(E_i^c + E_i^{WN}) + \sum_{i \notin \sigma_n} E_i^{WF} \tag{5.42}$$

and the total energy available for routing is

$$E(\sigma_n) = \sum_{i \in \sigma_n}(E_i^c + E_i^h). \tag{5.43}$$

Thus, the routing path is selected so that the total remaining energy after the routing is maximized as

$$\sigma^* = \arg \max_{\sigma_n \in \Omega}[E(\sigma_n) - C(\sigma_n)] \tag{5.44}$$

where Ω is the set of all possible routes and σ^* is the optimum route. This routing metric calculation takes the residual energy E_i, the harvested energy E_i^h and the wasted energy E_i^{WF} and E_i^{WN} into account. Simulation results were provided in Martinez et al. (2014) to show the performance of this protocol using solar energy. As expected, higher residual energy levels can be achieved using the proposed routing protocol.

The above studies only consider routing. In other studies, routing is also considered jointly with other functions in the upper layer. To this end, Avallone and Banchs (2016) considered a joint channel assignment and routing algorithm for mesh networks with extra constraints on the energy availability. This is applicable to systems with multiple channels so that both routing and channel assignment can increase the chance of success for multi-hop communications. In Hasenfratz et al. (2010), different routing protocols were investigated and compared in conjunction with the MAC protocols by maximizing the delivery ratio or the packet loss. Specifically, three routing algorithms of randomized Max-Flow, energy opportunistic weighted minimum energy and randomized minimum path recovery time have been studied.

5.3.2 Wireless Power Transfer

The studies in the previous subsection used ambient energy harvesting. Because of this, there is great uncertainty in the energy availability. Consequently, it is important to perform routing by taking the residual energy, the harvested energy and the wasted energy into account for maximum delivery ratio. In this subsection, wireless power transfer will be considered. As discussed before, this generates less uncertainty in the energy supply. However, since power transfer and data transmission may be performed in the same frequency band, coordination is required. Thus, studies on routing protocols in wireless powered systems mainly consider this coordination.

In Doose et al. (2010), wireless-charging-aware routing protocols were studied. Due to different distances and heights of the sensors to the power transmitter, the amount of energy received at different sensors is different. Hence, to reach a certain level of energy, the charging time for different sensors needs to be different too.

Denote t_i as the average charging time of the ith node and ϵ_i as the standard deviation of the charging time at the ith node. These values can be obtained experimentally. Also, denote $T_{max}(\sigma_n)$ and $\epsilon_{max}(\sigma_n)$ as the maximum charging time and the maximum standard deviation among all nodes on the path σ_n, respectively. These two maximum values will be included in the routing request packet and this packet will be forwarded to each node on the path and will be updated every time a sensor receives it by comparing its own charging time and standard deviation with them.

When the sensor receives the routing request packet, it waits for a delay of $t_i + \epsilon_i$ before forwarding it, so that the nodes with shorter charging times can forward it earlier. When the destination receives all these forwarded request packets from different nodes, it will choose the path with the minimum $T_{max}(\sigma_n)$ so that

$$\sigma^* = \min_{\sigma_n \in \Omega} \{ T_{max}(\sigma_n) \}. \tag{5.45}$$

After the destination chooses the path with the minimum maximum charging time, it sends back the route reply packet to inform the nodes on the chosen route or path. In this case, since charging and transmission use the same frequency band, the destination has to optimize the time allocation for charging and transmission with a fixed total of $T_c + T_x = T$, where T_c is the charging time, T_x is the transmission time, and T is the total packet time. The optimization problem proposed in Doose et al. (2010) hence becomes

$$\max_{T_c} \left\{ \frac{T_x R}{T} \right\} \tag{5.46}$$

$$(P_r - P_d)T_c - P_t T_x \geq 0 \tag{5.47}$$

$$N\left(T_c + \frac{P+H}{R}\right) \leq L_{max} \tag{5.48}$$

$$\frac{1}{S_0}[1 - kte^{\frac{-4700}{M+273}}] > \frac{1}{S_{max}} \tag{5.49}$$

$$T = T_c + T_x \tag{5.50}$$

where the throughput is maximized with respect to T_c, R is the data rate, P_r is the power harvested from the wireless charger during T_c, P_d is the idle power consumed during charging, P_t is the transmission power for data, N is the total number of nodes on the chosen path, P is the size of data in the packet, H is the size of overhead in the packet, L_{max} is the maximum end-to-end delay or latency allowed, S_0 is the initial equivalent series resistance of the supercapacitor, S_{max} is the maximum equivalent series resistance of the supercapacitor allowed, beyond which the capacitor will not work, k is a design constant, t is the operation time, and M is the absolute temperature. The first constraint makes sure that the consumed energy will be smaller than the harvested energy so that the node can stay alive after transmission. The second constraint limits the total latency caused by routing. The third constraint makes sure that the energy storage will work properly.

From the above optimization, the performance of this routing protocol was studied in Doose et al. (2010) in terms of throughput, latency, network lifetime, and residual energy. It was reported there that the latency decreases with the charging rate and increases with the size of data in the packet. Also, the maximum throughput decreases with the optimal charging time but the network lifetime increases with the optimal charging time.

In Tong et al. (2010) and Li et al. (2011), similar problems have been studied. Specifically, in Li et al. (2011), a joint charging and routing algorithm was proposed, where the sensor node is charged when needed to prolong the network lifetime, while in Tong et al. (2010), the sensor deployment was jointly designed with routing to make the best use of wireless charging.

5.4 Other Issues in the Upper Layers

The previous two sections have mainly examined the MAC protocols and the routing protocols for energy harvesting wireless communications. They are the two most important tasks of the upper layer. Next, we examine some other issues in the upper layer: scheduling; and effective capacity.

5.4.1 Scheduling

In transmission scheduling, the main problem is to adjust the transmission time and the transmission power for each time slot so that all data packets can be delivered by the minimum deadline, assuming randomly arriving data packets. Thus, one aims to minimize the delay with respect to transmission time and transmission power, under different traffic models. In energy harvesting wireless communications, this problem is further complicated by the fact that the energy arrives randomly too so that the transmission time and the transmission power must be adapted to the energy availability. Thus, the transmission scheduling problem in energy harvesting wireless communications requires the minimization of the delivery time with respect to the transmission time and the transmission power, for both randomly arriving data packets and randomly arriving energy. Figure 5.2 shows the transmission scheduling problem considered. This problem was studied in Yang and Ulukus (2012) for a single user case.

Assume that the initial amount of data packet is B_0 bits and the initial amount of energy is E_0 at the transmitter. During the packet delivery, the energy is harvested with

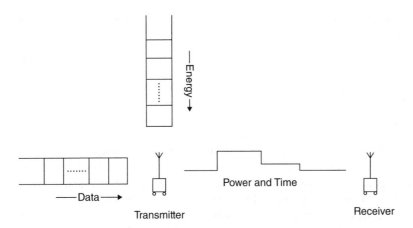

Figure 5.2 The transmission scheduling problem.

an amount of $E_1, E_2, \cdots, E_N$ at time instants of $t_1, t_2, \cdots, t_N$, respectively, where t_N is the last time instant that energy arrives within a total delivery time of T_d. This represents the energy arrival process with known amount and known arrival time, the deterministic model in Section 3.4. Note that this model has simplified the energy harvesting process and the energy storage process by assuming perfect energy conversion and perfect energy storage. In practice, there could be conversion loss or leakage. The data packet arrives with an amount of $B_1, B_2, \cdots, B_M$ at time instants of $T_1, T_2, \cdots, T_M$, respectively, where T_M is the last time instant that data packets arrive within the total delivery time of T_d. Again, the packet arrival time and amount are also known. Finally, within the total delivery time, assume that a sequence of transmission power $P_1, P_2, \cdots, P_K$ with transmission duration of $d_1, d_2, \cdots, d_K$ is adopted before the transmitter finishes the transmission. The transmission power and transmission duration are adapted to the energy arrival and packet arrival to minimize the delivery time. The energy arrival is independent of the packet arrival.

Using these assumptions, two scenarios were considered in Yang and Ulukus (2012). In the first scenario, a simpler case is considered, where the amount of data packet is fixed at B_0 and there is no random packet arrival during T_d. In this case, the total energy consumption and the total amount of bits transmitted before any time instant t are given by

$$E(t) = \sum_{k=1}^{\bar{K}} P_k d_k + P_{\bar{K}+1}\left(t - \sum_{k=1}^{\bar{K}} d_k\right) \tag{5.51}$$

$$B(t) = \sum_{k=1}^{\bar{K}} C(P_k)d_k + C(P_{\bar{K}+1})\left(t - \sum_{k=1}^{\bar{K}} d_k\right) \tag{5.52}$$

where $\bar{K} = \max\{k, \sum_{j=1}^{k} d_j \leq t\}$ is the largest time index that the transmission duration is changed before t, and $C(P_k)$ gives the transmission rate as a function of the transmission power. For example, in a static additive white Gaussian noise, $C(P_k) = \ln(1 + P_k)$ follows a logarithmic relationship. Based on the above equation, the transmission scheduling problem becomes

$$\min_{P_1, \cdots, P_K, d_1, \cdots, d_K} \{T_d\} \tag{5.53}$$

$$E(t) \le \sum_{n:t_n<t} E_n, \quad 0 \le t \le T_d \tag{5.54}$$

$$B(T_d) = B_0. \tag{5.55}$$

Here (5.53) shows the minimization of the total delivery time T_d with respect to K transmission power $P_1, \cdots, P_K$ and transmission time $d_1, \cdots, d_K$. Then (5.54) shows a constraint on the energy that the consumed energy must be smaller than the total harvested energy before time instant t, where the right-hand side of the inequality gives the total energy harvested before t. This energy constraint is similar to the one we used before in Section 4.2.5 and will be used in later chapters too. Finally, (5.55) shows a constraint on the data transmission that all the B_0 data bits must be delivered at the end of T_d.

It was reported in Yang and Ulukus (2012) that the optimum solutions to the above problem satisfy the following conditions as

$$P_k = \frac{\sum_{j=n_{k-1}}^{n_k-1} E_j}{t_{n_k} - t_{n_{k-1}}} \tag{5.56}$$

$$d_k = t_{n_k} - t_{n_{k-1}} \tag{5.57}$$

for $k = 1, 2, \cdots, K$, with $n_k = \arg\min_{n:t_n \le T_d, t_n > t_{n_{k-1}}} \left\{ \frac{\sum_{j=n_{k-1}}^{n-1} E_j}{t_n - t_{n_{k-1}}} \right\}$ and also $\sum_{k=1}^{K} C(P_k)d_k = B_0$ such that n_k is the time index of the energy arrival when the transmission power P_k switches to P_{k+1} or at time $t = t_{n_k}$ the power P_k switches to P_{k+1}. Details of the derivation can be found in Yang and Ulukus (2012).

In the second scenario, the data packet arrives randomly during the delivery time of T_d too but the time and amount of packet arrival are known. In this case, the optimization problem for packet scheduling becomes

$$\min_{P_1, \cdots, P_K, d_1, \cdots, d_K} \{T_d\} \tag{5.58}$$

$$E(t) \le \sum_{n:t_n<t} E_n, \quad 0 \le t \le T_d \tag{5.59}$$

$$B(t) \le \sum_{m:T_m<t} B_m, \quad 0 \le t \le T_d \tag{5.60}$$

$$B(T_d) = \sum_{m=0}^{M} B_m. \tag{5.61}$$

Thus, the two constraints are that the consumed energy must be smaller than the total energy harvested and that the delivered packet must be smaller than the total packet arriving, before any time instant t for $0 \le t \le T_d$. A procedure used to calculate the optimum solutions of P_k and d_k was also provided in Yang and Ulukus (2012).

This study provides some useful insights into the transmission scheduling problem with energy harvesting but to make them more practical, more studies need to be performed. For example, the power management needs to be considered. Also, the current study assumes perfect knowledge of the arrival time and amount of energy and packet but in reality such knowledge is not available or not perfectly available.

In Antepli et al. (2011), this problem was extended to a two-user Gaussian broadcast channel. In this work, assuming a fixed amount of data available at the transmitter that needs to be sent to user 1 and user 2 within a certain period of time, the effect of energy harvesting was examined. This corresponds to the first scenario in Yang and Ulukus (2012) but with B_1 and B_2 for user 1 and user 2, respectively, instead of B_0. The optimum solutions were obtained following numerical procedures.

5.4.2 Effective Capacity

Effective capacity is a concept proposed in Wu and Negi (2003) as a link layer channel model. Most existing channel models in the literature are physical layer channel models that determine the performance metrics of transmission and reception in the physical layer, such as symbol error rate, channel capacity, and outage probability. These models cannot be used to examine the link layer performance metrics, such as delay, delay violation probability, and network throughput, etc., directly. However, in systems that require quality of service (QoS) guarantees, such performance metrics are important to evaluate the link layer connection for admission control and resource allocation. To provide this QoS support, one needs an analysis of the queuing behavior of the connection, which cannot be extracted from the physical layer channel models. Thus, Wu and Negi (2003) aimed to obtain a link layer channel model that examines the link layer performance metrics directly without using the physical layer channel models. Figure 5.3 shows the link layer channel model with queuing considered.

Specifically, this model involves two random processes: the traffic process that describes the arrival of the data packet at the transmitter; and the service process that describes the departure of the data packet in the proposed channel. If the arriving data packet cannot be processed by the channel in time, the data packets will start to queue at the transmitter. Thus, the delay in the link layer is determined by the arriving rate in the traffic process and the departure rate in the service process.

Assume that the traffic process has a constant arriving rate of μ and that the queue has an infinite size. The service process is denoted as $S(t)$. The effective capacity that describes the service process is defined as

$$\alpha(u) = \frac{1}{u} \lim_{t \to \infty} \frac{1}{t} \log E\{e^{-uS(t)}\} \tag{5.62}$$

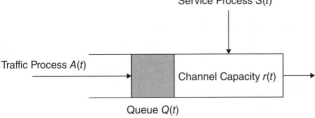

Figure 5.3 The link layer channel model.

where u is the argument of the function, $S(t) = \int_0^t r(\tau)d\tau$ is the service provided by the channel or the partial sum of $r(t)$ up to time instant t, $r(t)$ is the instantaneous rate of the channel, $\log E\{e^{-uS(t)}\}$ is the log-moment-generating-function of $S(t)$, and $u \geq 0$. Using the effective capacity, it can be shown that for a constant traffic arriving rate of μ, one has

$$\sup_t Pr\{D(t) \geq D_{max}\} \approx \gamma(\mu)e^{-\theta(\mu)D_{max}} \tag{5.63}$$

where $D(t)$ is the delay in the channel determined by the difference between the traffic and the service, D_{max} is the maximum delay allowed in the system to guarantee the QoS, $Pr\{D(t) \geq D_{max}\}$ is the probability that the delay is larger than the maximum delay that violates the QoS requirement or the delay-violation probability, $\gamma(\mu) = Pr\{D(t) \geq 0\}$ is the probability that the delay exists, and $\theta(\mu) = \mu\alpha^{-1}(\mu)$ is called the QoS exponent with $\alpha^{-1}(\cdot)$ being the inverse function of the effective capacity $\alpha(\cdot)$.

The values of $\gamma(\mu)$ and $\theta(\mu)$ are the parameters of the proposed link layer channel model that determine the quality of the connection. Thus, for a communications system that allows a maximum delay of D_{max} with delay-violation probability of ϵ, or $Pr\{D > D_{max}\} \leq \epsilon$, the maximum constant traffic rate supported by this channel to fulfill the delay requirement is μ, which can be solved using

$$\epsilon = \gamma(\mu)e^{-\theta(\mu)D_{max}}. \tag{5.64}$$

In this case, the service of the channel satisfies

$$Pr\{S(t) \leq \Phi(t)\} = Pr\{D > D_{max}\} \leq \epsilon \tag{5.65}$$

where $\Phi(t)$ is the upper bound given by $\Phi(t) = \mu[t - D_{max}]^+$ and $[\cdot]^+$ is the non-negative function defined as before. Thus, the link layer channel model using effective capacity links its channel parameters to the link layer performance metrics in a simple and direct relationship.

In Wu and Negi (2003), an example of the Rayleigh fading channel is given. It was shown that for a Rayleigh fading channel, the moment-generating-function of the service process can be approximated as

$$E\{e^{-uS(t)}\} \approx \frac{1}{|u\delta R + I|} \tag{5.66}$$

where $\delta = \frac{t}{N}$ is the time separation by dividing t into N slices, R is the covariance matrix of the Rayleigh channel gains, and I is the identity matrix.

For the additive white Gaussian noise channel, the instantaneous service rate is given as

$$r(t) = \ln\left[1 + \frac{a^2 P(t)}{N_0}\right] \tag{5.67}$$

where a is the static channel gain and $P(t)$ is the transmission power at time t. For fading channels, a becomes a random variable and hence the expectation in the moment-generating function needs to be derived. Thus, the power allocation can be studied to optimize the link layer delay using the effective capacity model. For example, in Yu et al. (2016), the effective capacity was used to optimize the spectral efficiency and the energy efficiency in green communications.

For energy harvesting wireless communications, especially for ambient energy harvesting, the power supply or the transmission power $P(t)$ may be random, as discussed in Chapter 4. Considering this randomness, new effective capacity can be calculated for variable-power transmission, such as in Gong et al. (2014). If fixed-power transmission is used instead, this randomness does not exist any more. Also, if wireless power is used, the harvested power may suffer from fading. Since power transfer and information transmission share the same channel, a and $P(t)$ may be correlated. Thus, the expression of the instantaneous rate will be more complicated due to energy harvesting.

5.5 Summary

In this chapter, the upper layer for energy harvesting wireless communications has been investigated. We have mainly focused on two functions in the upper layer: MAC protocols; and routing protocols. For the MAC protocols, only asynchronous protocols are viable in energy harvesting wireless communications, and for the routing protocols, only opportunistic routing algorithms are possible, due to the dynamic nature of the energy supply. The design objectives of these protocols are highly related to the energy sources used.

If ambient energy harvesting is used, where the sensors harvest energy from the ambient sources, such as the Sun and wind, the uncertainty in the energy supply is the main concern. In this case, the communications process follows a best-effort method, where the sensor transmits as many data as possible, subject to constraints on the energy availability. This energy availability is determined by the energy consumption (the traffic model) and the energy arrival (the harvesting model). In particular, for the MAC protocols, the duty cycle or the transmission power can be adjusted according to the residual energy. For transmission scheduling, both the transmission power and the transmission time are adapted to the traffic model and the harvesting model. For the routing protocols, the routing metric used to select the routing path should include the current energy level, the future energy level and possibly the energy wastage for energy-aware routing.

On the other hand, if wireless power is used, the sensors harvest energy from an intentional power transmitter. In this case, the uncertainty in energy supply has been greatly reduced, as power transfer is controllable. However, in some applications, power transfer and data transmission are performed in the same frequency band so that time multiplexing is necessary, as the sensor node cannot receive power and transmit data at the same time. Thus, coordination between energy harvesting and data transmission is important. For the MAC protocols, the design focuses on the fairness problem to prolong the network lifetime, as different sensors receive different amounts of power due to the different locations. For the routing protocol, the charging time needs to be considered in the routing path selection, as it affects the throughput and the latency. Alternatively, one can consider the effective capacity model that accounts for delay.

Up until now, we have covered the challenges brought by energy harvesting to wireless communications. In the following chapters, we will discuss several state-of-the-art energy harvesting communications systems as applications to cover the opportunities created by energy harvesting. These include wireless powered communications, energy harvesting cognitive radios, and energy harvesting relaying.

6

Wireless Powered Communications

6.1   Introduction

Wireless powered communications refer to the communications systems where the power of the communications device is supplied via wireless power transfer. This method is particularly useful for energy-constrained systems, such as a remote sensor or a mobile phone. In such systems, without wireless power transfer, when the battery is drained or the mains power is disconnected, the communications device will have to stop operations. Figure 6.1 shows some scenarios where wireless power can be used. In these cases, it is impossible or too costly to replace the battery of the device. Hence, wireless power transfer will be a good solution.

In general, energy harvesting wireless communications use different media to carry energy and information. For example, the communications device could harvest the solar power and then use the harvested solar power to send information via radio waves. One unique characteristic of wireless powered communications is that energy and information are often carried by the same wireless media. This saves the hardware cost.

There are many different sources of energy. There are also many different types of wireless systems. Thus, wireless powered communications can be implemented in different forms. For example, in Wang et al. (2015), wireless powered communications using lights have been discussed, where a solar panel was used to convert the solar power into electricity as an energy transducer. The same solar panel was also used as a photodetector to decode the modulated light signal for information. In Kisseleff et al. (2017), near-field magnetic induction was used to transmit data and wireless power at the same time. The inducted signal was split into several streams, one for data and the rest for power. By guaranteeing certain quality of service for data transmission, the total transferred power was maximized. One may also use sounds to transmit power and information simultaneously. However, the problem with lights and sounds is that they often require line-of-sight (LOS). The problem with near-field magnetic induction is that its efficiency drops very quickly as the distance increases. Due to these shortcomings, lights, sounds and magnetic induction have very limited applications. On the other hand, far-field radio frequency (RF) does not have these restrictions. It can penetrate walls or buildings through non-line-of-sight (NLOS) propagation, and it often covers distances of hundreds or thousands of meters. In this chapter, we will focus on wireless powered communications using far-field RF signals.

Energy Harvesting Communications: Principles and Theories, First Edition. Yunfei Chen.
© 2019 John Wiley & Sons Ltd. Published 2019 by John Wiley & Sons Ltd.

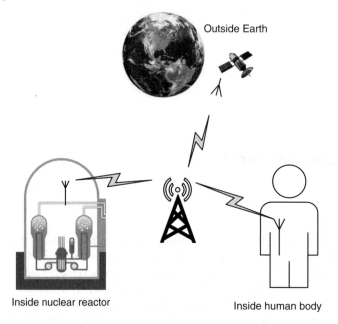

Figure 6.1 Some scenarios where wireless power is a promising alternative.

6.2 Types of Wireless Powered Communications

Since there are many different forms of wireless power in a communications system, it is useful to categorize them and then discuss them separately.

If the power transmitter has dedicated hardware for power transfer, this can be denoted as power beacon (PB)-based communications (Huang and Lau 2014). In this case, the PB is dedicated to power coverage in the network by broadcasting RF power, while the normal base stations still provide information coverage. If the power transmitter shares the hardware with the information transmitter, power transfer and information delivery can happen in different time slots. This can be denoted as hybrid access point (HAP)-based communications (Ju and Zhang 2014). In this case, the same transmitter will be used for power transfer during some periods of time and for information delivery during other periods of time. If the power transmitter and the information transmitter not only share the hardware but also operate in the same time slot, this can be denoted as simultaneous wireless information and power transfer (SWIPT). In SWIPT, the receiver splits the received signal into two parts in either the time domain or the power domain, one part for power reception and the other part for information reception. Figure 6.2 compares these three types. Note that they differ in the way wireless power is transferred. In the following sections, these three different forms will be discussed in detail.

Figure 6.2 Different types of wireless powered communications: (a) SWIPT; (b) HAP; and (c) PB.

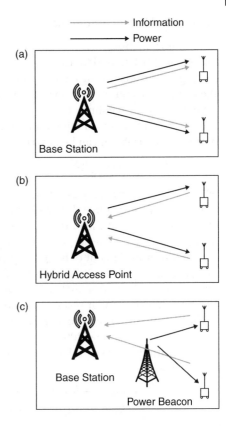

6.3 Simultaneous Wireless Information and Power Transfer

SWIPT is probably the earliest form of wireless powered communications that has been proposed in the literature. For a simple wireless power transfer circuit using magnetic induction, it can be proven that the power transfer efficiency is maximized at a specific frequency, or using a single sinusoid (Grover and Sahai 2010). However, according to Shannon's theorem, the information rate for a single sinusoid is zero, because its bandwidth is zero. For this reason, in the past few decades, messages and power have been transmitted using separate signals to maximize both the information rate and energy rate. Hence, we have power engineers working on power transfer and communications engineers working on message transmission. Two separate infrastructure networks have been designed: an electricity grid dedicated to power transfer; and communications networks dedicated to message transmission. It is reasonable to combine these two networks to save costs by using the same signal in the same network to transmit both power and information. In order to do this, a tradeoff between information rate and power transfer efficiency has to be made. This is the motivation of most recent studies on SWIPT.

6.3.1 Ideal Implementations

Several information theoretic studies were conducted in Varshney (2008) and Grover and Sahai (2010) to find for the first time the fundamental tradeoff between transmitting message and transmitting energy over a single noisy line. This tradeoff aims to optimize the energy rate and the information rate of the same signal at the same time.

Specifically, in Varshney (2008), a capacity-energy function is constructed by maximizing the mutual information under the constraint of a minimum received power. The optimization problem can be described as

$$C(B) = \max_{X : E[b(Y)] \geq B} I(X, Y) \tag{6.1}$$

where a discrete memoryless channel is considered, X is the input to the channel, Y is the output of the channel, $I(X, Y)$ is the mutual information between input and output, $b(Y)$ is the energy of the output Y, $E[b(Y)]$ is the average energy defined as $E[b(Y)] = \sum_{Y \in \mathbb{Y}} b(Y) p(Y)$, $\mathbb{Y}$ is the output alphabet, $p(Y)$ is the probability of Y, $0 \leq B \leq B_{max}$, B_{max} is the maximum element of $\mathbf{b}^T \mathbf{Q}$, $\mathbf{b}$ is a column vector of $b(Y)$ with all possible outputs in $\mathbb{Y}$, and $\mathbf{Q}$ is the transition probabilities between inputs in the input alphabet $\mathbb{X}$ and outputs in the output alphabet $\mathbb{Y}$. Several properties of $C(B)$ have been reported in Varshney (2008). Similar results are also applicable to a continuous Gaussian channel.

The optimum value can be found by using the above equation as a rate-energy tradeoff in most cases. Varshney (2008) has given an example for a particular binary symmetric channel with crossover probability of p. In this case, the capacity-energy function can be shown as

$$C(B) = \begin{cases} \log(2) - h_2(p), & 0 \leq B \leq \frac{1}{2} \\ h_2(B) - h_2(p) & \frac{1}{2} \leq B \leq 1 - p \end{cases} \tag{6.2}$$

where $h_2(\cdot)$ is the binary entropy function. Figure 6.3 shows $C(B)$ versus B in (6.2). This function is not a continuous function. Several other examples have also been given in Varshney (2008). Interested readers are referred to Varshney (2008) for more details.

Grover and Sahai (2010) studied another tradeoff between power transfer and information rate by using the same signal. Recognizing the fact that the power transfer circuit is a frequency-selective channel, this study aims to maximize the total information rate

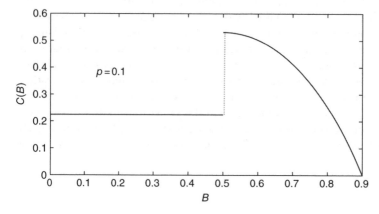

Figure 6.3 $C(B)$ versus B in (6.2).

subject to one constraint on the total power available at the transmitter and another constraint on the total transferred power at the receiver. Specifically, the output voltage of the receiver can be expressed as

$$V_{o,i} = h_i \sqrt{R_r} I_{s,i} + Z_i \tag{6.3}$$

where i indexes the ith frequency band by dividing the total bandwidth into N parts such that each band is narrow-band frequency-non-selective channel (the channel is frequency-selective across the whole bandwidth of $\sum_{i=1}^{N} f_i$), h_i is the transfer function of the ith band at frequency f_i, R_r is the load at the receiver, $I_{s,i}$ is the amplitude of the input current at the ith band so that the current through the receiver load is $h_i I_{s,i}$, and Z_i is the additive white Gaussian noise.

The average power consumed over the internal resistance R_s at the transmitter is $P_{s,i} = E[|I_{s,i}|^2]R_s$, where $E[|I_{s,i}|^2]$ is the average over the random messages. The average power consumed at the load in the receiver is $P_{r,i} = E[|I_{s,i}|^2]|h_i|^2 R_r$. Thus, the total power in the ith band available at the transmitter is $P_i = E[|I_{s,i}|^2](R_s + |h_i|^2 R_r)$, including both power consumed at the transmitter and the power delivered to the receiver. Using P_i, one has $P_{r,i} = \eta_i P_i$, where $\eta_i = \frac{|h_i|^2 R_r}{R_s + |h_i|^2 R_r}$ is the power transfer efficiency of the ith band.

Denote P_A as the total power available at the transmitter and P_B as the total power that is required to be delivered to the receiver. The optimization problem can be described as

$$C(P_A, P_B) = \max_{P_i : \sum_{i=1}^{N} P_i \leq P_A, \sum_{i=1}^{N} \eta_i P_i \geq P_B} \left\{ \sum_{i=1}^{N} \log_2 \left(1 + \frac{\eta_i P_i}{W} \right) \right\} \tag{6.4}$$

where W is the noise power.

Thus, from (6.4), the sum rate is maximized by optimizing the power allocation across different bands, subject to the total transmitted power and the total delivered power constraints.

The maximum sum rate follows the water-filling algorithm. The optimal solution to (6.4) is given by

$$C(P_A, P_B) = \sum_{i=1}^{N} \log_2 \left(1 + \frac{\eta_i P_i^{opt}}{W} \right) \tag{6.5}$$

where $P_i^{opt} = \max\{ \frac{\log_2(e)}{\lambda^{opt} - \eta_i \mu^{opt}} - \frac{W}{\eta_i}, 0 \}$ is the optimum power that should be allocated to the ith band, $\max\{\cdot, \cdot\}$ takes the maximum of two values, and the parameters λ^{opt} and μ^{opt} satisfy the two equations

$$\sum_{i=1}^{N} \max \left\{ \left(\frac{\log_2(e)}{\lambda^{opt} - \eta_i \mu^{opt}} - \frac{N}{\eta_i} \right), 0 \right\} = P_A \tag{6.6}$$

$$\sum_{i=1}^{N} \eta_i \max \left\{ \left(\frac{\log_2(e)}{\lambda^{opt} - \eta_i \mu^{opt}} - \frac{N}{\eta_i} \right), 0 \right\} = P_B. \tag{6.7}$$

A continuous version of this optimization was also discussed in Grover and Sahai (2010), where η_i is replaced by $\eta(f)$, the power transfer efficiency as a function of frequency, and summation is replaced by integration over frequency.

Comparing (6.1) with (6.4), one sees that (6.1) imposes a constraint on the average transferred power, while (6.4) imposes a constraint on the total transmitted power and a constraint on the total transferred power. Thus, they lead to different solutions. Another difference between them is that (6.4) considers the frequency selectivity of the channel while (6.1) does not. However, both of them aim to maximize the information rate. Alternatively, one can also maximize the transferred power subject to constraints on the information rate. Other variants are also possible and may be useful in different applications.

6.3.2 Practical Implementations

The information theoretic bounds give the limiting performances of SWIPT. However, these bounds are only theoretically possible. In practice, it is difficult to use the same signal for both power transfer and information decoding at the same time, as the signal used for information decoding cannot be harvested for energy again, and vice versa. The results in Grover and Sahai (2010) and Varshney (2008) have assumed that one can use the signal for energy harvesting and then use the same signal again for information decoding. Thus, practical schemes for SWIPT are needed.

Such practical designs were first discussed in Zhang and Ho (2013) and Zhou et al. (2013b). In the following, Zhou et al. (2013b) will be used to explain how these practical designs work. We will focus on the separated information and energy receiver case in Zhou et al. (2013b), as the integrated information and energy receiver case in Zhou et al. (2013b) has certain limitations.

Since energy and information carried by the same signal cannot be processed at the same time, practical SWIPT implementations solve this problem in a simple way by splitting the signal into two parts: one part for power transfer or energy harvesting; and the other part for message transmission or information decoding. Thus, one has two main methods of SWIPT: time switching (TS) and power splitting (PS). Figure 6.4 compares these two methods. Generally speaking, TS splits the signal in the time domain, while PS splits the signal in the power domain. Their principles are explained in detail in the following.

6.3.2.1 Time Switching

For TS, the signal is split in the time domain. A switch is used such that during one part of the transmission time the received signal is connected to an energy harvester for power transfer and during the other part of the transmission time the received signal is switched to an information decoder for message delivery. Thus, the most important design parameter in this method is called the time-switching coefficient. The switching can happen within one packet between its data part and energy part, or between packets, depending on whether the whole packet or only part of the packet is used for power transfer. Without loss of generality, one can assume that the total transmission time is T seconds, of which αT seconds are used for power transfer and $(1 - \alpha)T$ seconds are used for information delivery, where $0 \le \alpha \le 1$ is the TS coefficient. The time T could be the symbol, frame or packet interval. Thus, the signal received for processing can be expressed as

$$y(t) = \begin{cases} h\sqrt{P_s}s(t) + n(t), & 0 \le t \le \alpha T \\ h\sqrt{P_s}s(t) + n(t) & \alpha T \le t \le T \end{cases} \tag{6.8}$$

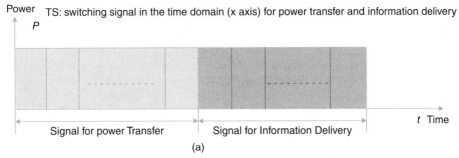

(a)

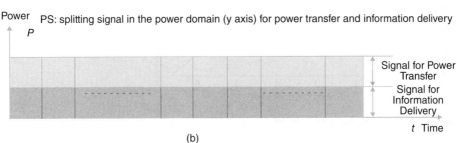

(b)

Figure 6.4 Comparison of (a) TS and (b) PS.

where P_s is the transmission power, $s(t)$ is the transmitted data symbol or any symbol, h is the complex channel gain between the transmitter and the receiver and $n(t)$ is the complex additive white Gaussian noise (AWGN) with mean zero and variance $2\sigma^2$. Note that this model is a baseband model where the signal has already been down-converted and the in-phase and quadrature components have been combined. From (6.8), the first part is used for energy harvesting to give the total harvested energy as

$$E = \eta \alpha P_s |h|^2 T \tag{6.9}$$

where η is the conversion efficiency of the harvester, which has been discussed in Chapter 3. Also, $E[|s(t)|^2] = 1$ has been assumed so that all the transmitted symbols have unit power. The second part is used for information decoding, and its achievable rate is given by

$$C = (1 - \alpha)\log_2\left(1 + \frac{P_s|h|^2}{2\sigma^2}\right) \tag{6.10}$$

where $(1 - \alpha)$ takes the penalty of power transfer into account (αT could have been used for data transmission for higher information rates). As can be seen from (6.9) and (6.10), a larger value of α gives more harvested energy but at the same time reduces the information rate. This tradeoff can be best described by using the rate-energy function defined in Zhou et al. (2013b) as

$$C(R, Q) = \left\{ Q \le \alpha \eta P_s |h|^2, R \le (1 - \alpha)\log_2\left(1 + \frac{P_s|h|^2}{2\sigma^2}\right) \right\} \tag{6.11}$$

where $\alpha \eta P_s |h|^2$ is the harvested energy (the term power and the term energy are used interchangeably in this chapter). Figure 6.5 shows the rate-energy function for TS. Note

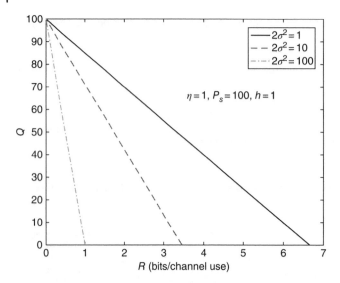

Figure 6.5 The rate-energy function for TS.

that the straight line is only the boundary and the actual rate-energy region is below the straight line. One sees that the rate-energy region increases when the noise power decreases, or when the signal-to-noise ratio (SNR) increases. The maximum rate or the maximum power can be achieved at the two ends of this straight line. On the other hand, if there is any constraint on rate or power, the optimum point will be somewhere between these two end points. This function is essentially the same as $C(B)$ studied in Varshney (2008) and $C(P_A, P_B)$ studied in Grover and Sahai (2010), only in a different form. Remember that we started from the fact that power transfer efficiency and information rate cannot be maximized at the same time and hence, we look for a tradeoff between them.

6.3.2.2 Power Splitting

For PS, the signal is split in the power domain. A power splitter is used in the circuit so that the received signal as an input will be divided into two parts: one part is fed into the energy harvester for energy harvesting and the other part is fed into the decoder for information decoding. Thus, unlike TS, energy harvesting and information decoding can be processed simultaneously but separately. The most important design parameter in this case is the PS factor, similar to the TS coefficient. Thus, the received signals for power transfer and for information delivery, respectively, can be written as

$$y_1(t) = \sqrt{\rho P_s} hs(t) + \sqrt{\rho} n_a(t) + n_d(t), \ 0 \le t \le T \tag{6.12a}$$

$$y_2(t) = \sqrt{(1-\rho)P_s} hs(t) + \sqrt{1-\rho} n_a(t) + n_d(t), \ 0 \le t \le T \tag{6.12b}$$

where ρ is the PS factor, $n_a(t)$ is the complex AWGN introduced by the antenna with mean zero and variance $2\sigma_a^2$, and $n_d(t)$ is the complex AWGN introduced by the conversion with mean zero and variance $2\sigma_d^2$. Again, this is a baseband signal model.

Several observations can be made. First, $n(t)$ in (6.8) is related to $n_a(t)$ and $n_d(t)$ in (6.12a,b) as $n(t) = n_a(t) + n_d(t)$. In other words, the AWGN in (6.8) actually consists of noise from the antenna as well as noise from processing. Thus, $2\sigma^2 = 2\sigma_a^2 + 2\sigma_d^2$. Secondly, the power splitting is applied to the received RF signal. Thus, one has $\sqrt{\rho}n_a(t)$ and $\sqrt{1-\rho}n_a(t)$, as $n_a(t)$ is part of the RF signal and is also split.

Using (6.12a,b), the total energy harvested from the energy signal is given by

$$E = \eta\rho P_s|h|^2 T \tag{6.13}$$

and the achievable rate of the information signal is given by

$$C = \log_2\left[1 + \frac{(1-\rho)P_s|h|^2}{(1-\rho)2\sigma_a^2 + 2\sigma_d^2}\right]. \tag{6.14}$$

Unlike TS, there is no penalty of power transfer in terms of information rate, as energy harvesting and information decoding are processed simultaneously. Thus, it has been shown in many studies that PS can provide a higher data rate than TS, under the same other conditions. As can be seen from (6.13) and (6.14), when ρ increases, more power can be harvested but the information rate decreases. Using the rate-energy function defined in Zhou et al. (2013b), this tradeoff is given as

$$C(R, Q) = \left\{Q \leq \eta\rho P_s|h|^2, R \leq \log_2\left[1 + \frac{(1-\rho)P_s|h|^2}{(1-\rho)2\sigma_a^2 + 2\sigma_d^2}\right]\right\}. \tag{6.15}$$

Figure 6.6 gives an example of the rate-energy function for PS. Again, the line represents the boundary of the rate-energy region. Unlike TS, in this case, the boundary for PS is not a straight line and it is actually a convex function. Thus, PS has a larger rate-energy region. It can be shown that PS and TS have the same maximum power and maximum rate or the same end points for the curves, under the same conditions, but for a tradeoff between the two end points, PS normally has a larger power or rate than TS. This tradeoff is the same tradeoff as what we are looking for in TS and in Grover and Sahai (2010)

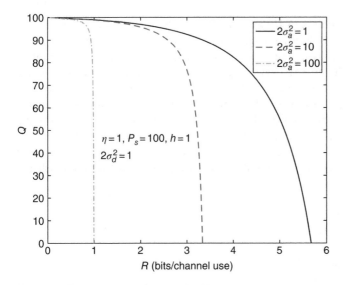

Figure 6.6 The rate-energy function for PS.

and Varshney (2008). In fact, most studies on SWIPT have been conducted to find this tradeoff under different conditions, such as multiple antennas, joint designs of power allocation and scheduling, etc. Some examples will be given later.

6.3.2.3 General Scheme

Although not necessary, it is useful to unify TS and PS to have a better view of how they operate. In Zhou et al. (2013b), TS and PS were unified under dynamic power splitting (DPS). Specifically, denote ρ_k as the PS factor for the kth symbol, frame, or packet. Following a similar procedure as before, the harvested energy can be derived as

$$E_k = \eta \rho_k P_s |h|^2 T \tag{6.16}$$

and the achievable rate is

$$C_k = \log_2 \left[1 + \frac{(1 - \rho_k) P_s |h|^2}{(1 - \rho_k) 2\sigma_a^2 + 2\sigma_d^2} \right]. \tag{6.17}$$

Considering the average harvested power and the average rate, for K symbols, frames, or packets, the rate-energy function is

$$C(R, Q) = \left\{ Q \leq \frac{\eta P_s |h|^2}{K} \sum_{k=1}^{K} \rho_k, R \leq \frac{1}{K} \sum_{k=1}^{K} \log_2 \left[1 + \frac{(1 - \rho_k) P_s |h|^2}{(1 - \rho_k) 2\sigma_a^2 + 2\sigma_d^2} \right] \right\}. \tag{6.18}$$

For TS discussed before, using the model of DPS, one has

$$\rho_k = \begin{cases} 1, & k = 1, 2, \cdots, \alpha K \\ 0, & k = \alpha K + 1, \alpha K + 2, \cdots, K \end{cases} \tag{6.19}$$

and for PS discussed before, one has

$$\rho_k = \rho, \quad k = 1, 2, \cdots, K. \tag{6.20}$$

It can be verified that, if one uses (6.19) and (6.20) in (6.18), one can obtain (6.11) and (6.15), respectively. Figure 6.7 shows a diagram that unifies TS and PS under the same structure. In this figure, $n(t)$ is $n_a(t)$ for PS, as $n_d(t)$ will only occur inside the information decoding box or the energy harvesting box. There are other forms of SWIPT implementation. For example, one can combine TS with PS so that a time switcher is first used to obtain dedicated power transfer time and then a power splitter is used during the information delivery time to split the signal into two parts for energy harvesting and information decoding. These variants are not discussed here. It was further proved in Zhou et al. (2013b) that the PS scheme provides the best tradeoff between energy and information among these schemes. The only shortcoming of PS is that the hardware

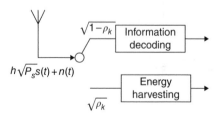

Figure 6.7 A unified structure for TS and PS.

implementation of a power splitter is much more complicated than a time switcher. Hence, one has to take both performance and complexity into account when choosing these SWIPT schemes.

Another issue that often causes confusion is that, in the above discussion and in most literature, the noise power is ignored in energy harvesting. Thus, in (6.9), (6.13), and (6.16), only the energy of the signal part is accounted for. This is based on the assumption that the noise is normally much weaker than the signal. If the noise is comparable with the signal, then the noise power can be added back in (6.9), (6.13), and (6.16), as it is part of the received energy. In addition to PS and TS, other SWIPT techniques can also be used. For example, antenna switching that splits a subset of antennas for information and another subset of antennas for energy was proposed in Ben Khelifa and Alouini (2017a) and Ben Khelifa et al. (2017). This splits the received signal in the space domain.

6.4 Hybrid Access Point

HAP is another useful form of wireless powered communications (Ju and Zhang 2014). The definition of HAP comes from the fact that the access point or the base station in this network serves as both a power transmitter and an information transmitter and so is hybrid. In most applications, the function of power transmitter and the function of information transmitter are performed at the HAP in a time-division-duplex (TDD) manner, as frequency-division-duplex would lead to extra costs at both the HAP and the remote device. In this case, as will be shown later, HAP is theoretically very similar to the TS scheme in SWIPT. Figure 6.8 shows a HAP wireless powered communications system with one HAP and K nodes. SWIPT is suitable for point-to-point wireless powered communications, while HAP is suitable for point-to-multi-point wireless powered communications.

6.4.1 Rate-Energy Tradeoff

Consider a HAP wireless powered communications system, where one HAP serves K remote devices. The TDD protocol proposed in Ju and Zhang (2014) is adopted,

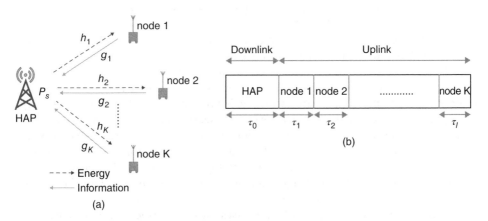

Figure 6.8 HAP wireless powered communications: (a) HAP structure; and (b) HAP link time.

where the HAP broadcasts wireless power in the downlink for τ_0 seconds, followed by information transmission from the nodes to the HAP in the TDD uplink using the harvested energy. Denote τ_k as the transmission time of node k, where $k = 1, 2, \cdots, K$. The fundamental tradeoff between energy and information occurs when the total time $\tau_0 + \sum_{k=1}^{K} \tau_k = T$ is fixed. Thus, if one increases τ_0 or the power transfer time, more energy can be harvested by the nodes such that they can use a larger transmission power for the information transmission in the uplink, while the information transmission time will be reduced such that less data symbols can be sent. Thus, unlike SWIPT where the tradeoff between energy and information is considered for the use of the same signal, HAP considers this tradeoff for the use of the same link. If one lets $K = 1$, one can easily see that this problem is very similar to the TS scheme in SWIPT. Next, assume that the received signal in the downlink is given by

$$y_k = h_k \sqrt{P_s} s + n_k \tag{6.21}$$

where $k = 1, 2, \cdots, K$, h_k is the complex channel gain from the HAP to the kth node, P_s is the transmission power of the HAP, s is the transmitted symbol and n_k is the complex AWGN at the kth node with mean zero and variance $2\sigma^2$. This signal is used for energy harvesting to give the harvested energy as

$$E_k = \eta P_s |h_k|^2 \tau_0. \tag{6.22}$$

The harvested energy is then used to transmit the information from the kth node to the HAP during τ_k as

$$r_k = g_k \sqrt{P_k} s_k + z_k \tag{6.23}$$

where g_k is the complex channel gain from the kth node to the HAP, s_k is the data symbol transmitted, z_k is the complex AWGN at the HAP during τ_k with mean zero and variance $2\sigma^2$, and P_k is the transmission power of the kth node given by

$$P_k = \frac{E_k}{\tau_k} = \eta P_s |h_k|^2 \frac{\tau_0}{\tau_k}. \tag{6.24}$$

Thus, the achievable information rate for the kth node is

$$R_k = \tau_k \log_2 \left(1 + \frac{P_k |g_k|^2}{2\sigma^2} \right) = \tau_k \log_2 \left(1 + \frac{\eta |g_k|^2 |h_k|^2 P_s \tau_0}{2\sigma^2 \tau_k} \right). \tag{6.25}$$

Note that the penalty τ_k must be added to account for the fact that only τ_k seconds are used by the kth node for information transmission.

In the derivation of (6.25), several assumptions have been made. First, it is assumed that all the harvested energy is used for transmission. This is the "harvest-then-transmit" protocol. It is possible to use only part of the harvested energy, if there is some energy storage device at the remote device. However, for sensor networks, this may not be necessary or possible. Secondly, all the energy loss has been ignored during the processing, except the conversion loss at the energy harvester represented by η. Thirdly, perfect synchronization is essential for this scheme, in fact, for all TDD protocols. Finally, h_k and g_k are random variables but only change from one transmission to the other. They remain constant during one transmission. Thus, block fading is assumed.

It can be shown that the rate in (6.25) increases with τ_0 for fixed τ_k, as more energy will be harvested for information transmission. It can also be shown that the rate in (6.25)

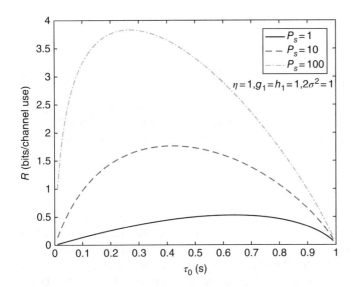

Figure 6.9 R_1 versus τ_0 when there is one node.

increases with τ_k for fixed τ_0, as more transmission time will be provided for information delivery. However, when the total time is fixed such that τ_0 and τ_k cannot increase at the same time, an optimal tradeoff exists. Figure 6.9 shows this tradeoff when there is only one node. It can be seen that an optimum value of τ_0 exists that maximizes the information rate. This optimization changes when P_s or the SNR changes.

The sum rate is obtained by adding the rates of all K devices to give

$$R(\tau_0, \cdots, \tau_K) = \sum_{k=1}^{K} R_k = \sum_{i=1}^{K} \tau_k \log_2 \left(1 + \frac{\eta |g_k|^2 |h_k|^2 P_s \tau_0}{2\sigma^2 \tau_k} \right). \tag{6.26}$$

Thus, the rate-energy function to be optimized in this system is

$$\max_{\tau_0, \cdots, \tau_K} \{R(\tau_0, \cdots, \tau_K)\}$$

$$s.t. \sum_{k=1}^{K} \tau_k \leq T$$

$$\tau_k \geq 0, k = 1, 2, \cdots, K. \tag{6.27}$$

Using the Lagrange multiplier, this optimization problem can be solved to give the optimum time intervals as

$$\tau_k^{opt} = \begin{cases} \dfrac{a_{opt} - 1}{A + a_{opt} - 1}, & k = 0 \\[2ex] \dfrac{\gamma_k}{A + a_{opt} - 1} & k = 1, 2, \cdots, K \end{cases} \tag{6.28}$$

where $A = \sum_{k=1}^{K} \gamma_k$, T has been set to 1 for convenience,

$$\gamma_k = \frac{\eta |h_k|^2 |g_k|^2 P_s}{2\sigma^2} \tag{6.29}$$

is the SNR of the kth node, and a_{opt} is the solution to the equation $A = a \ln a - a + 1$. Thus, the maximum sum rate is given by

$$R_{max} = \sum_{i=1}^{K} \tau_k^{opt} \log_2 \left(1 + \frac{\eta |g_k|^2 |h_k|^2 P_s \tau_0^{opt}}{2\sigma^2 \tau_k^{opt}} \right). \tag{6.30}$$

One sees that, in order to achieve this optimum rate, knowledge of the channel gains h_k and g_k is required at the beginning of each transmission, based on which the time intervals will be allocated. Thus, this scheme is suitable for quasi-fading channels where h_k and g_k change slowly.

6.4.2   Fairness Issue

The time allocation in Section 6.4.1 gives the best tradeoff between energy and information for the link. However, it is not fair to all devices. To see this, one notes from (6.28) that the length of time interval allocated to the kth device is proportional to the SNR of the device γ_k. From (6.29), the SNR increases with the channel gains h_k and g_k. In a wireless communication channel, the channel gains increase when the distance between the transmitter and the receiver decreases (Stuber 2001). Thus, nodes closer to the HAP have larger channel gains for both h_k and g_k. Consequently, they will be allocated a longer transmission time, which is not fair to other nodes. This is called the doubly near-far problem in Ju and Zhang (2014).

To tackle this fairness problem, another sum rate optimization problem can be formulated as

$$\max_{\tau_0, \cdots, \tau_K, R_0} \{R(\tau_0, \cdots, \tau_K)\}$$

$$s.t. \sum_{k=1}^{K} \tau_K \leq T$$

$$\tau_k \geq 0, k = 1, 2, \cdots, K.$$

$$R_k \geq R_0, k = 1, 2, \cdots, K. \tag{6.31}$$

where an additional constraint has been added to make sure each node has at least a guaranteed minimum rate of transmission. This optimization does not have any analytical solution but can be solved iteratively. Interested readers are referred to Ju and Zhang (2014) for the detailed procedure.

6.4.3   Channel Knowledge Issue

Another issue related to the energy-rate tradeoff in the HAP wireless powered communications is the channel knowledge. As mentioned before, one needs to predict h_k and g_k for the next transmission in order to allocate the time intervals beforehand. This prediction seems not to be reliable in a dynamic wireless channel. Also, this knowledge needs to be updated frequently if the channels change fast. The overhead incurred by the prediction may not be tolerable in this case.

To avoid these issues, instead of using the rate in (6.25), one can average it over the channel gains h_k and g_k and then try to optimize the average sum rate. In the following, the average rate is obtained, which can be used in the optimization of (6.27) or (6.31).

It has already been shown in Section 6.4.1 that the achievable rate of the kth device is

$$R_k = \tau_k \log_2(1 + \gamma_k) = \tau_k \log_2\left(1 + \frac{\eta P_s \tau_0}{2\sigma^2 \tau_k}|g_k|^2|h_k|^2\right). \tag{6.32}$$

Similarly, the bit error rate (BER) of the kth device is

$$B_k = \frac{1}{2}\text{erfc}(\sqrt{\gamma_k}) = \frac{1}{2}\text{erfc}\left(\sqrt{\frac{\eta P_s \tau_0}{2\tau_k \sigma^2}}|g_k||h_k|\right) \tag{6.33}$$

by using the instantaneous SNR, where erfc($\cdot$) is the complementary error function of the Gaussian distribution.

Assume Nakagami-m fading channels. The joint probability density function (PDF) of $|g_k|$ and $|h_k|$ is given by Simon and Alouini (2005)

$$f_{|g_k|,|h_k|}(x_1, x_2) = \frac{4m^{m+1}(x_1 x_2)^m e^{-\frac{m}{1-\rho_k}\left(\frac{x_1^2}{\Omega_1} + \frac{x_2^2}{\Omega_2}\right)}}{\Gamma(m)\Omega_1\Omega_2(1-\rho_k)(\Omega_1\Omega_2\rho_k)^{\frac{m-1}{2}}}I_{m-1}\left(\frac{2m\sqrt{\rho_k}x_1 x_2}{\sqrt{\Omega_1\Omega_2}(1-\rho_k)}\right) \tag{6.34}$$

where $I_{m-1}(\cdot)$ is the $(m-1)$th order modified Bessel function of the first kind (Gradshteyn and Ryzhik 2000), $\Gamma(\cdot)$ is the Gamma function, m is the Nakagami-m parameter, Ω_1 is the average fading power in the uplink, Ω_2 is the average fading power in the downlink, and ρ_k is the correlation coefficient between $|g_k|$ and $|h_k|$. If the Jakes' model applies, using Simon and Alouini (2005), the correlation coefficient is determined by

$$\rho_k = J_0^2\left(2\pi f_m T \sum_{i=0}^{k-1}\tau_i\right) \tag{6.35}$$

where $J_0(\cdot)$ is the zeroth order Bessel function of the first kind and f_m is the maximum Doppler shift. The time difference between g_k and h_k in the fading process is given by $\sum_{i=0}^{k-1}\tau_i$, as the uplink data transmission from the users to the access point takes place sequentially.

On the other hand, the special case occurs when g_k and h_k are independent, which is the case often used in the literature and is an approximation to (6.34) when $f_m T$ or τ_k are large such that $\rho_k \to 0$. In this case, one has

$$f_{|g_k|,|h_k|}(x_1, x_2) = \frac{4m^{m+1}(x_1 x_2)^{2m-1}}{\Gamma^2(m)(\Omega_1\Omega_2)^m}e^{-\frac{m}{\Omega_1}x_1^2 - \frac{m}{\Omega_2}x_2^2}. \tag{6.36}$$

In the following, we are going to use (6.34) and (6.36) to derive the average achievable rate and the average BER of the kth device.

6.4.3.1 Average Achievable Rate

Using (6.34), the average achievable rate can be calculated as

$$\bar{R}_k = \int_0^\infty \int_0^\infty \tau_k \log_2\left(1 + \frac{\eta P_s \tau_0}{2\sigma^2 \tau_k}x_1^2 x_2^2\right)f_{|g_k|,|h_k|}(x_1, x_2)dx_1 dx_2 \tag{6.37}$$

which requires the solution to a two-dimensional integral. We perform a two-dimensional variable transformation as $x = x_1$ and $z = x_1 x_2$. Then, using the Jacobian and its determinant, one has

$$\bar{R}_k = \int_0^\infty \int_0^\infty \tau_k \log_2\left(1 + \frac{\eta P_s \tau_0}{2\sigma^2 \tau_k}z^2\right)f_{|g_k|,|h_k|}\left(x, \frac{z}{x}\right)\frac{1}{x}dxdz \tag{6.38}$$

where the integrations over x_1 and x_2 have been transformed into the integrations over x and z.

For correlated links, using (6.34), the integral becomes

$$\bar{R}_k = \frac{4m^{m+1}\tau_k}{\Gamma(m)\Omega_1\Omega_2(1-\rho_k)(\Omega_1\Omega_2\rho_k)^{\frac{m-1}{2}}} \int_0^\infty \int_0^\infty \log_2\left(1 + \frac{\eta P_s\tau_0}{2\sigma^2\tau_k}z^2\right)z^m$$

$$I_{m-1}\left(\frac{2m\sqrt{\rho_k}z}{\sqrt{\Omega_1\Omega_2}(1-\rho_k)}\right)\frac{1}{x}e^{-\frac{m}{1-\rho_k}\left(\frac{x^2}{\Omega_1}+\frac{z^2}{x^2\Omega_2}\right)}dxdz. \tag{6.39}$$

The integration over x can be solved by using an equation in Gradshteyn and Ryzhik (2000, eq. (3.478.4)) to give (Gao et al. 2018)

$$\bar{R}_k = \frac{4m^{m+1}\tau_k}{\Gamma(m)\Omega_1\Omega_2(1-\rho_k)(\Omega_1\Omega_2\rho_k)^{\frac{m-1}{2}}} \int_0^\infty \log_2\left(1 + \frac{\eta P_s\tau_0}{2\sigma^2\tau_k}z^2\right)z^m$$

$$I_{m-1}\left(\frac{2m\sqrt{\rho_k}z}{\sqrt{\Omega_1\Omega_2}(1-\rho_k)}\right)K_0\left(\frac{2mz}{(1-\rho_k)\sqrt{\Omega_1\Omega_2}}\right)dz \tag{6.40}$$

where $K_0(\cdot)$ is the zeroth order modified Bessel function of the second kind (Gradshteyn and Ryzhik 2000). This integral cannot be simplified further unless approximations are applied. However, such a one-dimensional integral is very easy to calculate using standard mathematical software, such as MATLAB and MATHEMATICA. On the other hand, if one does need an approximation, one has (Gao et al. 2018)

$$\bar{R}_k = \frac{4m^{m+1}\tau_k}{\Gamma(m)\Omega_1\Omega_2(1-\rho_k)(\Omega_1\Omega_2\rho_k)^{\frac{m-1}{2}}} \sum_{i=0}^\infty \frac{\{m\sqrt{\rho_k}/[\sqrt{\Omega_1\Omega_2}(1-\rho_k)]\}^{m+2i-1}}{i!\Gamma(m+i)}$$

$$\int_0^\infty \log_2\left(1 + \frac{\eta P_s\tau_0}{2\sigma^2\tau_k}z^2\right)z^{2m+2i-1}K_0\left[\frac{2mz}{(1-\rho_k)\sqrt{\Omega_1\Omega_2}}\right]dz \tag{6.41}$$

where the series expansion of $I_{m-1}(\cdot)$ in Gradshteyn and Ryzhik (2000, eq. (8.445)) has been used. The function of $K_0(x)$ decays very fast with x. It can be shown that the integrand in (6.41) is very small when $x > 10$. Thus, we can perform a least-squares curve-fitting on $K_0(x)$ for $0 < x < 10$, which gives us $K_0(x) \approx 2.7e^{-1.9x}$. Using this approximation in (6.41), one has (Gao et al. 2018)

$$\bar{R}_k \approx \frac{10.8m^{m+1}\tau_k}{\Gamma(m)\Omega_1\Omega_2(1-\rho_k)(\Omega_1\Omega_2\rho_k)^{\frac{m-1}{2}}} \sum_{i=0}^\infty \frac{\{m\sqrt{\rho_k}/[\sqrt{\Omega_1\Omega_2}(1-\rho_k)]\}^{m+2i-1}}{i!\Gamma(m+i)\ln 2}$$

$$\left[\frac{(1-\rho_k)\sqrt{\Omega_1\Omega_2}}{3.8m}\right]^{2m+2i}\int_0^\infty \ln\left[1 + \frac{\eta P_s\tau_0}{2\sigma^2\tau_k}\frac{(1-\rho_k)^2\Omega_1\Omega_2}{3.8m^2}t^2\right]t^{2m+2i-1}e^{-t}dt$$

$$= \frac{10.8m^{m+1}\tau_k}{\Gamma(m)\Omega_1\Omega_2(1-\rho_k)(\Omega_1\Omega_2\rho_k)^{\frac{m-1}{2}}} \sum_{i=0}^\infty \frac{\{m\sqrt{\rho_k}/[\sqrt{\Omega_1\Omega_2}(1-\rho_k)]\}^{m+2i-1}}{i!\Gamma(m+i)\ln 2}$$

$$\left[\frac{(1-\rho_i)\sqrt{\Omega_1\Omega_2}}{3.8m}\right]^{2m+2i}F\left[\frac{\eta P_s\tau_0}{2\sigma^2\tau_k}\frac{(1-\rho_i)^2\Omega_1\Omega_2}{3.8m^2}, 2m+2i-1\right] \tag{6.42}$$

where we have defined

$$F(a,n) = \int_0^\infty \ln(1 + at^2)t^ne^{-t}dt \tag{6.43}$$

with n being an integer, and it can be solved iteratively using integration by parts as (Gao et al. 2018)

$$
\begin{aligned}
F(a, n) &= \int_0^\infty \ln(1 + at^2) t^n e^{-t} dt \\
&= -\int_0^\infty \ln(1 + at^2) t^n de^{-t} \\
&= n \int_0^\infty \ln(1 + at^2) t^{n-1} e^{-t} dt + 2 \int_0^\infty \frac{t^{n+1}}{1/a + t^2} e^{-t} dt \\
&= nF(a, n-1) + G(a, n)
\end{aligned}
\tag{6.44}
$$

where $G(a, n) = (-1)^{\frac{n+1}{2}} (\frac{1}{\sqrt{a}})^n [ci(\frac{1}{\sqrt{a}}) \sin(\frac{1}{\sqrt{a}}) - si(\frac{1}{\sqrt{a}}) \cos(\frac{1}{\sqrt{a}})] + \sum_{j=1}^{\frac{n+1}{2}}(n + 1 - 2j)!$ $(-\frac{1}{a})^{j-1}$ for odd values of n, $G(a, n) = (-1)^{\frac{n}{2}-1}(\frac{1}{\sqrt{a}})^n [ci(\frac{1}{\sqrt{a}}) \cos(\frac{1}{\sqrt{a}}) + si(\frac{1}{\sqrt{a}}) \sin(\frac{1}{\sqrt{a}})] +$ $\sum_{j=1}^{\frac{n}{2}}(n + 1 - 2j)!(-\frac{1}{a})^{j-1}$ for even values of n, and $F(a, 0) = \ln a + 2[\ln \frac{1}{\sqrt{a}} - ci(\frac{1}{\sqrt{a}})$ $\cos(\frac{1}{\sqrt{a}}) - si(\frac{1}{\sqrt{a}}) \sin(\frac{1}{\sqrt{a}})]$ by using Gradshteyn and Ryzhik (2000, eq. (4.338.1)), and $ci(\cdot)$ and $si(\cdot)$ are the cosine integral and the sine integral, respectively, using Gradshteyn and Ryzhik (2000, eq. (3.356.1) and eq. (3.356.2)).

Similarly, if the links are independent, using (6.36), one has

$$
\bar{R}_k = \frac{4m^{2m}\tau_k}{\Gamma^2(m)(\Omega_1\Omega_2)^m} \int_0^\infty \log_2\left(1 + \frac{\eta P_s \tau_0}{2\sigma^2 \tau_k} z^2\right) z^{2m-1} K_0\left(\frac{2mz}{\sqrt{\Omega_1\Omega_2}}\right) dz.
\tag{6.45}
$$

Equation (6.45) can also be obtained from (6.42) by using the series expansion of the Bessel function $I_{m-1}(\cdot)$ (Gradshteyn and Ryzhik 2000, eq. (8.445)) and letting $\rho_k \to 0$ in the expanded result, assuming that the integration and the limiting operations can exchange orders. Using $K_0(x) \approx 2.7e^{-1.9x}$, a simpler approximation can also be derived as

$$
\bar{R}_k \approx \frac{10.8m^{2m}\tau_k}{\Gamma^2(m)(\Omega_1\Omega_2)^m \ln 2} \left(\frac{\sqrt{\Omega_1\Omega_2}}{3.8m}\right)^{2m} F\left(\frac{\eta P_s \tau_0}{2\sigma^2 \tau_k} \frac{\Omega_1\Omega_2}{3.8^2 m^2}, 2m - 1\right).
\tag{6.46}
$$

6.4.3.2 Average BER

Using (6.34), the average BER can be obtained as

$$
\bar{B}_k = \frac{1}{2} \int_0^\infty \int_0^\infty erfc\left(\sqrt{\frac{\eta P_s \tau_0}{2\tau_k \sigma^2}} x_1 x_2\right) f_{|g_k|,|h_k|}(x_1, x_2) dx_1 dx_2.
\tag{6.47}
$$

By using the same method as that in the previous section, similar integral expressions for the average BER can be obtained. However, unlike the logarithm function, since the erfc function is convergent, integration by parts can be used to solve all the integrals. To do this, the average BER can be rewritten as

$$
\bar{B}_i = \frac{1}{2} \int_0^\infty erfc\left(\sqrt{\frac{\eta P_s \tau_0}{2\tau_k \sigma^2}} z\right) f_{|g_k||h_k|}(z) dz
\tag{6.48}
$$

where $f_{|g_k||h_k|}(z)$ is the PDF of the random variable $|g_k||h_k|$. From (6.48), one has

$$
\bar{B}_k = \frac{1}{2} \int_0^\infty erfc\left(\sqrt{\frac{\eta P_s \tau_0}{\tau_k \sigma^2}} z\right) dF_{|g_k||h_k|}(z)
\tag{6.49}
$$

where $F_{|g_k||h_k|}(z)$ is the cumulative distribution function (CDF) of $|g_k||h_k|$. Using integration by parts, one has from (6.49)

$$\bar{B}_k = \sqrt{\frac{\eta P_s \tau_0}{2\pi\sigma^2 \tau_k}} \int_0^\infty e^{-\frac{\eta P_s \tau_0}{2\tau_k \sigma^2}z^2} F_{|g_k||h_k|}(z)dz. \tag{6.50}$$

Note that the logarithm function is not convergent so that the integration by parts will lead to divergence in the calculation. Thus, this method cannot be used in the calculation of the average achievable rate. Note also that the calculation of (6.50) requires the CDF of $|g_k||h_k|$, which will be derived next.

The CDF of $|g_k||h_k|$ is defined as

$$F_{|g_k||h_k|}(z) = Pr\{x_1 x_2 < z, 0 < x_1 < \infty, 0 < x_2 < \infty\}. \tag{6.51}$$

Using the joint PDF in (6.34), this gives

$$F_{|g_k||h_k|}(z) = \frac{4m^{m+1}}{\Gamma(m)\Omega_1\Omega_2(1-\rho_k)(\Omega_1\Omega_2\rho_k)^{\frac{m-1}{2}}} \int_0^\infty x_2^m e^{-\frac{m}{1-\rho_k}\frac{x_2^2}{\Omega_2}}$$
$$\times \int_0^{z/x_2} x_1^m e^{-\frac{m}{1-\rho_k}\frac{x_1^2}{\Omega_1}} I_{m-1}\left(\frac{2m\sqrt{\rho_k}x_1 x_2}{\sqrt{\Omega_1\Omega_2}(1-\rho_k)}\right) dx_1 dx_2. \tag{6.52}$$

Using Gradshteyn and Ryzhik (2000, eq. (8.445)), one further has (Gao et al. 2018)

$$F_{|g_k||h_k|}(z) = \frac{4m^{m+1}}{\Gamma(m)\Omega_1\Omega_2(1-\rho_k)(\Omega_1\Omega_2\rho_k)^{\frac{m-1}{2}}} \sum_{i=0}^\infty \frac{\left[\frac{m\sqrt{\rho_k}}{\sqrt{\Omega_1\Omega_2(1-\rho_k)}}\right]^{m-1+2i}}{i!\Gamma(m+i)}$$
$$\times \int_0^\infty x_2^{2m-1+2i} e^{-\frac{m}{1-\rho_k}\frac{x_2^2}{\Omega_2}} \int_0^{z/x_2} x_1^{2m+2i-1} e^{-\frac{m}{1-\rho_k}\frac{x_1^2}{\Omega_1}} dx_1 dx_2. \tag{6.53}$$

We solve the inner integral first, which is a function of x_2. By letting $t = x_1^2$ and using Gradshteyn and Ryzhik (2000, eq. (3.351.1)), one has

$$\int_0^{z/x_2} x_1^{2m+2i-1} e^{-\frac{m}{1-\rho_k}\frac{x_1^2}{\Omega_1}} dx_1$$
$$= \frac{(m+i-1)!}{2(\frac{m}{(1-\rho_k)\Omega_1})^{m+i}} \left[1 - e^{-\frac{m}{1-\rho_k}\frac{z^2}{\Omega_1 x_2^2}} \sum_{j=0}^{m+i-1} \frac{(z^2/x_2^2)^j}{j!} \left[\frac{(1-\rho_k)\Omega_1}{m}\right]^j\right]. \tag{6.54}$$

Using (6.54) in (6.53), the integral becomes (Gao et al. 2018)

$$F_{|g_k||h_k|}(z) = \frac{4m^{m+1}}{\Gamma(m)\Omega_1\Omega_2(1-\rho_k)(\Omega_1\Omega_2\rho_k)^{\frac{m-1}{2}}} \sum_{i=0}^\infty \frac{\left[\frac{m\sqrt{\rho_k}}{\sqrt{\Omega_1\Omega_2(1-\rho_k)}}\right]^{m-1+2i}}{2i!\left[\frac{m}{(1-\rho_k)\Omega_1}\right]^{m+i}}$$
$$\times \left\{\int_0^\infty x_2^{2m-1+2i} e^{-\frac{m}{1-\rho_k}\frac{x_2^2}{\Omega_2}} dx_2 - \sum_{j=0}^{m+i-1} \frac{z^{2j}\left[\frac{(1-\rho_k)\Omega_1}{m}\right]^j}{j!}\right.$$
$$\left.\times \int_0^\infty x_2^{2m+2i-2j-1} e^{-\frac{m}{1-\rho_k}\frac{x_2^2}{\Omega_2}-\frac{m}{1-\rho_k}\frac{z^2}{\Omega_1 x_2^2}} dx_2\right\}. \tag{6.55}$$

Using equations from Gradshteyn and Ryzhik (2000, eq. (3.461.3) and eq. (3.478.4)) in (6.55), the CDF is (Gao et al. 2018)

$$
F_{|g_k||h_k|}(z) = \sum_{i=0}^{\infty} \frac{2\rho_k^i m^{2m+2i}}{i! \Gamma(m)(\Omega_1\Omega_2)^{m+i}(1-\rho_k)^{m+2i}}
$$
$$
\times \left\{ \frac{(m+i-1)!(\Omega_1\Omega_2)^{m+i}(1-\rho_i)^{2m+2i}}{2m^{2m+2i}} \right.
$$
$$
\left. - \sum_{j=0}^{m+i-1} \frac{1}{j!} \left[\frac{z(1-\rho_k)\sqrt{\Omega_1\Omega_2}}{m} \right]^{m+j-i} K_{m+j-i}\left[\frac{2mz}{(1-\rho_k)\sqrt{\Omega_1\Omega_2}} \right] \right\}. \quad (6.56)
$$

Then, the average BER can be derived by substituting (6.56) into (6.50), which gives

$$
\bar{B}_k = \sqrt{\frac{\eta P_s \tau_0}{2\pi\sigma^2\tau_k}} \sum_{i=0}^{\infty} \frac{2\rho_k^i m^{2m+2i}}{i!\Gamma(m)(\Omega_1\Omega_2)^{m+i}(1-\rho_k)^{m+2i}}
$$
$$
\left\{ \frac{(m+i-1)!(\Omega_1\Omega_2)^{m+i}(1-\rho_k)^{2m+2i}}{2m^{2m+2i}} \right.
$$
$$
\times \int_0^{\infty} e^{-\frac{\eta P_s \tau_0}{2\tau_k\sigma^2}z^2} dz - \sum_{j=0}^{m+i-1} \frac{1}{j!} \left[\frac{(1-\rho_k)\sqrt{\Omega_1\Omega_2}}{m} \right]^{m+j-i}
$$
$$
\left. \times \int_0^{\infty} z^{m+i+j} e^{-\frac{\eta P_s \tau_0}{2\tau_k\sigma^2}z^2} K_{m+j-i}\left[\frac{2mz}{(1-\rho_k)\sqrt{\Omega_1\Omega_2}} \right] dz \right\}. \quad (6.57)
$$

Solving the two one-dimensional integrals using equations in Gradshteyn and Ryzhik (2000, eq. (3.461.3) and eq. (6.631.3)), one can derive the average BER as (Gao et al. 2018)

$$
\bar{B}_k = \sqrt{\frac{\eta P_s \tau_0}{2\pi\sigma^2\tau_k}} \sum_{i=0}^{\infty} \frac{2\rho_k^i m^{2m+2i}}{i!\Gamma(m)(\Omega_1\Omega_2)^{m+i}(1-\rho_k)^{m+2i}}
$$
$$
\times \left\{ \frac{(m+i-1)!(\Omega_1\Omega_2)^{m+i}(1-\rho_k)^{2m+2i}}{4m^{2m+2i}\sqrt{\eta P_s\tau_0/(2\pi\tau_k\sigma^2)}} \right.
$$
$$
- \sum_{j=0}^{m+i-1} \frac{1}{j!} \left[\frac{(1-\rho_k)\sqrt{\Omega_1\Omega_2}}{m} \right]^{m+i-j+1} \frac{\Gamma(m+i+0.5)\Gamma(j+0.5)}{4\left(\sqrt{\frac{\eta P_s\tau_0}{2\tau_k\sigma^2}}\right)^{m+j+i}}
$$
$$
\left. \times e^{\frac{m^2\tau_k\sigma^2}{2(1-\rho_k)^2\Omega_1\Omega_2\eta P_s\tau_0}} W_{-\frac{m+j+i}{2},\frac{m+j-i}{2}}\left[\frac{2m^2\tau_k\sigma^2}{(1-\rho_k)^2\Omega_1\Omega_2\eta P_s\tau_0} \right] \right\} \quad (6.58)
$$

where $W(\cdot)$ is the Whittaker function (Gradshteyn and Ryzhik 2000).

When the links are independent, using a similar method, the average BER in this case is given by

$$
\bar{B}_k = \frac{1}{2} - \frac{1}{2} \frac{m^{m-1}}{\Gamma(m)\sqrt{\pi}(\Omega_1\Omega_2)^{\frac{m-1}{2}}} \left(\sqrt{\frac{\eta P_s\tau_0}{2\tau_i\sigma^2}} \right)^{1-m} \sum_{i=0}^{m-1} \frac{\Gamma(m+0.5)\Gamma(i+0.5)}{i!}
$$
$$
\times \left(m\sqrt{\frac{2\tau_k\sigma^2}{\Omega_1\Omega_2\eta P_s\tau_0}} \right)^i e^{\frac{m^2\tau_k\sigma^2}{2\eta P_s\tau_0\Omega_1\Omega_2}} W_{-\frac{m+i}{2},\frac{m-i}{2}}\left(\frac{2m^2\tau_k\sigma^2}{\eta P_s\tau_0\Omega_1\Omega_2} \right). \quad (6.59)
$$

One can also let $m = 1$ to obtain the results for Rayleigh fading channels. This will simplify the results further.

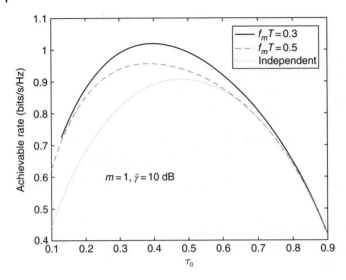

Figure 6.10 Average rate versus τ_0 for different values of $f_m T$ in the Jakes' model.

6.4.3.3 Numerical Examples

In this section, the performance of wireless powered communications (WPC) will be examined through numerical examples using the average rate and the average BER. In the examination, we set $\eta = 0.5$, and $\Omega_1 = \Omega_2 = 1$, while we vary the values of $\tau_0, f_m T, m$, and $\bar{\gamma} = \frac{P_s}{2\sigma^2}$. For simplicity, we only consider the case of one user such that $\tau_0 + \tau_1 = 1$, except in Figure 6.20.

Figures 6.10–6.16 examine the average rate performance of the WPC system under different conditions. In particular, Figure 6.10 shows the achievable rate versus τ_0 for different values of $f_m T$. The value of $f_m T$ determines the correlation of the links. The smaller the value of $f_m T$ is, the more correlated the links will be. Several observations can be made from Figure 6.10. First, there exists an optimum τ_0 in all the cases considered, as expected, as a larger τ_0 generates more harvested power and higher SNR in the received signal at the access point, but it also reduces the effective time for data transmission. Secondly, different values of $f_m T$ give different achievable rates. For example, the optimum τ_0 for $f_m T = 0.5$ is around 0.4, while the optimum τ_0 for independent links is around 0.5. Their maximum achievable rates are different too. Thus, the link correlation does affect the system performance. On the other hand, when $\tau_0 > 0.6$, their performances are very similar. From (6.35), $\rho_1 = J_0^2(2\pi f_m T \tau_0)$. Thus, the correlation coefficient in general decreases when τ_0 increases. When $\tau_0 = 0.6$ and $f_m T = 0.5$, one can find that $\rho_1 \approx 0.08$, which is very close to 0. Thus, although the correlation affects the performance, this effect may be ignored when the correlation coefficient is small. Finally, if one considers the independent links as the case when $f_m T \to \infty$, one sees that the achievable rate increases and the optimum τ_0 decreases when $f_m T$ decreases, as the link correlation benefits the performance and for smaller correlation a larger value of τ_0 is needed to harvest more power.

Figures 6.11 and 6.12 show the average rate versus τ_0 for different values of $\bar{\gamma}$ and m, respectively. One sees that the achievable rate performance of the WPC system can be significantly affected by the values of $\bar{\gamma}$ and m. Specifically, the achievable rate increases

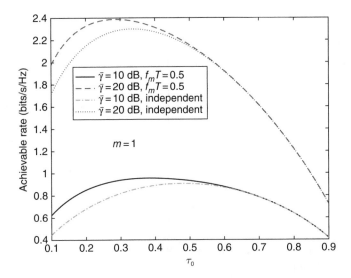

Figure 6.11 Average rate versus τ_0 for different values of $\bar{\gamma}$ in the channel.

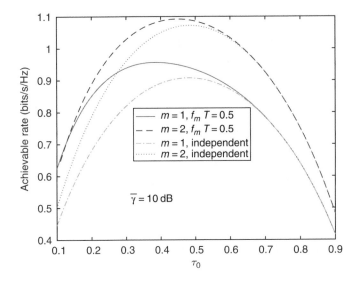

Figure 6.12 Average rate versus τ_0 for different values of the Nakagami-m parameter in the channel.

when $\bar{\gamma}$ and m increase, as expected, as the channel conditions become better for larger values of $\bar{\gamma}$ and m. In Figure 6.11, the optimum τ_0 decreases when $\bar{\gamma}$ increases, while in Figure 6.12, the optimum τ_0 increases when m increases. Similar to Figure 6.10, comparing the performance for correlated links with that for independent links, one sees that their performances are similar when τ_0 is large due to low correlation but are significantly different when τ_0 is small with large correlation.

Figures 6.13 and 6.14 show the maximum achievable rate and the optimum τ_0 versus $\bar{\gamma}$ for different values of $f_m T$, respectively. One sees from Figure 6.13 that the difference caused by link correlation increases when $\bar{\gamma}$ increases or $f_m T$ decreases, as expected. For

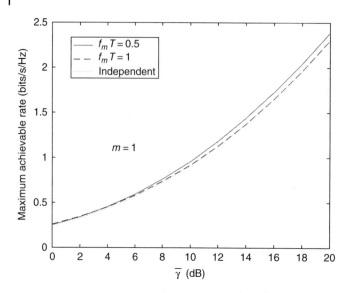

Figure 6.13 Maximum average rate versus $\bar{\gamma}$ for different values of $f_m T$ in the Jakes' model.

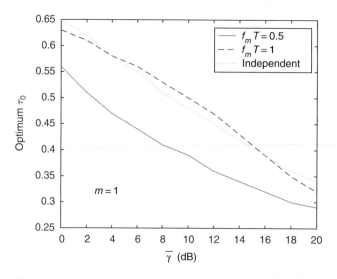

Figure 6.14 Optimum τ_0 versus $\bar{\gamma}$ for different values of $f_m T$ in the Jakes' model.

small $\bar{\gamma}$ and large $f_m T$, the difference is negligible. On the other hand, from Figure 6.14, the optimum values of τ_0 that reach the maximum achievable rates are significantly different between correlated links and independent links. For example, when $\bar{\gamma} = 10$ dB, the optimum τ_0 for independent links is 0.5, while the optimum τ_0 for correlated links is around 0.38. Since the maximum value of τ_0 is 1, this represents a 12% difference.

Figures 6.15 and 6.16 show the maximum achievable rate and the optimum τ_0 versus $\bar{\gamma}$ for different values of m, respectively. From Figure 6.15, the maximum achievable rate increases when m increases. However, the difference between correlated links and

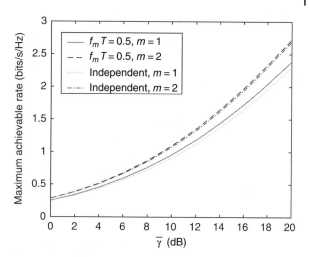

Figure 6.15 Maximum average rate versus $\bar{\gamma}$ for different values of the Nakagami-m parameter in the channel.

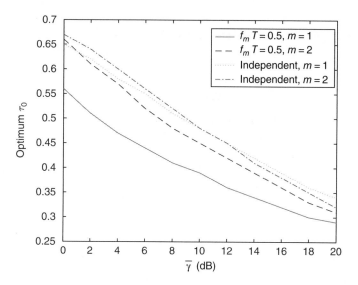

Figure 6.16 Optimum τ_0 versus $\bar{\gamma}$ for different values of the Nakagami-m parameter in the channel.

independent links decreases when m increases. This implies that the link correlation is not important when the channel condition is good enough, as it can compensate the performance degradation caused by low link correlation. Similarly, from Figure 6.16, the optimum τ_0 increases when m increases but the difference between correlated links and independent links also decreases when m increases. For example, when $\bar{\gamma} = 10$ dB, $m = 1$ gives the optimum τ_0 of 0.38 for $f_m T = 0.5$ and the optimum τ_0 of 0.5 for independent links, while $m = 2$ gives the optimum τ_0 of 0.45 for $f_m T = 0.5$ and the optimum τ_0 of 0.5 for independent links.

Figures 6.17–6.19 show the average BER performance of the system for different parameters. One sees from these figures that the BER decreases when $\bar{\gamma}$ increases or m increases. However, the BER changes little when $f_m T$ changes in the cases considered,

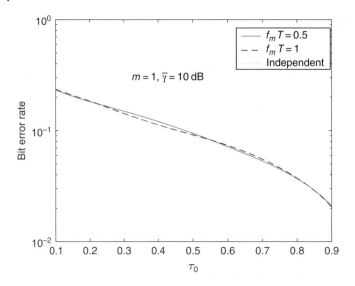

Figure 6.17 Average BER versus τ_0 for different values of $f_m T$ in the Jakes' model.

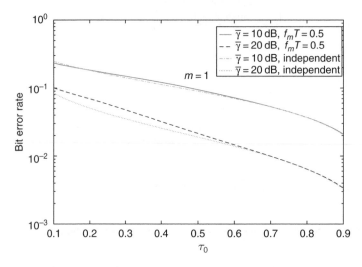

Figure 6.18 Average BER versus τ_0 for different values of $\bar{\gamma}$ in the channel.

implying that the link correlation does not affect the BER performance of the system much. Also, the BER always decreases when τ_0 increases, as a larger τ_0 will produce more harvested power and hence higher SNR in the received signal.

Figure 6.20 shows the average sum rate of two users versus τ_0, when $\bar{\gamma} = 10$ dB, $f_m T = 0.5$, and $m = 1$ for different values of τ_1. In this case, $\tau_0 + \tau_1 + \tau_2 = 1$. One sees that there exists an optimum τ_0 that maximizes the average sum rate for a fixed value of τ_1. When the value of τ_1 increases from 0.05 to 0.30 with a step size of 0.05, the maximum rate increases and the corresponding optimum τ_0 decreases. When τ_1 is

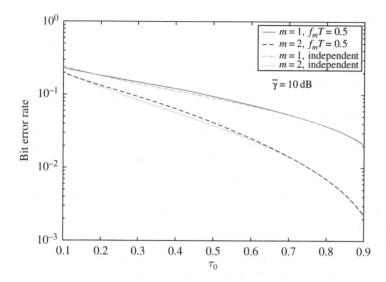

Figure 6.19 Average BER versus τ_0 for different values of the Nakagami-m parameter in the channel.

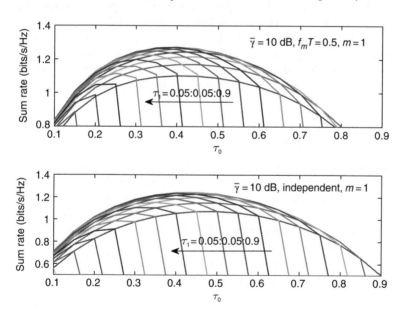

Figure 6.20 Average sum rate versus τ_0 for two devices.

larger than 0.30 and keeps increasing, the maximum rate decreases again, implying that there is a global maximum sum rate.

In summary, the tradeoff between energy and information for the HAP wireless powered communications is considered for the use of the same link. The balance between the power transfer time and the information delivery time needs to be studied. This leads to many optimization problems, including the basic ones discussed in this section. More details on the derivation can be found in Gao et al. (2018).

6.5 Power Beacon

PB is the last form of wireless powered communications that will be discussed here. Compared with SWIPT and HAP, in PB-based wireless powered communications, wireless power transfer and wireless information transmission are more independent of each other. For SWIPT, the same RF signal has to be used to deliver both power and information. For HAP, the same link has to be shared in a TDD manner for power and information. However, for PB, a separate power network composed of multiple PBs is implemented to deliver power. To avoid interfering with the base station that delivers information, PB often operates at a different frequency band. Also, at the remote nodes, two sets of antennas and RF fronts are required, one for information and one for power. Thus, PB-based wireless powered communications simplifies the tradeoff between energy and information, at the cost of additional infrastructure.

Since power transfer and information delivery are relatively independent of each other in PB, not much research work needs to be done in terms of their tradeoff. Compared with conventional wireless communications systems, the only change here is that the power is delivered by RF signals instead of battery. Thus, most research on PB focuses on how the RF power can be delivered efficiently and on time, for example, by using full-duplex radios (Chalise et al. 2017), by designing waveforms for power transfer (Ku et al. 2016), and by combining it with HAP or SWIPT (Ma et al. 2015).

The only resource that has to be shared by both energy and information in PB is perhaps the space, as the power network overlays with the information network. Hence, the spatial distribution of PBs and base stations need to be coordinated. Another constraint is the energy causality, as energy has to be harvested before it can be used. These aspects are discussed in the following sections.

6.5.1 System and Design Problem

Consider a hybrid communications–power network where multiple base stations, multiple PBs, and multiple mobile nodes are randomly distributed with densities of λ_b, λ_p, and λ_m, respectively, in a certain area. The transmission power of the mobile is p, and the transmission power of the PB is q. Figure 6.21 shows the structure of a mobile receiver that can be used in such a PB-based wireless powered communications system. Since the energy storage unit cannot charge and supply power at the same time, two energy storage units are needed, one for charging and one for power supply for the data transmission. Their roles will be switched, depending on the status of the power receiver and the data transceiver.

There are two performance metrics. The first one is the signal outage at the base station that serves a specific mobile. It can be given as

$$P_O = Pr\left\{\frac{p/d_1^\alpha}{I + \sigma^2} < \gamma_0\right\} \tag{6.60}$$

where d_1 is the random distance between the base station and the mobile, α is the path loss exponent, $I + \sigma^2$ is the interference-plus-noise power, and γ_0 is the threshold. Thus, if the signal from the mobile to the base station has a signal-to-interference-plus-noise ratio (SINR) below a threshold, the base station will not be able to detect this signal and hence a signal outage occurs. The second one is the power outage at the mobile. If

Figure 6.21 A mobile receiver used in PB-based wireless powered communications.

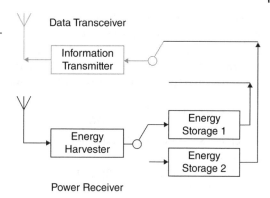

the mobile has a limited storage and the instantaneous power it receives from the PB is not enough for the next transmission, a power outage occurs. This only happens when the energy storage is limited. If the energy storage is unlimited, the power outage may not occur.

The power transfer can be performed in two ways: isotropic power transfer; or directed power transfer. Isotropic transfer uses an omnidirectional antenna, while directed transfer uses a directional antenna or array. In the case of isotropic transfer, the mobile receives power from all PBs within the range so that the received instantaneous power is

$$P = \sum_{i \in \Phi} \frac{qd}{[\max\{d_i, v\}]^{\beta}} \tag{6.61}$$

where Φ represents the set of PBs within the range, d_i is the distance between the mobile and the ith PB in Φ, $v \geq 1$ is a constant, d is the reference path loss, and β is the path loss exponent between mobiles and PBs. Here, the non-singular path loss model is used to avoid the issue in the traditional Friis formula that the path loss decreases when the distance increases for distances smaller than 1 m. In the case of directed transfer, the mobile will receive power from the mainlobe of the nearest PB serving it as well as side-lobes from other PBs. The received instantaneous power is

$$P = \frac{c_m qd}{[\max\{d_2, v\}]^{\beta}} + \sum_{j \in \Psi} \frac{c_s qd}{[\max\{d_j, v\}]^{\beta}} \tag{6.62}$$

where c_m is a constant representing the mainlobe response, d_2 is the distance between the mobile and its nearest PB, Ψ is the set of all PBs that affect the mobile, $c_s < c_m$ is the sidelobe response of all other PBs, d_j is the distance between the mobile and the jth PB in Ψ, and all other symbols are defined as before.

The design goals here are to find all possible combinations of p, q, λ_p, and λ_b that allow the signal outage to be smaller than a threshold ϵ and the power outage to be smaller than a threshold δ. These parameters are all larger than 0. It has been derived in Huang and Lau (2014) that, for isotropic transfer, these system parameters must satisfy (Huang and Lau 2014)

$$q\lambda_p \lambda_b^{\frac{\alpha}{2}} \geq (1 - \frac{2}{\beta}) \frac{\omega \sigma^2 v^{\beta - 2}}{\pi e d \mu} \tag{6.63}$$

where ω is the duty cycle scheduled by the base station for the mobile to transmit information, σ^2 is the noise power at the base station, e and d are the reference path losses in the communications network and the power network, respectively, and μ is a constant determined by ϵ. A curve between μ and ϵ has been provided in Huang and Lau (2014).

For directed transfer, they also satisfy (Huang and Lau 2014)

$$c_m q \lambda_b^{\frac{\alpha}{2}} (1 - e^{-\pi \lambda_p v^2}) \geq \frac{\omega \sigma^2 v^\beta}{ed\mu} \tag{6.64}$$

which is a subset of (6.42). Equations (6.42) and (6.43) are for unlimited energy storage at the mobile.

In the case when the energy storage is limited, for isotropic transfer, one has (Huang and Lau 2014)

$$q \lambda_p^{\frac{\ell}{2}} \lambda_b^{\frac{\alpha}{2}} \geq \left(\frac{-\log \delta}{\pi} \right)^{\frac{\ell}{2}} \frac{\sigma^2}{ed\mu} \tag{6.65a}$$

$$q \lambda_b^{\frac{\alpha}{2}} \geq \frac{\sigma^2 v^\beta}{ed\mu} \tag{6.65b}$$

and for directed transfer, one has (Huang and Lau 2014)

$$z_m q \lambda_p^{\frac{\ell}{2}} \lambda_b^{\frac{\alpha}{2}} \geq \left(\frac{-\log \delta}{\pi} \right)^{\frac{\ell}{2}} \frac{\sigma^2}{ed\mu} \tag{6.66a}$$

$$z_m q \lambda_b^{\frac{\alpha}{2}} \geq \frac{\sigma^2 v^\beta}{ed\mu}. \tag{6.66b}$$

A detailed discussion of these equations can be found in Huang and Lau (2014). Note that all these results are based on stochastic geometry models with randomly distributed base stations, PBs and mobiles. There has been no system level simulation of these results. The actual implementation is likely to be more complicated; especially the coordination between the wireless power network and the wireless communications network requires considerable effort.

6.5.2 More Notes

In summary, PB-based wireless powered communications require more infrastructure support in terms of extra PBs and extra RF fronts than SWIPT and HAP, but with a simpler tradeoff between energy and information. Thus, most research in PB-based wireless powered communications focuses on the RF power transfer and the PB network deployment, as the communications side of this system has been well studied. For SWIPT and HAP, power transfer and information delivery have to share the same signal or the same link. Hence, the tradeoff in these wireless powered communications is more complicated, and most of the research focuses on this tradeoff in different applications, such as multi-input-multiple-output (Zhang and Ho 2013), physical layer security (Liu et al. 2014a), joint time and power allocation (Luo et al. 2013), and even joint resource allocation and power splitting (Shi et al. 2014). A common point is that all these works study the fundamental tradeoff between energy and information. In the following, some of these interesting studies will be discussed. To have a complete view of all these studies, a cited reference search of Grover and Sahai (2010), Zhou et al. (2013b), or Ju and Zhang (2014) will be helpful.

6.6   Other Issues

6.6.1   Effect of Interference on Wireless Power

The interference in a wireless network normally degrades the system performance, as it reduces the SINR. There have been quite a few studies on the effect of interference on wireless powered communications. For example, Diamantoulakis et al. (2017) examined the effect of interference on HAP. Zhong et al. (2015a) studied the effect of interference on the throughput of a PB scheme. In Liu et al. (2013), the effect of interference on the scheduling of energy and information at the transmitter was studied. In Shen et al. (2014), a multi-user interference channel was considered by maximizing the sum rate subject to energy harvesting constraints. In Chen et al. (2017a), the effect of interference generated by sending wireless power was studied.

On the other hand, interference is also a good source of energy. In particular, for wireless powered communications, interference is some ambient RF energy that can actually be harvested for power. Consequently, this may improve the data transmission later due to a higher transmission power. In Zhao et al. (2017b), a detailed survey on the use of interference in energy harvesting has been provided.

Thus, it is interesting to study these two contradicting effects of interference on wireless powered systems. In the following, the HAP system is considered. In this system, the interference can increase the amount of the power received at the nodes in the downlink. This increased power could be used to send the information back to the HAP using a higher transmission power, to compensate the reduced SINR due to interference in the uplink. Rayleigh fading channels are used. The average SINR and the average rate will be derived next.

6.6.1.1   System and Assumptions

As discussed before, in HAP, power transfer and information transmission are performed sequentially. The base station first transmits energy to the nodes for τ_0 seconds. The nodes harvest this energy and uses it to transmit data to the base station for τ_1 seconds, where $\tau_0 + \tau_1 = T$ is the total time.

Using this assumption, the signal received by the node in the downlink is

$$y = \sqrt{P_s}h + \sum_{i=1}^{N} \sqrt{Q_i}u_ix_i + n \qquad (6.67)$$

where P_s is the transmission power of the base station, h is the fading channel gain and is a complex Gaussian random variable with mean zero and variance $2\alpha_h^2$, N is the number of interferes in the downlink, Q_i is the transmission power of the ith interferer, u_i is the fading gain from the ith interferer and is a complex Gaussian random variable with mean zero and variance $2\alpha_{u_i}^2$, x_i is the transmitted symbol of the ith interferer with $E\{|x_i|^2\} = 1$, $E\{\cdot\}$ is the expectation operator, and n is the AWGN with mean zero and variance $2\sigma^2$. This signal is harvested, and the harvested energy is given by

$$E_h = \eta(P_s|h|^2 + \sum_{i=1}^{N} Q_i|u_i|^2)\tau_0 \qquad (6.68)$$

where η is the conversion efficiency of the energy harvester. From this equation, the interference increases the amount of energy harvested.

The harvested energy is then used to transmit the information to the base station. The signal received by the base station is

$$z = \sqrt{P_r} g x + \sum_{j=1}^{M} \sqrt{T_j} v_j x_j + n \tag{6.69}$$

where $P_r = \frac{\tau_0}{\tau_1} \eta (P_s |h|^2 + \sum_{i=1}^{N} Q_i |u_i|^2)$ is the node transmission power using the harvested energy, g is the complex Gaussian fading gain with mean zero and variance $2\alpha_g^2$, x is the transmitted signal with $E\{|x|^2\} = 1$, M is the number of interfering nodes, T_j is the jth interfering node's transmission power, v_j is the fading gain from the jth interfering node and is a complex Gaussian random variable with mean zero and variance $2\alpha_{v_j}^2$, x_j is the transmitted symbol with $E\{|x_j|^2\} = 1$, and n is again the noise with mean zero and variance $2\sigma^2$. Thus, the SINR in this case is

$$\gamma = \frac{\tau_0}{\tau_1} \eta |g|^2 \frac{P_s |h|^2 + \sum_{i=1}^{N} Q_i |u_i|^2}{\sum_{j=1}^{M} T_j |v_j|^2 + 2\sigma^2}. \tag{6.70}$$

It can be seen that the interference increases the SINR in the numerator while decreasing the SINR in the denominator. The actual effect of the interference cannot be observed from this equation, and has to be analyzed.

6.6.1.2 Performances with Interference

First, the probability density functions (PDFs) of two random variables are needed. In the numerator, let $S = P_s |h|^2 + \sum_{i=1}^{N} Q_i |u_i|^2$. Since $|h|^2, |u_i|^2, i = 1, 2, \cdots, N$ are independent exponential random variables, one has (Johnson et al. 1994)

$$f_S(s) = \sum_{i=0}^{N} \frac{\prod_{k=0}^{N} \frac{1}{2\alpha_{u_k}^2 Q_k}}{\prod_{k=0, k \neq i}^{N} (\frac{1}{2\alpha_{u_k}^2 Q_k} - \frac{1}{2\alpha_{u_i}^2 Q_i})} e^{-\frac{s}{2\alpha_{u_i}^2 Q_i}}, \quad s > 0 \tag{6.71}$$

where $\alpha_{u_0} = \alpha_h$ and $Q_0 = P_s$. In the denominator, defining $\Omega = \sum_{j=1}^{M} T_j |v_j|^2$, one has

$$f_\Omega(w) = \sum_{j=1}^{M} \frac{\prod_{k=1}^{M} \frac{1}{2\alpha_{v_k}^2 T_k}}{\prod_{k=1, k \neq j}^{M} (\frac{1}{2\alpha_{v_k}^2 T_k} - \frac{1}{2\alpha_{v_j}^2 T_j})} e^{-\frac{w}{2\alpha_{v_j}^2 T_j}}, \quad w > 0. \tag{6.72}$$

The average SINR of the signal received by the base station is obtained as

$$\bar{\gamma} = \frac{\tau_0}{\tau_1} \eta 2\alpha_g^2 (2\alpha_h^2 P_s + \sum_{i=1}^{N} 2\alpha_{u_i}^2 Q_i) E \left\{ \frac{1}{\Omega + 2\sigma^2} \right\} \tag{6.73}$$

where

$$E \left\{ \frac{1}{\Omega + 2\sigma^2} \right\} = -\sum_{j=1}^{M} \frac{\prod_{k=1}^{M} \frac{1}{2\alpha_{v_k}^2 T_k} e^{\frac{\sigma^2}{\alpha_{v_j}^2 T_j}} Ei(-\frac{\sigma^2}{\alpha_{v_j}^2 T_j})}{\prod_{k=1, k \neq j}^{M} (\frac{1}{2\alpha_{v_k}^2 T_k} - \frac{1}{2\alpha_{v_j}^2 T_j})} \tag{6.74}$$

using an equation in Gradshteyn and Ryzhik (2000, eq. (3.352.4)), and $Ei(\cdot)$ is the exponential integral (Gradshteyn and Ryzhik 2000, eq. (8.211.1)).

The average rate is derived as

$$\bar{\tau} = E\{\tau_1 \log_2(1+\gamma)\} = I_1 - I_2 \tag{6.75}$$

where, using an equation in Gradshteyn and Ryzhik (2000, eq. (4.337.1)), one has

$$I_1 = \sum_{j=1}^{M} \frac{\tau_1 \prod_{k=1}^{M} \frac{1}{2\alpha_{v_k}^2 Q_k} 2\alpha_{v_j}^2 Q_j}{\ln 2 \prod_{k=1,k\neq j}^{M} (\frac{1}{2\alpha_{v_k}^2 Q_k} - \frac{1}{2\alpha_{v_j}^2 Q_j})} \int_0^\infty f_Z(z)$$

$$\times \left[\ln(2\sigma^2 + \frac{c\eta z}{1-c}) - e^{\frac{2\sigma^2 + \frac{\tau_0 \eta z}{\tau_1}}{2\alpha_{v_j}^2 Q_j}} Ei\left(-\frac{2\sigma^2 + \frac{c\eta z}{1-c}}{2\alpha_{v_j}^2 Q_j} \right) \right] dz. \tag{6.76}$$

and

$$I_2 = \frac{\tau_1}{\ln 2} \sum_{j=1}^{M} \frac{\prod_{k=1}^{M} \frac{1}{2\alpha_{v_k}^2 Q_k} 2\alpha_{v_j}^2 Q_j}{\prod_{k=1,k\neq j}^{M} (\frac{1}{2\alpha_{v_k}^2 Q_k} - \frac{1}{2\alpha_{v_j}^2 Q_j})}$$

$$\times \left[\ln(2\sigma^2) - e^{\frac{\sigma^2}{\alpha_{v_j}^2 Q_j}} Ei\left(-\frac{\sigma^2}{\alpha_{v_j}^2 Q_j} \right) \right]. \tag{6.77}$$

with $f_Z(z) = \frac{1}{\alpha_g^2} \sum_{i=0}^{N} \frac{\prod_{k=0}^{N} \frac{1}{2\alpha_{u_k}^2 W_k} K_0(2\sqrt{\frac{z}{2\alpha_{u_i}^2 W_i 2\alpha_g^2}})}{\prod_{k=0,k\neq i}^{N} (\frac{1}{2\alpha_{u_k}^2 W_k} - \frac{1}{2\alpha_{u_i}^2 W_i})}$, by using $Z = |g|^2 S$ and an equation in Gradshteyn and Ryzhik (2000, eq. (3.471.9)), $K_0(\cdot)$ is the zeroth order modified Bessel function of the second type (Gradshteyn and Ryzhik Gradshteyn and Ryzhik, (8.407.1)).

6.6.1.3 Performances without Interference

For comparison, the case of no interference is also discussed below. In this case, the two interference terms in (6.70) can be removed. Thus, the SINR becomes

$$\gamma = \frac{\tau_0}{\tau_1} \eta |g|^2 \frac{P_s|h|^2}{2\sigma^2}. \tag{6.78}$$

Then, it is quite straightforward to derive the average SINR as

$$\bar{\gamma} = \frac{2\tau_0}{\tau_1} \eta \alpha_g^2 \alpha_h^2 \frac{P_s}{\sigma^2}. \tag{6.79}$$

The average rate is derived as

$$\bar{\tau} = \tau_1 \int_0^\infty \log_2(1 + \frac{\tau_0}{\tau_1} \frac{\eta P_s}{2\sigma^2} y) f_Y(y) dy \tag{6.80}$$

where $Y = |g|^2|h|^2$ and $f_Y(y) = \frac{1}{2\alpha_g^2 \alpha_h^2} K_0(2\sqrt{\frac{y}{4\alpha_g^2 \alpha_h^2}})$.

6.6.1.4 Numerical Examples

Next, the performance of HAP with interference is compared with that without interference to examine the overall effect of interference. In the comparison, $\tau_0 = 0.5$, $T = 1$, $\eta = 0.5$, $\gamma_0 = 3$, $P_s = 1$, $W_i = 1$, and $Q_j = 1$. Also, let $SNRh = \frac{\alpha_h^2}{\sigma^2}$, $SIRh = \frac{\alpha_h^2}{A_u^2}$, and $EI = \frac{A_u^2}{A_v^2}$.

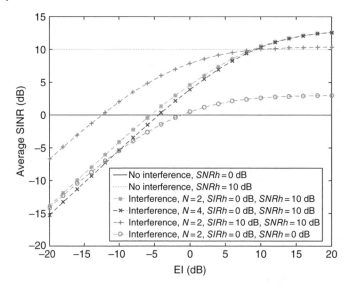

Figure 6.22 Average SINR versus *EI* for different values of *SNRh*, *SIRh* and *N*.

be the SNR of the downlink, the signal-to-interference ratio of the downlink, and the ratio of the interference power in the downlink to that in the uplink, respectively, where $A_u^2 = \sum_{i=1}^{N} \alpha_{u_i}^2 W_i$ and $A_v^2 = \sum_{j=1}^{M} \alpha_{v_j}^2 Q_j$. Also, set $M = N$, $[\alpha_{u_1}^2, \alpha_{u_2}^2, \cdots, \alpha_{u_N}^2] = [2^1, 2^2, \cdots, 2^N] \frac{A_u^2}{\sum_{i=1}^{N} 2^i}$ and $[\alpha_{v_1}^2, \alpha_{v_2}^2, \cdots, \alpha_{v_M}^2] = [2^1, 2^2, \cdots, 2^M] \frac{A_v^2}{\sum_{j=1}^{M} 2^j}$. Further, $\alpha_h^2 = \alpha_g^2 = 1$ so that A_u^2 changes with *SIRh*, A_v^2 changes with *EI*, and σ^2 changes with *SNRh*. The parameter of *EI* determines the relative strength of the interference in the downlink that contributes to energy harvesting to the strength of the interference in the uplink that reduces the SINR. It reflects the tradeoff between harm and benefit done by interference.

Figure 6.22 shows the average SINR versus *EI*. One sees that the average SINR with interference is larger than that without interference when *EI* is larger than around 9 dB when *SNRh* = 10 dB. Thus, the interference in the downlink needs to be at least 10 times as much as that in the uplink in order to make it beneficial. For *SNRh* = 0 dB, the average SINR with interference is larger than that without interference when *EI* is larger than around −1 dB. Thus, as *SNRh* decreases, it is easier to benefit from interference. Also, when *SIRh* increases or *N* decreases, the performance difference decreases, as expected. In this case, the threshold beyond which the average SINR with interference is larger than that without interference remains approximately the same. Finally, when *EI* increases, an upper limit occurs, as for large *EI*, the term $\sum_{j=1}^{M} \alpha_{v_j}^2 T_j$ in (6.70) can be ignored such that the performance is determined by *SNRh* and *SIRh*, which are fixed for each curve in our figures. Figure 6.23 shows the average rate versus *EI*. Similar insights can be obtained. Note that the thresholds between interference and no interference are larger than that in Figure 6.22.

In conclusion, the overall effect of interference, beneficial or harmful, is largely determined by the SNR and the ratio of the interference powers but it is always beneficial beyond a certain threshold. A more detailed discussion on this issue can be found in Chen et al. (2016a).

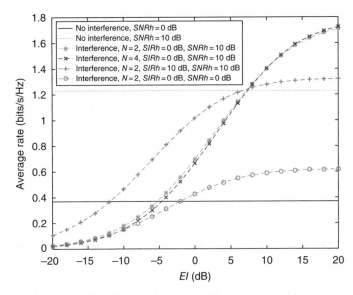

Figure 6.23 Average rate versus *EI* for different values of *SNRh*, *SIRh* and *N*.

6.6.2 Effect of Interference by Wireless Power

In the previous subsection, the effect of interference caused by information transmission on the performance of wireless powered communications has been investigated. In this subsection, we will change the angle of view by examining the effect of interference caused by wireless power on the performance of information transmission.

Interference is well known as one of the most fundamental limits in wireless networks. When wireless powered communications is performed, this issue becomes even more severe, as the use of wireless power will generate more and larger interference, which degrades the performances of information users in the same network. Effectively, due to wireless power, all the battery power and mains connection in the conventional networks become interference. For instance, in the PB networks, wireless power is broadcast by the PB to the whole network. This will introduce extra interference to all devices operating at the same frequency band, even though they may not need wireless power. In SWIPT, in order to deliver power and information by using the same signal, the transmission power needs to be raised so that there is enough energy to be split either in the time domain or the power domain at the receiver for both energy and information, causing great interference to other nodes. In HAP, the downlink wireless power transfer may become interference to devices that do not need wireless power. Also, due to the hardware constraints, the energy receiver often has a lower sensitivity than the information receiver. For example, the information receiver could have a sensitivity of −50 or −40 dBm, while the energy receiver often has a sensitivity of −20 or −10 dBm. This means that a larger transmission power has to be used if power is to be delivered efficiently. Thus, wireless powered networks are likely to suffer from more and stronger interference.

Although there have been quite a few studies on the effect of interference on wireless powered communications, including the previous subsection, very few studies have considered the interference generated by wireless power. In the following, the effect of interference generated by wireless power will be studied using the information rate for the PB, HAP, and SWIPT systems.

6.6.2.1 System and Assumptions

Consider the downlink of a multi-node multi-cell or a multi-cluster network. In this network, the desired node receives information from its nearest base station. If PB is used for wireless power transfer, it generates interference to this node as

$$I_B = \sum_{i=1}^{I} \sqrt{\frac{P_i}{d_i^v}} h_i x_i \alpha_i + \sum_{j=1}^{J} \sqrt{\frac{Q_j}{d_j^v}} g_j y_j \beta_j, \tag{6.81}$$

where I is the number of undesired data-transmitting base stations, P_i is their transmission power, d_i is their distance to the desired node, h_i is the fading channel gain between them, x_i is their transmitted symbol, α_i represents the asynchronization with data-transmitting base stations (they could start transmission earlier or later than the desired node), J is the number of undesired power-transmitting base stations, Q_j is their transmission power, d_j is their distance, g_j is the fading channel gain between them, w_j is their transmitted symbol, β_j represents the asynchronization with power-transferring base stations, and v is the path loss exponent. The first term in (6.81) is interference from the normal data transmission, while the second term in (6.81) is the extra interference incurred by wireless power transfer.

Without loss of generality, assume that all distances are uniformly distributed over a ring such that their PDF is

$$f_d(x) = \frac{2x}{r_h^2 - r_l^2}, \ r_l \leq x \leq r_h, \tag{6.82}$$

where r_h denotes the radius of the outer ring and r_l stands for the radius of the inner ring. The fading gains h_i and g_j are complex Gaussian random variables with zero-mean and variance $2\alpha^2$. M-ary phase shift keying (PSK) modulation is used such that $|x_i| = |y_j| = 1$. Assume that the frame length is T. The interfering base stations start their transmission randomly at t_0, where t_0 is uniformly distributed between 0 and T. Thus, their interfering power over the period of T is determined by $\frac{t_0}{T}$, which is uniformly distributed between 0 and 1 and represented by α_i and β_j.

Similarly, if HAP is adopted, the total interference at the desired node is

$$I_H = \sum_{i=1}^{I_1} \sqrt{\frac{P_i}{d_i^v}} h_i x_i \alpha_i + \sum_{i=I_1+1}^{I} \sqrt{\frac{Q_i}{d_i^v}} g_i y_i \beta_j, \tag{6.83}$$

where I_1 denotes the number of data-transmitting base stations among all I base stations and $I - I_1$ is the number of power-transmitting base stations.

Finally, using SWIPT, the interference seen by the desired node is

$$I_S = \sum_{i=1}^{I} \sqrt{\frac{P_i + Q_i}{d_i^v}} h_i x_i \alpha_i, \tag{6.84}$$

where the ith AP has a transmission power of $P_i + Q_i$, with P_i for information decoding and Q_i for energy harvesting. Essentially, a power-splitting receiver for SWIPT is considered.

Using these expressions, the received signal at the desired node is given by

$$y = \sqrt{\frac{P_0}{d_0^v}} h_0 x_0 + z,$$

(6.85)

where z can be replaced by I_B, I_H and I_S depending on the scheme used, P_0 is the transmission power, d_0 is the distance, h_0 is the fading gain, and x_0 is the transmitted symbol of the nearest base station. Finally, an interference-limited scenario is considered such that noise is ignored in the received signal. Also, the distance d_0 is assumed to be constant.

6.6.2.2 Average Interference Power

For the PB scheme, the average interference power can be derived from (6.67) as (Chen et al. 2017a)

$$E\{|I_B|^2\} = \frac{2\alpha^2}{3} d \left[\sum_{i=1}^{I} P_i + \sum_{j=1}^{J} Q_j \right].$$

(6.86)

where $d = E\{\frac{1}{d_i^v}\} = E\{\frac{1}{d_j^v}\} = \frac{2(\ln r_h - \ln r_l)}{r_h^2 - r_l^2}$ for $v = 2$ and $d = \frac{2(r_l^{2-v} - r_h^{2-v})}{(r_h^2 - r_l^2)(v-2)}$ for $v > 2$.

Similarly, for the HAP scheme (Chen et al. 2017a),

$$E\{|I_H|^2\} = \frac{2\alpha^2}{3} d \left[\sum_{i=1}^{I_1} P_i + \sum_{i=I_1+1}^{I} Q_i \right].$$

(6.87)

Finally, for the SWIPT scheme (Chen et al. 2017a),

$$E\{|I_S|^2\} = \frac{2\alpha^2}{3} d \left[\sum_{i=1}^{I} P_i + \sum_{i=1}^{I} Q_i \right].$$

(6.88)

From the above, two remarks can be made. First, by subtracting $\frac{2\alpha^2}{3} d \sum_{i=1}^{I} P_i$ from (6.86)–(6.88), the average interference power generated by PB is $\frac{2\alpha^2}{3} d \sum_{j=1}^{J} Q_j$, by HAP is $\frac{2\alpha^2}{3} d \sum_{i=I_1+1}^{I} (Q_i - P_i)$, and by SWIPT is $\frac{2\alpha^2}{3} d \sum_{i=1}^{I} Q_i$. Secondly, PB and SWIPT have the same average interference power when $J = I$. Also, PB and HAP have the same average interference power when $\sum_{i=I_1+1}^{I} P_i + \sum_{j=1}^{J} Q_j = \sum_{i=I_1+1}^{I} Q_i$. On the other hand, SWIPT always has larger average interference power than HAP.

6.6.2.3 Rate

To find the rate, the PDF of the instantaneous signal-to-interference ratio (SIR) is required. The SIR for the PB scheme can be written as $\gamma_p = \frac{W|h_0|^2}{|I_B|^2}$, where $W = \frac{P_0}{d_0^v} |x_0|^2$. Also, let $u_p = \sum_{i=1}^{I} \frac{P_i}{d_i^v} |x_i|^2 \alpha_i^2 + \sum_{j=1}^{J} \frac{Q_j}{d_j^v} |w_j|^2 \beta_j^2$. It is easy to show $|h_0|^2$ is an exponential random variable and $|I_B|^2$ is also an exponential random variable conditioned on u_p. Thus, the conditional PDF of γ_p given u_p can be expressed as

$$f_{\gamma_p | u_p}(z) = \frac{W u_p}{(u_p z + W)^2} = \frac{W \frac{1}{u_p}}{(z + W \frac{1}{u_p})^2},$$

(6.89)

which holds from Gradshteyn and Ryzhik (2000, eq. (3.351.3)). Using (6.89) and an equation in Gradshteyn and Ryzhik (2000, eq. (4.291.15)), the conditional rate is

$$C(u_p) = \frac{W}{\ln 2} \frac{\ln u_p - \ln W}{u_p - W} = \frac{W}{\ln 2} \frac{\frac{1}{u_p} \ln \frac{W}{u_p}}{W \frac{1}{u_p} - 1}. \tag{6.90}$$

The next step is to find the PDF of u_p or $\frac{1}{u_p}$ for the unconditional rate. Approximations can be used. It is found that the inverse Gaussian distribution fits u_p the best for large values of I and J. For $\frac{1}{u_p}$, the Gamma and Weibull distributions provide good approximations. Using the Gamma approximation, the PDF of $\frac{1}{u_p}$ is

$$f_{\frac{1}{u_p}G}(t) \approx \frac{t^{k-1}}{\Gamma(k)k^\theta} e^{-\frac{t}{\theta}} \tag{6.91}$$

where $\Gamma(\cdot)$ denotes the Gamma function (Gradshteyn and Ryzhik 2000, eq. (8.310.1)), k stands for the shape parameter, and θ is the scale parameter of the Gamma distribution, and using the Weibull approximation, the PDF of $\frac{1}{u_p}$ is

$$f_{\frac{1}{u_p}W}(t) \approx \frac{p}{q}\left(\frac{t}{q}\right)^{p-1} e^{-\left(\frac{t}{q}\right)^p}, \tag{6.92}$$

where p and q are the shape parameter and scale parameter of the Weibull distribution, respectively. Moment-matching gives $k\theta = E\{\frac{1}{u_p}\}$, $k^2\theta^2 + k\theta^2 = E\{\frac{1}{u_p^2}\}$, $q\Gamma(1 + \frac{1}{p}) = E\{\frac{1}{u_p}\}$, $q^2\Gamma(1 + \frac{2}{p}) = E\{\frac{1}{u_p^2}\}$. The moments of $\frac{1}{u_p}$ can be simulated.

Using the PDF of $\frac{1}{u_p}$, the unconditional rate for the PB scheme can be derived as (Chen et al. 2017a)

$$C_1 = \frac{1}{\Gamma(k)(\theta W)^k \ln 2} \int_0^\infty \frac{\ln t}{t-1} t^k e^{-\frac{t}{\theta W}} dt, \tag{6.93}$$

$$C_2 = \frac{p}{\ln 2(qW)^p} \int_0^\infty \frac{\ln t}{t-1} t^p e^{-\left(\frac{t}{qW}\right)^p} dt, \tag{6.94}$$

which hold for the Gamma and Weibull approximations, respectively. Further, $\frac{\ln t}{1-t} \approx at^b$, where $a = 1.3$ and $b = -0.7$ from curve-fitting. Thus, using an equation from Gradshteyn and Ryzhik (2000, eq. (3.381.4)) for (6.96) and another equation from Gradshteyn and Ryzhik (2000, eq. (3.478.1)) for (6.97), they can be further simplified to (Chen et al. 2017a)

$$C_1 = \frac{a(\theta W)^{b+1}\Gamma(k+b+1)}{\Gamma(k) \ln 2}, \tag{6.95}$$

$$C_2 = \frac{a}{\ln 2}(qW)^{b+1}\Gamma\left(\frac{p+b+1}{p}\right). \tag{6.96}$$

Similar results for HAP and SWIPT schemes can also be obtained after replacing u_p by $u_h = \sum_{i=1}^{I_1} \frac{P_i}{d_i^v} |x_i|^2 \alpha_i^2 + \sum_{i=I_1+1}^{I} \frac{Q_i}{d_i^v} |w_i|^2 \beta_i^2$ for HAP, and $u_s = \sum_{i=1}^{I} \frac{P_i+Q_i}{d_i^v} |x_i|^2 \alpha_i^2$ for SWIPT.

6.6.2.4 Numerical Examples

Next, some numerical examples are presented to illustrate the effects of interference caused by wireless power transfer. In the figures, $P_i = P = 20\ W$, $Q_j = Q = 40\ W$, $r_h = 100\ m$, $r_l = 1\ m$, and $v = 3$. Also, let $\hat{\gamma} = \frac{W}{\frac{d}{3}IP}$ be the benchmark average SIR when there is no wireless power. In SWIPT, PS is assumed.

Fig. 6.24 shows the rate versus $\hat{\gamma}$ using different approximations for the HAP. Figure 6.24a employs $I_1 = I - I_1 = 10$, while Figure 6.24b uses $I_1 = I - I_1 = 1000$. One sees from Figure 6.24 that the Gamma and Weibull approximations are very close to the simulated results, while the inverse Gaussian approximation only works when the values of I_1 and $I - I_1$ are large. It can be shown that similar conclusions hold for the PB and SWIPT schemes. They are not presented here to save space.

Figures 6.25 and 6.26 plot the rate versus $\hat{\gamma}$ for different wireless powered networks, when $I = J = 10$, $I_1 = 5$, and assuming the Gamma approximation. The case of no wireless power is obtained by setting $I = I_1$. One sees that the case without wireless power has the highest rate, as expected. However, this case has to be supported by conventional battery power for operation. One also sees that wireless power in general degrades the rate performance. In particular, PB and SWIPT have the smallest rate due to the largest interference caused, while HAP has the smallest interference and thus higher rate than PB and SWIPT. Nevertheless, the rate increases with $\hat{\gamma}$ and v. Figure 6.27 shows the rate performance of PB for different J. One sees that, when $J = 2$, PB has a similar throughput

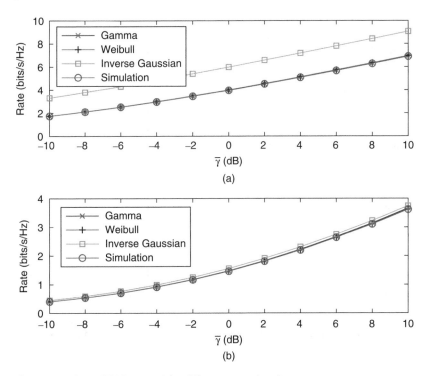

Figure 6.24 Rate of HAP versus $\hat{\gamma}$ for different approximations.

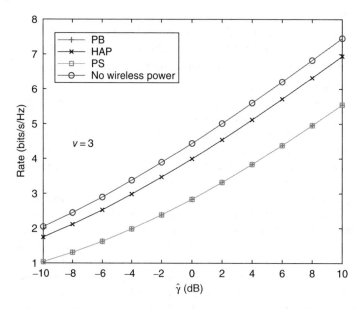

Figure 6.25 Rate versus $\hat{\gamma}$ for different wireless powered networks when $I = J = 10$ and $I_1 = 5$ for $v = 3$.

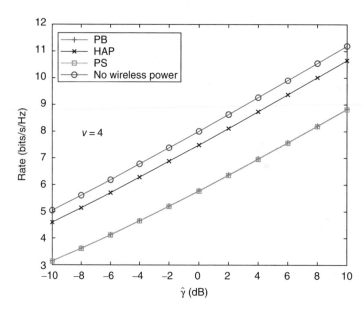

Figure 6.26 Rate versus $\hat{\gamma}$ for different wireless powered networks when $I = J = 10$ and $I_1 = 5$ for $v = 4$.

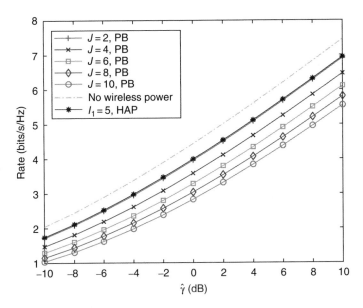

Figure 6.27 Comparison of PB using different *J* and HAP.

to HAP. Figure 6.28 compares different values of I_1 for HAP. A larger I_1 gives higher rate. More results on the effect of interference can be found in Chen et al. (2017a).

6.6.3 Exploitation of Interference

In the previous two subsections, the effect of interference on and caused by the wireless power has been studied. In this subsection, the interference is used as an energy source to power up the devices. In particular, nodes in the idle mode can harvest energy from peer nodes in the network that are transmitting data. Most RF energy harvesting works consider either energy harvesting from dedicated sources or energy harvesting from ambient sources. The ambient sources have great uncertainty, while the dedicated sources require extra infrastructure. In Xie et al. (2017), peer harvesting was studied, where idle nodes harvest energy from other transmitting peer nodes. This method offers a good alternative. Compared with dedicated sources, peer harvesting reuses the RF power transmitted by peer nodes to increase the lifetime of the operation. Compared with ambient sources, the transmitted signals from peer nodes in the same network are more controllable and hence have less uncertainty.

In this subsection, we analyze how long peer harvesting for a wireless network can last. In this network, multiple nodes share the same channel and transmit data in turn. When nodes are not transmitting, they will harvest energy from the transmitting nodes. The harvested energy can then be used to perform extra transmission, which can then be harvested by other nodes again. We are interested in finding out the number of extra transmissions each node can make by using the extra energy from peer harvesting.

Specifically, we consider a wireless network that has N independent nodes. These nodes access the same channel in a time-division-duplex manner by using the round-robin algorithm, where each node is assigned a time slot of T seconds so that all

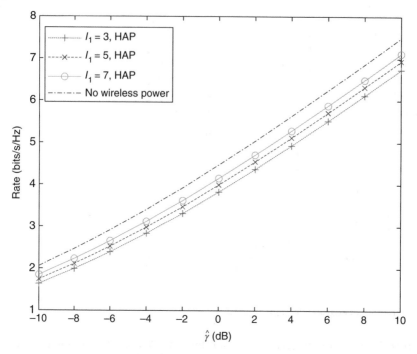

Figure 6.28 Comparison of different l_1 for HAP.

nodes finish their transmission in NT seconds. Without loss of generality, we assume that the nodes transmit in the order of $1, 2, \cdots , N$. Also, assume that all the nodes have data to transmit during their assigned time slots and only transmit during their assigned slots. Each transmission has a fixed transmission power of P. The initial energy available at each node before any transmission starts is assumed to be E_0. Essentially, it is interesting to find out how many transmissions the total energy of NE_0 can be used for until all nodes die down due to insufficient energy.

Denote E_{ki} as the initial energy at the ith node before the kth round of transmission starts. Using the above assumptions, one can derive the energy matrix and present it as

$$
\mathbf{E}_k = \begin{bmatrix} -PT\alpha_{k1} & \frac{\eta PT|h_{12k}|^2}{d_{12k}^v}\alpha_{k1} & \cdots & \frac{\eta PT|h_{1Nk}|^2}{d_{1Nk}^v}\alpha_{k1} \\ \frac{\eta PT|h_{21k}|^2}{d_{21k}^v}\alpha_{k2} & -PT\alpha_{k2} & \cdots & \frac{\eta PT|h_{2Nk}|^2}{d_{2Nk}^v}\alpha_{k2} \\ \vdots & \vdots & \ddots & \vdots \\ \frac{\eta PT|h_{N1k}|^2}{d_{N1k}^v}\alpha_{kN} & \frac{\eta PT|h_{N2k}|^2}{d_{N2k}^v}\alpha_{kN} & \cdots & -PT\alpha_{kN} \end{bmatrix}
\tag{6.97}
$$

where a plus sign of the element indicates energy harvesting, a minus sign of the element indicates energy consumption, and P and T, respectively, are the transmission power and the transmission time of the node giving the energy consumption of PT. Importantly, the transmission indicator is given by

$$
\alpha_{ki} = \begin{cases} 1, & E_{ki} > PT, \\ 0, & E_{ki} < PT, \end{cases}
\tag{6.98}
$$

so that the ith node only transmits in the kth round if its energy is larger than PT and otherwise it does not transmit, η is the conversion efficiency of the energy harvester as discussed in Chapter 3, h_{ijk} is the fading coefficient between the ith node and the jth node in the kth round, d_{ijk} is their distance, v is the path loss exponent, and $i = 1, 2, \cdots, N$ represent different nodes. The fading coefficients are assumed independent for different values of i and j. Rayleigh fading channels are used so that the fading coefficient is a circularly symmetric complex Gaussian random variable with mean 0 and variance $2\sigma^2$.

In the energy matrix, the first row gives the energy change after node 1 finishes its time slot. If node 1 has enough energy to transmit, node 1 will have an energy consumption of PT while other nodes will harvest an energy of $\frac{\eta PT|h_{1ik}|^2}{d_{1ik}^v}$ from node 1's transmission, $i = 2, 3, \cdots, N$. If node 1 does not transmit, the energy changes will be zero at all nodes in this case. Assume $E_0 = K * PT$. After K rounds, the extra energy harvested at the ith node is given by

$$E_{(K+1)i} = \sum_{l=1}^{K} \sum_{j=1, j\neq i}^{N} \frac{\eta PT|h_{jil}|^2}{d_{jil}^v}. \tag{6.99}$$

This energy harvested from peer nodes during their transmissions in the K rounds can be used to perform extra data transmissions. The larger the value of $E_{(K+1)i}$, the more extra transmissions the ith node can perform. Thus, the number of extra transmissions each node makes indicates how long the lifetime of the network can be.

Assume that m extra transmissions can be made. The energy matrix of the $(K + m - 1)$th round can be derived from (6.97) by replacing k with $K + m - 1$ to give

$$\mathbf{E}_{(K+m-1)} = \begin{bmatrix} -PT\alpha_{(K+m-1)1} & \cdots & \frac{\eta PT|h_{1N(K+m-1)}|^2}{d_{1N(K+m-1)}^v}\alpha_{(K+m-1)1} \\ \frac{\eta PT|h_{21(K+m-1)}|^2}{d_{21(K+m-1)}^v}\alpha_{(K+m-1)2} & \cdots & \frac{\eta PT|h_{2N(K+m-1)}|^2}{d_{2N(K+m-1)}^v}\alpha_{(K+m-1)2} \\ \vdots & \ddots & \vdots \\ \frac{\eta PT|h_{N1(K+m-1)}|^2}{d_{N1(K+m-1)}^v}\alpha_{(K+m-1)N} & \cdots & -PT\alpha_{(K+m-1)N} \end{bmatrix}. \tag{6.100}$$

From (6.100), by summing up its ith column and using iteration, the residual energy after the $(K + m - 1)$th round or the initial energy of the $(K + m)$th round can be derived as

$$E_{(k+m)i} = E_{(K+1)i} - PT\sum_{p=1}^{m-1}\alpha_{(K+p)i}$$

$$+ \sum_{p=1}^{m-1}\sum_{j=1, j\neq i}^{N} \frac{\eta PT|h_{ji(K+p)}|^2}{d_{ji(K+p)}^v}\alpha_{(K+p)j}, \tag{6.101}$$

where $i = 1, 2, \cdots, N$. To allow for the transmission of m extra rounds, one must have $E_{(k+p)i} > PT$ for $p = 1, 2, \cdots, m$ and $i = 1, 2, \cdots, N$. Thus, the probability that m extra rounds of transmission can be made is determined by

$$P_m = Pr\{E_{(k+p)i} > PT, p = 1, 2, \cdots, m, i = 1, 2, \cdots, N\}. \tag{6.102}$$

In (6.102), the random fading coefficient makes the event uncertain. One has from (6.102)

$$P_m = \prod_{i=1}^{N} Pr\{G_{(K+1)i} > PT,$$

(6.103)

$$G_{(K+1)i} + G_{(K+2)i} > 2PT, \cdots, \sum_{p=1}^{m} G_{(K+p)i} > mPT\},$$

where

$$G_{(K+1)i} = \sum_{l=1}^{K} \sum_{j=1, j\neq i}^{N} \frac{\eta PT |h_{jil}|^2}{d_{jil}^v}$$

(6.104)

$$G_{(K+p)i} = \sum_{j=1, j\neq i}^{N} \frac{\eta PT |h_{ji(K+p-1)}|^2}{d_{ji(K+p-1)}^v}, p = 2, \cdots, m.$$

(6.105)

Thus, m is a discrete random variable with probability mass function given by (6.103). We are interested in the distribution of this random variable. A general expression of the distribution is not possible but for small values of m, it is possible. In particular, when $m = 1$, one has (Chen et al. 2018)

$$P_1 = \prod_{i=1}^{N} \left[\sum_{j_1=1, j_1\notin\phi_i}^{NK} \beta_{j_1 i(K+1)} e^{-\lambda_{j_1 i(K+1)} PT} \right].$$

(6.106)

where $\lambda_{j_1 i(K+1)} = \frac{d_{j_1 i}^v}{2\sigma^2 \eta PT}$, $\beta_{j_1 i(K+1)} = \prod_{l_1=1, l_1\neq j_1}^{NK} \frac{\lambda_{l_1 i(K+1)}}{\lambda_{l_1 i(K+1)} - \lambda_{j_1 i(K+1)}}$, and the distances $d_{1i}, d_{2i}, \cdots, d_{(NK)i}$ correspond to $d_{1i1}, d_{2i1}, \cdots, d_{NiK}$, respectively, and we have stacked the two summations in (6.106) into one summation and in the stacked summation there are $(N-1)K$ terms, as ϕ_i contains the N terms excluded in the second summation of (6.106).

When $m = 2$, one has (Chen et al. 2018)

$$P_2 = \prod_{i=1}^{N} \{ \sum_{j_1=1, j_1\notin\phi_i}^{NK} \beta_{j_1 i(K+1)} e^{-\lambda_{j_1 i(K+1)} 2PT}$$

$$\times [1 + \sum_{j_2=1, j_2\neq i}^{N} \frac{\beta_{j_2 i(K+2)}(1 - e^{PT(\lambda_{j_1 i(K+1)} - \lambda_{j_2 i(K+2)})})}{\frac{\lambda_{j_2 i(K+2)}}{\lambda_{j_1 i(K+1)}} - 1}]\}.$$

(6.107)

where $\lambda_{j_2 i(K+2)} = \frac{d_{j_2 i(K+1)}^v}{2\sigma^2 \eta PT}$, $\beta_{j_2 i(K+2)} = \prod_{l_2=1, l_2\neq j_2}^{N} \frac{\lambda_{l_2 i(K+2)}}{\lambda_{l_2 i(K+2)} - \lambda_{j_2 i(K+2)}}$, $\lambda_{j_3 i(K+3)} = \frac{d_{j_3 i(K+2)}^v}{2\sigma^2 \eta PT}$, and $\beta_{j_3 i(K+3)} = \prod_{l_3=1, l_3\neq j_3}^{N} \frac{\lambda_{l_3 i(K+3)}}{\lambda_{l_3 i(K+3)} - \lambda_{j_3 i(K+3)}}$.

When $m = 3$, one has (Chen et al. 2018)

$$P_3 = \prod_{i=1}^{N} [I_1 + I_2 + I_3 + I_4]$$

(6.108)

where

$$I_1 = \sum_{j_1=1, j_1\notin\phi_i}^{NK} \beta_{j_1 i(K+1)} e^{-\lambda_{j_1 i(K+1)} 3PT}$$

$$I_2 = \sum_{j_1=1, j_1 \notin \phi_i}^{NK} \sum_{j_2=1, j_2 \neq i}^{N} \frac{\beta_{j_1 i(K+1)} \beta_{j_2 i(K+2)}}{\frac{\lambda_{j_2 i(K+2)}}{\lambda_{j_1 i(K+1)}} - 1} (e^{-\lambda_{j_1 i(K+1)} 3PT} - e^{-\lambda_{j_2 i(K+2)} 3PT})$$

$$I_3 = \sum_{j_1=1, j_1 \notin \phi_i}^{NK} \sum_{j_2=1, j_2 \neq i}^{N} \sum_{j_3=1, j_3 \neq i}^{N} \beta_{j_1 i(K+1)} \beta_{j_2 i(K+2)} \beta_{j_3 i(K+3)}$$

$$\times \left[\frac{e^{-\lambda_{j_1 i(K+1)} 3PT} - e^{-(\lambda_{j_1 i(K+1)} 2PT + \lambda_{j_2 i(K+2)} PT)}}{\lambda_{j_2 i(K+2)} - \lambda_{j_1 i(K+1)}} - \frac{e^{-\lambda_{j_1 i(K+1)} 3PT} - e^{-(\lambda_{j_1 i(K+1)} 2PT + \lambda_{j_3 i(K+3)} PT)}}{\lambda_{j_3 i(K+3)} - \lambda_{j_1 i(K+1)}} \right]$$

$$I_4 = \sum_{j_1=1, j_1 \notin \phi_i}^{NK} \sum_{j_2=1, j_2 \neq i}^{N} \sum_{j_3=1, j_3 \neq i}^{N} \beta_{j_1 i(K+1)} \beta_{j_2 i(K+2)} \beta_{j_3 i(K+3)}$$

$$\times \frac{e^{-\lambda_{j_2 i(K+2)} 3PT} - e^{-(\lambda_{j_2 i(K+2)} 2PT + \lambda_{j_3 i(K+3)} PT)}}{\lambda_{j_3 i(K+3)} - \lambda_{j_1 i(K+1)}} \frac{e^{\lambda_{j_2 i(K+2)} - \lambda_{j_1 i(K+1)} 2PT} - 1}{\lambda_{j_2 i(K+2)} - \lambda_{j_1 i(K+1)}}.$$

A similar method can be used to derive results for $m = 4$ and so on. However, the expressions are becoming more and more complicated as m increases. Also, the CDF of m can be calculated as $F_m(x) = \sum_{m=1}^{x} P_m$. The above results assume that there is only one node transmitting in each time slot. If more than one node is transmitting at the same time, such as two-way relaying or non-orthogonal multiple access, similar results can be obtained by replacing $j \neq i$ with $j \neq \psi_i$, where ψ_i is the set of nodes that transmit at the same time as the ith node. In this case, there will be less harvested energy due to simultaneous transmissions but the total time of each round will be reduced to increase the throughput.

Next, numerical examples are presented to show the number of extra transmissions. In these examples, three typical network topologies are considered: all nodes are evenly distributed over a line with length L and fixed positions; all nodes are randomly distributed over a line with length L; and all nodes are randomly distributed over a disc with radius R. The line models may be found in vehicular communications on a highway, while the disc model may be found in cellular communications. Since it is difficult to obtain analytical results for large values of m, we use computer simulation. In the simulation, we set $\eta = 0.5$, $PT = 1$, $2\sigma^2 = 1$, and other parameters are given in the figures.

Figures 6.29 and 6.30 show P_m versus m for different system settings. One sees that the probability of having extra rounds of transmission decreases when m increases. This is expected, as all the energy will eventually be consumed such that it becomes less and less possible to perform extra transmissions. Comparing the three different topologies in Figure 6.29, one sees that the random line has the highest probability, followed by the fixed line and then the random disc. This implies that a network with nodes randomly distributed over a line can harvest more energy and thus can have a longer operation time than those randomly distributed over a disc or evenly distributed over a line. Comparing Figure 6.29 with Figure 6.30, one also sees that a larger value of K allows for higher probability of extra transmission. This means that one can increase the operational time by allocating more initial energy.

Figure 6.31 shows P_m versus m when two nodes transmit at the same time in each time slot. In this case, the fixed line becomes the worst case, while the random line is still the best case. The data rate increases by using simultaneous transmission and the energy that can be harvested in each transmission also increases but the number

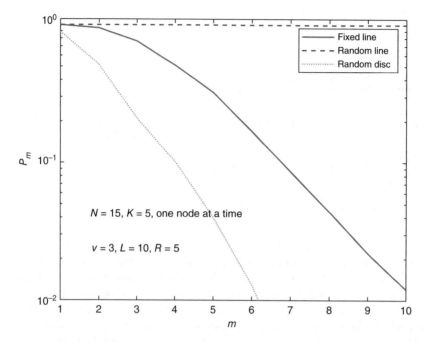

Figure 6.29 P_m versus m for different topologies when only one node transmits in each time slot using round-robin.

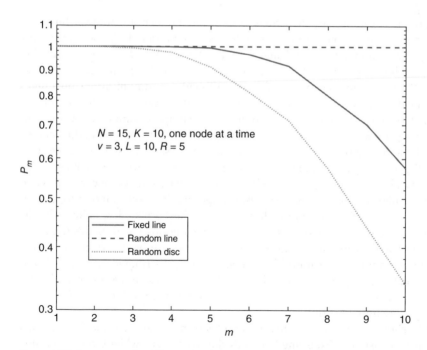

Figure 6.30 P_m versus m for different topologies when only one node transmits in each time slot using round-robin.

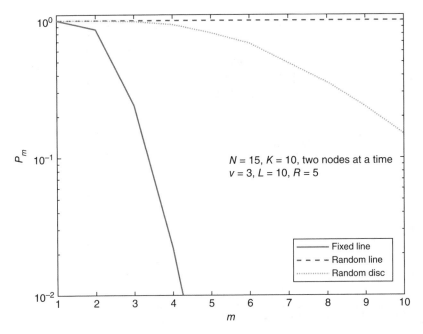

Figure 6.31 P_m versus m for different topologies when two nodes transmit data simultaneously in each time slot using round-robin.

of transmissions in each round will be reduced. This is different from the single-node transmission case. Figure 6.32 shows the average value of m versus the number of nodes N. Indeed, on average the number of extra transmissions increases with the number of nodes. This increase is very slow at the beginning but then dramatic after a threshold before it slows down again. The random line has the highest average, followed by the random disc and the fixed line.

Figure 6.33 focuses on the random line case for different system parameters. In this case, the length of the line has been increased to $L = 30\,m$. One sees that the probability of extra transmission is very sensitive to K. A decrease of K from 10 to 5 leads to a dramatic drop in probability. The probability also decreases when N decreases from 15 to 10, as less energy can be harvested in each round due to less transmitting nodes. This is also observed from Figure 6.32. Finally, comparing one-node transmission with two-node transmission, the probability of extra transmission is smaller in two-node transmission. This implies that the effect of having less transmissions is larger than the effect of having more energy harvested in each transmission. More details on the derivation can be found in Chen et al. (2018).

6.6.4 Multiple Antennas

In wireless communications, multiple-input-multiple-output (MIMO) has been proven as an effective means of achieving high data rate. Similar ideas can also be applied to wireless powered communications. Consider the use of multiple antennas in SWIPT (Zhang and Ho 2013). In this case, the transmitter uses M antennas, and the receiver uses N antennas. The received signal is used for both energy harvesting and information

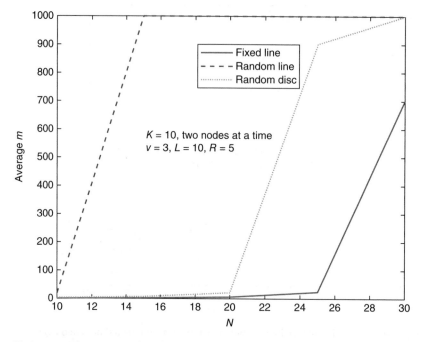

Figure 6.32 The average value of *m* for different topologies when two nodes transmit data simultaneously in each time slot using round-robin.

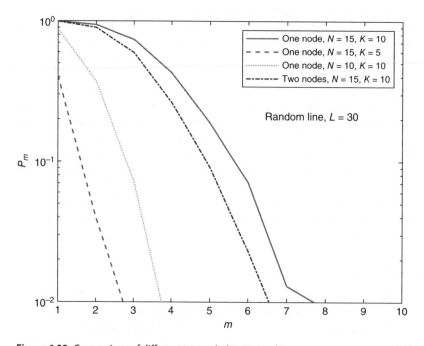

Figure 6.33 Comparison of different transmission strategies.

decoding. The channel matrix between the transmitter and the receiver is denoted by **H**, which is a $N \times M$ matrix. The received signal is thus denoted as

$$\mathbf{y} = \mathbf{Hs} + \mathbf{z} \tag{6.109}$$

where **y** is a $N \times 1$ vector representing received signals on all antennas, **s** is a $M \times 1$ vector representing the transmitted signals on all antennas, and **z** is a $N \times 1$ vector representing AWGN. Denote $\mathbf{S} = E[\mathbf{ss}^H]$ as the covariance matrix of **s**, where $E[\cdot]$ is the expectation operation and H is the conjugate transpose operation.

If the received signal in (6.109) is used for energy harvesting, the harvested power (energy normalized by the symbol interval) is given by

$$Q = \eta E[||\mathbf{Hs}||^2] = \eta tr(\mathbf{HSH}^H) \tag{6.110}$$

where $tr(\cdot)$ is the trace operation and η is the conversion efficiency.

On the other hand, if the received signal in (6.109) is used for information decoding, the achievable information rate is

$$R = \log|\mathbf{I} + \mathbf{HSH}^H| \tag{6.111}$$

where **I** is the $M \times M$ identity matrix and $|\cdot|$ is the determinant operation.

Similar to the single antenna case, the optimization problem here is to maximize both Q and R, which leads to a tradeoff between energy and information. If TS is used, this tradeoff is represented by the rate-energy function as

$$C(R,Q) = \{Q \le \eta\alpha \cdot tr(\mathbf{HS}_E\mathbf{H}^H), R \le (1-\alpha)\log|\mathbf{I} + \mathbf{HS}_I\mathbf{H}^H|,$$
$$tr(\mathbf{S}_I) < P_s, tr(\mathbf{S}_E) < P_s, \mathbf{S}_I \ge 0, \mathbf{S}_E \ge 0\} \tag{6.112}$$

where $tr(\mathbf{S}_I)$ is the transmission power during information decoding and $tr\{\mathbf{S}_E\}$ is the transmission power during energy harvesting. Comparing (6.112) and (6.11), one sees that they are very similar to each other.

If PS is used, the rate-energy function becomes

$$C(R,Q) = \{Q \le \eta \cdot tr(\Lambda_1\mathbf{HSH}^H), R \le \log|\mathbf{I} + \Lambda_2^{1/2}\mathbf{HSH}^H\Lambda_2^{1/2}|,$$
$$tr(\mathbf{S}) < P_s, \mathbf{S} \ge 0\} \tag{6.113}$$

where $\Lambda_1 = diag(\rho_1, \cdots, \rho_N)$ is a diagonal matrix that has all the PS factors of the receiving antennas in the diagonal line, and $\Lambda_2 = diag(\upsilon_1, \cdots, \upsilon_N)$, $\upsilon_n = \frac{1-\rho_n}{(1-\rho_n)2\sigma_a^2+2\sigma_d^2}$, and $2\sigma_a^2$ and $2\sigma_d^2$ are the noise power for the RF antenna and the RF-to-baseband conversion defined as before.

Using (6.112) and (6.113), further optimizations are possible. For example, one can maximize Q subject to a constraint on R, or maximize R subject to a constraint on Q. The parameters to be optimized could be P_s, α, ρ, M, N, and so on. Alternatively, energy beamforming and information beamforming can also be used for MIMO systems (Liu et al. 2014b). In this case, **s** will be replaced by **Us** and **H** will be replaced by **VH**, where **U** and **V** are the precoding and decoding matrices at the transmitter and the receiver, respectively. Then, **U** and **V** can be jointly maximized with P_s, α, ρ, M, and N, leading to even more variants.

These two rate-energy functions are the basis of all these optimization studies in the literature. In fact, the tradeoff proposed in Varshney (2008) and Grover and Sahai (2010) are the starting points of all these studies.

6.7   An Example: Wireless Powered Sensor Networks

Wireless sensor networks (WSNs) have been widely used in our daily lives, such as in smart cities, asset monitoring, home automation, and logistics. When Internet of Things networks are deployed, they will become more popular, as many applications will have embedded sensing devices installed for various purposes (Raza et al. 2017).

As WSNs spread and scale up, one issue is becoming more and more serious. Most existing sensors rely on either battery or mains connection for power supply. For mobile applications, battery is actually the only choice. However, all batteries have a limited lifetime. When the battery runs out, it has to be replaced, or the whole network could break down. In a sensor network with hundreds or even thousands of sensors, this is almost a mission impossible, not to mention that some sensors could be embedded in buildings or human bodies such that their replacement incurs a huge amount of extra cost or pain.

Meanwhile most sensor networks are used for low-power applications, such as environment monitoring, where sensors only need to send data occasionally with very low duty cycles. Their power consumption is often on the scale of milliwatts or even microwatts. On the other hand, studies have shown that even the ambient RF energy can provide a harvested power on the scale of milliwatts or microwatts (Azmat et al. 2016). Thus, wireless power and wireless sensor networks are a perfect match (Xie et al. 2013). Indeed, most current wireless power designs are for sensor networks or other low-power applications.

For example, in Tong et al. (2010), commercially available wireless power products were used to test the efficiency of wireless power transfer for sensor networks. The results revealed that the efficiency is quite low so that custom-made hardware may be necessary to supply wireless power for sensor networks. In Peng et al. (2010) a mobile robot was used to charge a sensor network. The optimal scheduling and the optimal path of this robot were studied to prolong the lifetime of the network. Their results showed that wireless power can improve the network lifetime, but the power transfer efficiency is still an issue that needs to be resolved for better results. The power transfer efficiency of different harvesters has been extensively studied in Chapter 3. In a wider context, Erol-Kantarci and Touftah (2012) studied a rechargeable WSN used for smart grid monitoring. This scheme is similar to the PB wireless powered networks, where a few dedicated power transmitters are installed to supply power for the sensor network. Kamalinejad et al. (2015) reviewed and studied the use of wireless power in the Internet of Things by focusing on the extended network lifetime. Some future challenges have also been discussed.

There are many other applications of sensor networks using wireless power. Interested readers are referred to Sudevalayam and Kulkarni (2011) and Lu et al. (2015) for a more complete review of wireless powered sensing.

6.8   Summary

This chapter has mainly focused on wireless powered communications. Three different types of wireless powered communications have been discussed: SWIPT; HAP; and PB. They have different pros and cons. For PB, it simplifies the fundamental tradeoff between energy and information, but it incurs a large amount of extra infrastructure cost. This extra infrastructure cost may not be desirable for most existing applications and hence, it is more suitable for future networks. For SWIPT and HAP, they are relatively easy to implement with little infrastructure upgrade but the implementation involves complicated protocol design problems. They are more suitable for existing networks that require additional wireless power capability, as the changes will be mainly on protocols rather than network infrastructure.

Several important research issues on wireless powered communications have also been reviewed. It has been emphasized that most of these research problems start from the tradeoff between energy and information. In most cases, the amount of harvested energy and the information rate can be derived, based on which different system parameters can be optimized. However, in these cases, the power that can be supplied by far-field wireless transfer is still low. Thus, an example of using low wireless power in sensor networks has been examined.

7

Energy Harvesting Cognitive Radios

7.1 Introduction

7.1.1 Cognitive Radio

Cognitive radio (CR) was proposed in 2005 to deal with the "spectrum scarcity" problem (Haykin 2005). Since Marconi's first trans-Atlantic radio transmission, more and more radio systems have been deployed over the years. All of these systems require a frequency band in the radio spectrum for operation. However, the radio spectrum is a scarce resource. In particular, the portion of the radio spectrum that is suitable for existing wireless technologies is very limited, mainly from 30 MHz to 3 GHz. If the frequency is too high, the propagation distance will decrease dramatically due to blockage. If the frequency is too low, the bandwidth will not be large enough to meet the quality of service (QoS) requirement. As more and better wireless services are being deployed, this portion of the radio spectrum is becoming more and more crowded. This is the so-called "spectrum scarcity" problem.

On the other hand, various studies in different countries have reported that the radio spectrum is in fact seriously under-utilized, although it looks very crowded (Chen and Oh 2016a). For example, studies commissioned by the Federal Communications Commission (FCC) in the US and the Office of Communications in the UK revealed that the percentage of the radio spectrum that is being used at any time in any location is usually low, between 15% and 80% in most cases (FCC Spectrum Policy Task 2002; QinetiQ 2007). Even in the city center of London, the frequency bands occupied by some very popular radio systems are rarely in operation (QinetiQ 2007). This creates the "spectrum under-utilization" problem.

Thus, we have a serious dilemma of insufficient spectrum resource and inefficient spectrum usage at the same time. This dilemma is largely caused by the regulators' current practices that impose a fixed spectrum access policy. Under this policy, a wireless system will normally be assigned an exclusive part of the radio spectrum as a license for operation to avoid interference from other users. This part of the radio spectrum cannot be used by any other users for any other purposes. For example, a TV band cannot be used to transmit mobile signals, and vice versa. This policy was made decades ago when there were not so many radio systems and when technologies were not so advanced as to be able to handle interference. Such a policy is not sustainable and is outdated given the dramatic development of wireless communications systems and the associated advances in software and hardware in recent years.

Energy Harvesting Communications: Principles and Theories, First Edition. Yunfei Chen.
© 2019 John Wiley & Sons Ltd. Published 2019 by John Wiley & Sons Ltd.

To solve the problems of both "spectrum scarcity" and "spectrum under-utilization", dynamic spectrum access can be employed. One promising solution to the "spectrum scarcity" problem is to improve the efficiency of current spectrum utilization by allowing unlicensed "non-interfering" access to the licensed bands. In 2004, the FCC issued a notice to propose "no-harm" use of the licensed TV broadcasting bands in the USA (FCC 2004), as most US families use either satellite TV or cable TV. In the UK, Ofcom also proposed to increase the percentage of market-based bands to 71.5%. Owners of the market-based bands are allowed to trade and lease their bands to whoever wants to re-use their bands to maximize the social benefit of the spectrum. These governmental actions represent a new shift in the spectrum management policy. CR is a technological enabler of this new dynamic access policy. Figure 7.1 shows how the CR systems can use the spatial and temporal opportunities for operation. If the spatial opportunities are explored, CR and the primary user (PU) can operate at the same time but are far away from each other to avoid interference. If the temporal opportunities are explored, CR and PU operate at different time slots but in the same geographical area.

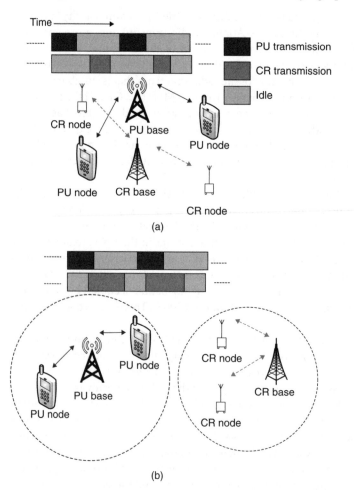

Figure 7.1 CR systems using (a) temporal and (b) spatial opportunities.

Figure 7.2 The frame structure of a conventional CR.

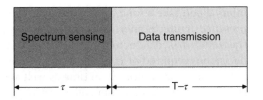

7.1.2 Cognitive Radio Functions

CR is a radio device that has cognition. The cognition is acquired by learning from and adapting to the radio environment during the operation. Therefore, CR is a radio-environment-aware device. To be more specific, CR first finds the parts of the radio spectrum that are not being occupied at some specific times in some specific locations, or the under-utilized spectrum, and then moves its operation to these parts called "spectrum holes" for opportunistic access, so that it does not need a fixed frequency band or license. Due to this attractive property, CR has already been adopted in several standardization works (Sherman et al. 2008).

In the physical layer, CR has two main functions: spectrum sensing; and data transmission. Spectrum sensing is probably more important than data transmission in the context of CR, as an inaccurate sensing will lead to interference with other systems, including the primary systems that own the license of the under-utilized spectrum, because they operate at the same frequency band in the radio spectrum. This will violate the non-interfering access rule. Consequently, it is of paramount importance to obtain spectrum sensing results as accurately as possible. Figure 7.2 shows the frame structure of a CR system with two main functions. When the total duration of the frame is fixed, there is a tradeoff between sensing and transmission (Liang et al. 2008).

7.1.3 Spectrum Sensing

The key function in CR is to find the empty frequency bands in the radio environment. This is accomplished by spectrum sensing.

Energy detection is probably one of the simplest methods for spectrum sensing (Urkowitz 1967). It measures the energy of the signal received from the interested band and compares this measurement with a predetermined threshold. The energy detector is very simple and easy to implement. However, it suffers from the noise uncertainty at a low signal-to-noise-ratio (SNR) (Tandra and Sahai 2008). This is caused by the error in the estimation of the noise power that is required in setting the detection threshold (Urkowitz 1967). Also, in a wireless environment with shadowing and/or fading, the random variation in the signal strength could reduce the SNR significantly and hence the energy detector may not be able to distinguish a heavily shadowed and/or faded signal from a zero signal very well. Consequently, it will think the channel is free to cause interference to the PU, the owner of the frequency band, who is actually operating.

To overcome this problem, feature-based detectors can be used. The feature-based detectors differentiate between signal and noise by using features of the signal, such as covariance and cyclostationarity, instead of the signal power (Chen et al. 2011b). They do not need the noise power in the detection threshold so they do not have the noise uncertainty problem. However, calculations of these features often require a large observation interval and take a long time. Moreover, their performances deteriorate quickly in the presence of interference, as interference may have the same features as the signal. Thus, the feature-based detectors may be less desirable in applications where a quick detection decision is required or in applications where interference occurs.

To avoid long computation time as well as high computational cost, another method is to use energy detection in collaborative spectrum sensing (Chen and Beaulieu 2009). In collaborative spectrum sensing, energy measurements from several CR users are combined to reduce the noise uncertainty. Since collaborative spectrum sensing only involves linear or quadratic combination of the measurements, it provides simple, quick yet reliable detectors for spectrum sensing, even at a fairly low SNR. However, in order to perform collaboration, measurements have to be transmitted from the CR users to the fusion center through the CR links. This adds overheads to the CR network. Moreover, a strict network control has to be implemented to coordinate the transmission of measurements from all CR users.

7.1.4 Energy Harvesting Cognitive Radio

An energy harvesting CR is a radio device that has both cognition and energy harvesting capability. The cognition allows the unlicensed use of licensed frequency bands. Hence, it exploits the spectrum opportunities in the radio environment to help improve the spectral efficiency. On the other hand, energy harvesting is an important technology to achieve battery-free operations by replacing batteries with ambient energies from the surrounding environment. In particular, recent advances in electronics have made it possible to harvest the ambient electromagnetic waves in the radio environment for wireless communications, such as radio frequency identification and wireless body area networks. Thus, energy harvesting exploits the energy opportunities in the environment to save energy cost.

Two main costs of a wireless communications system are spectrum cost and energy cost. The spectrum cost can be considerably reduced by using CR with "free" bandwidth, while the energy cost can be significantly reduced by using energy harvesting with "free" energy. Thus, to provide a viable solution to reliable and sustainable wireless communications, it is imperative to design efficient energy harvesting CRs to reduce both spectrum and energy costs. For some applications, such as sensor networks, where a low-cost device is key to large-scale deployment, this will be very useful.

In the conventional CR, the main design constraint is collision avoidance. The CR needs to avoid collision with the PUs as much as possible, as it has lower priority on the use of the licensed band. This is often translated into a constraint on the transmission power, the transmission location, or the sensing period, etc. The conventional CR has two main functions: spectrum sensing; and data transmission. Hence, most design problems for the conventional CR involve the tradeoff between the accuracy of spectrum sensing and the throughput of data transmission, or the sensing-throughput tradeoff (Liang et al. 2008). The main design goal is to maximize the throughput or information rate subject to constraints on collision avoidance.

In the energy harvesting CR, the new feature of energy harvesting brings new challenges into CR designs. First, CR will have three functions to perform in the physical layer: spectrum sensing; data transmission; and energy harvesting. Thus, instead of the tradeoff between sensing and transmission, one has the tradeoff between sensing, transmission, and harvesting. This makes the designs much more complicated. Secondly, in addition to the constraint on collision avoidance, there is an additional energy causality constraint. This constraint basically imposes a limitation that the energy must be harvested before it can be used for spectrum sensing or data transmission, or CR cannot use any future energy. In this case, unlike the conventional CR, the energy in the energy harvesting CR is dynamic and unstable. Hence, there is a possibility that the licensed channel is free but the CR user cannot use it due to insufficient energy. It is also possible that the licensed channel is busy but instead of remaining idle the CR user can harvest energy from the PU. The energy harvesting and spectrum sensing functions are performed at a receiver, while the data transmission function is performed at a transmitter. Their main design goals can be the maximization of throughput but can also be the maximization of the harvested energy subject to constraints on collision avoidance and energy causality.

Figure 7.3 compares the conventional CR with the energy harvesting CR in terms of their design problems. If the energy is harvested from the PU or the secondary

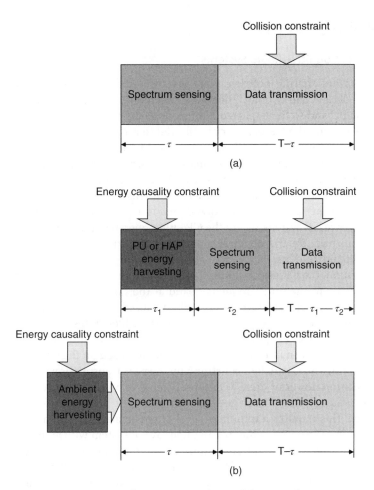

Figure 7.3 Comparison of (a) conventional CR and (b) energy harvesting CR.

base station, it requires extra energy harvesting time in the data packet. If the energy is harvested from the ambient sources, harvesting can be performed at the same time as sensing and transmission so that no extra harvesting time is required but the energy causality constraint still applies with greater uncertainty. Thus, most studies on energy harvesting CR focus on the tradeoff between sensing, transmission, and harvesting, subject to constraints on collision avoidance and energy causality. The collision avoidance is related to the dynamics of the PU traffic. The energy causality is related to the dynamics of the energy arrival process. If the CR harvests energy from the PU, the problem will be even more complicated, as the PU traffic and the harvested energy will be correlated.

Next, we will discuss several important technical challenges in energy harvesting CR. Before doing this, it is necessary to explain how the conventional CR and its spectrum sensing function work, as they are also part of the principles based on which energy harvesting CR works.

7.2 Conventional Cognitive Radio

7.2.1 Different Types of Cognitive Radio Systems

Depending on the spectrum sharing policies, there are three different types of CR systems: interweave; underlay; and overlay. They have different levels of cooperation between the CR and the PU and so their spectrum utilization efficiency also varies.

In the interweave CR system, the CR user is only allowed to access the licensed frequency band when the PU is not using it. This is also called opportunistic spectrum access. This policy imposes the strictest limitation on the CR user and provides the best protection for the PU. It is adopted in applications where the CR user and the PU do not have any interaction or cooperation. In this system, spectrum sensing is vital. It helps to find the "spectrum holes" in the spatial and temporal domains. Figure 7.1 actually gives the diagram of an interweave CR system, where the CR either operates at different time slots or different locations from the PU, depending on whether it exploits the temporal opportunities or spatial opportunities.

In the underlay CR system, the CR user is allowed to coexist with the PU, that is, the CR and the PU can operate in the same frequency band at the same time. From the PU's point of view, there is an increased level of noise floor caused by the CR that may affect its performance but this effect is negligible as long as the CR operates within the limit. This is very similar to the spread spectrum idea or the ultra-wide bandwidth technique. To protect the PU, there is an interference temperature imposed on the CR user. This interference temperature is essentially a constraint on the peak transmission power and the average transmission power of the CR, or a probability of power outage of the PU. Thus, in this case, although the CR can coexist with the PU, it must operate at a low transmission power. This requires a minimum level of cooperation or interaction between the CR and the PU. Figure 7.4 compares an underlay CR system with an interweave CR system, both of which utilize the spectrum holes in the time domain. Their main difference is how they behave when the PU is detected. In interweave CR, it stops transmission when the PU is detected, while in underlay CR, it reduces its transmission power when the PU is detected.

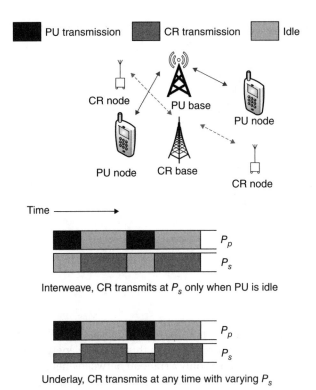

Figure 7.4 Comparison of interweave CR and underlay CR when the temporal opportunities are exploited.

Compared with the interweave system, the underlay system does not stop transmission when the PU is operating, as it is allowed to coexist with the PU, but it does need power adjustment. In other words, the transmission power of the interweave system can be considered as a discrete set of either zero or P_s, while the transmission power of the underlay system is within a continuous range depending on the interference limitation. The interweave system is suitable for high data rate or high-power applications, while the underlay system is suitable for low data rate or low-power applications.

In the overlay CR system, the CR user has full cooperation with the PU. For example, the PU transmitter can forward its signal to the CR and ask the CR to relay this signal to the PU receiver. The CR can combine its own signal with the relayed PU signal so that it can make use of this opportunity to deliver information to the CR receiver. In other words, the CR user helps the PU in exchange for the use of the licensed frequency band. In most cases, the CR needs to have important information about the PU network, such as its modulation scheme, its coding scheme, etc. Thus, in the overlay system, the CR and PU must trust each other. Figure 7.5 shows a diagram of an overlay CR system, where the CR cooperates with the PU for PU transmission in exchange for spectrum access. In this case, spectrum sensing is not needed, as the CR can get the spectrum occupancy information from the PU network directly.

Among these three systems, overlay has the full cooperation, underlay has the minimum cooperation, while interweave has no cooperation between the CR and the PU.

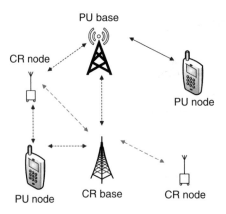

PU base

CR node

PU node

PU node   CR base   CR node

Figure 7.5 Diagram of an overlay CR system.

PU to PU link

CR to CR link

PU to CR or CR to PU link

Consequently, interweave has the widest application. In an energy harvesting CR, if the overlay principle is used, spectrum sensing is not required. In this case, the energy harvesting CR designs will be much simpler, as there is only a tradeoff between energy harvesting and data transmission, which is very similar to the tradeoff between energy and rate discussed in Chapter 6. What makes CR unique is its spectrum sensing function. Thus, this chapter focuses on interweave and underlay CRs. Next, several commonly used spectrum sensing methods will be reviewed.

7.2.2 Spectrum Sensing Methods

7.2.2.1 Energy Detection
The most commonly used spectrum sensing method is energy detection (Urkowitz 1967). It stems from the detection theory in statistics.

In a static additive white Gaussian noise (AWGN) channel, the energy detection can be described by a binary hypothesis testing problem as

$$H_0(\text{channel free}) : \mathbf{y} = \mathbf{n} \tag{7.1a}$$

$$H_1(\text{channel occupied}) : \mathbf{y} = \mathbf{s} + \mathbf{n} \tag{7.1b}$$

where $\mathbf{y} = [y_1 \quad y_2 \cdots y_K]$ is the received sample, $\mathbf{n} = [n_1 \quad n_2 \cdots n_K]$ is the AWGN with mean zero and variance σ^2, and $\mathbf{s} = [s_1 \quad s_2 \cdots s_K]$ is the PU signal. This model can describe the case when a single CR performs spectrum sensing. In this case, K denotes the number of samples taken at different time instants at the CR. In this model, the elements in $\mathbf{s}$ are deterministic values.

There are many different detection rules. A commonly used rule is the maximum *a posteriori* rule. If the maximum *a posteriori* rule is used, the optimum detector can be designed as

$$P[H_0]f(\mathbf{y}|H_0) \overset{H_0}{\underset{H_1}{\gtrless}} P[H_1]f(\mathbf{y}|H_1) \tag{7.2}$$

where $P[H_0]$ and $P[H_1]$ are the *a priori* probabilities of H_0 and H_1, respectively, $f(\mathbf{y}|H_0) = \frac{1}{(2\pi\sigma^2)^{K/2}}e^{-\frac{\mathbf{y}\mathbf{y}^T}{2\sigma^2}}$ and $f(\mathbf{y}|H_1) = \frac{1}{(2\pi\sigma^2)^{K/2}}e^{-\frac{(\mathbf{y}-\mathbf{s})(\mathbf{y}-\mathbf{s})^T}{2\sigma^2}}$ are the likely functions, and $()^T$ represents the transpose operation. Thus, using the likelihood functions in the likelihood ratio test, one has

$$\mathbf{y}\mathbf{s}^T \overset{H_1}{\underset{H_0}{\gtrless}} \sigma^2 \ln \frac{P[H_0]}{P[H_1]} + \frac{1}{2}\mathbf{s}\mathbf{s}^T. \tag{7.3}$$

This is a coherent detector that requires knowledge of $\mathbf{s}$, which is not realistic in an interweave system. In fact, had $\mathbf{s}$ been known, there is no need for spectrum sensing, because the PU is there.

In a fading channel or in a static AWGN channel with random PU signals, which is normally the case in a wireless communications system, the values of $\mathbf{s}$ become random. Denote $\mathbf{s}$ as Gaussian random variables with mean zero and covariance matrix $\alpha^2 \mathbf{I}$, where $\mathbf{I}$ is the $K \times K$ identity matrix, that is, the PU signal samples are independent and identically distributed with a common variance of α^2. Then, one has $f(\mathbf{y}|H_0) = \frac{1}{(2\pi\sigma^2)^{K/2}}e^{-\frac{\mathbf{y}\mathbf{y}^T}{2\sigma^2}}$ and $f(\mathbf{y}|H_1) = \frac{1}{[2\pi(\sigma^2+\alpha^2)]^{K/2}}e^{-\frac{\mathbf{y}\mathbf{y}^T}{2(\sigma^2+\alpha^2)}}$. Using them in (7.2) and calculating the likelihood ratio, one has the detector as

$$\mathbf{y}\mathbf{y}^T \overset{H_1}{\underset{H_0}{\gtrless}} D. \tag{7.4}$$

This is an energy detector, because $\mathbf{y}\mathbf{y}^T$ gives the energy of the received samples. If the maximum *a posteriori* rule is used, it can be shown that the detection threshold is given by $D = \frac{\sigma^2(2\sigma^2+2\alpha^2)}{\alpha^2} \ln \left[\frac{P[H_0]}{P[H_1]}(\frac{\sigma^2+\alpha^2}{\sigma^2})^{K/2} \right]$.

One sees that, if the energy is larger than the threshold D, the decision is that the channel is occupied, while if the energy is smaller than the threshold D, the decision is that the channel is free. This agrees with intuition. One also sees that this detector needs knowledge of the average fading power and the noise power in the calculation of D. This leads to the noise uncertainty problem (Tandra and Sahai 2008), as the actual values of these parameters are not known in practice and they have to be estimated. The estimation error has a lower limit so that noise uncertainty will occur.

The model in (7.1) can also be adapted to describe the case when multiple CRs perform collaborative spectrum sensing. In this case, each CR takes one sample and K denotes the number of CRs. All these K samples are sent to the fusion center for a decision. If all the CR users take more than one sample in the time domain, the vectors will become matrices in (7.1) but the detection method will be similar.

The energy detector in (7.4) can be used to detect the spectrum holes in the time domain. For applications that exploit the spectrum opportunities in the spatial domain, one can use the distance-dependent path loss (Visotsky et al. 2005; Chen and Beaulieu

(2009). In this case, collaborative spectrum sensing has to be performed, where CR users at different locations sample the signal received from the licensed channel and then these samples are sent to a fusion center for a final decision. In this case, one has the binary hypothesis testing problem as

$$H_0(\text{channel free}) : \mathbf{z} \sim \mathcal{N}(\mu(R+\delta) \times \mathbf{1}, \sigma^2 \Sigma) \tag{7.5a}$$

$$H_1(\text{channel occupied}) : \mathbf{z} \sim \mathcal{N}(\mu(R) \times \mathbf{1}, \sigma^2 \Sigma) \tag{7.5b}$$

where $i = 1, 2, \cdots, N$ index samples taken at different CR users, $\mathcal{N}(\cdot, \cdot)$ represents a normal distribution, $\mathbf{z} = [z_1 \quad z_2 \cdots z_N]$ are the samples used for sensing, $\delta > 0$ and $\sigma^2 \Sigma$ is the covariance matrix of $\mathbf{z}$, $\mathbf{z} = 10\log_{10}\mathbf{P}$ is the dB value of the average power received from the licensed channel by the N CR users, $\mathbf{P}$ is the actual value of the average power representing shadowing and is lognormally distributed with mean $\mu(r) \times \mathbf{1}$, $\mu(r)$ is the distance-dependent path loss, R is the safe distance so that the CR users only transmit data in the licensed frequency band of the PU when r is larger than R.

From (7.5), the detector can be derived as (Visotsky et al. 2005)

$$\frac{1 \times \Sigma^{-1} \times \mathbf{z}^T}{1 \times \Sigma^{-1} \times \mathbf{1}^T} \overset{H_1}{\underset{H_0}{\gtrless}} D \tag{7.6}$$

where D is the detection threshold. Using the maximum *a posteriori* detection rule, one has $D = \frac{\sigma^2}{(\mu(R)-\mu(R+\delta))1 \times \Sigma^{-1} \times 1^T} \ln \frac{P[H_0]}{P[H_1]} + \frac{\mu(R)+\mu(R+\delta)}{21 \times \Sigma^{-1} \times 1^T} N.$

Using these detectors, two important performance measures can be defined: the probability of false alarm; and the probability of detection. The probability of false alarm is defined as

$$P_f = Pr\{H_1|H_0\}. \tag{7.7}$$

It represents the opportunity that the CR user has lost due to inaccurate sensing. Sometimes it is also called the probability of missed opportunity.

The probability of detection is defined as

$$P_d = Pr\{H_1|H_1\}. \tag{7.8}$$

It represents the protection the CR user can provide for the PU.

One sees that, in order to have a good sensing result, spectrum sensing should be designed in a way such that P_d is as large as possible to maximize the protection for the PU, while P_f is as small as possible to minimize the missed opportunity. Statistical theories have shown that the maximum *a posteriori* rule aims to minimize the overall error of $P[H_0]Pr\{H_1|H_0\} + P[H_1]Pr\{H_0|H_1\}$. However, within this overall error, it can be seen that, for CR systems, $Pr\{H_0|H_1\} = 1 - Pr\{H_1|H_1\}$ is more important than $Pr\{H_1|H_0\}$, as $Pr\{H_0|H_1\}$ determines the interference to the PU from the CR and for interweave systems the opportunistic spectrum access of the CR must be "non-interfering" according to the regulations. Thus, the maximum *a posteriori* rule cannot fulfill the regulatory requirements.

In most CR studies, the Neyman–Pearson detection rule is adopted, where one fixes the probability of false alarm $Pr\{H_1|H_0\}$ as β, while minimizing $Pr\{H_0|H_1\}$ or maximizing P_d. Using the Neyman–Pearson rule, one needs to find the detection threshold in (7.4) and (7.6) by calculating P_f and fixing it at β.

In (7.4), under the hypothesis of H_0, $\mathbf{yy}^T = \mathbf{nn}^T$ follows a Gamma distribution with shape parameter $\frac{K}{2}$ and scale parameter $2\sigma^2$. Thus, one has the probability of false alarm as

$$\beta = 1 - G\left(D, \frac{K}{2}, 2\sigma^2\right) \qquad (7.9)$$

where $G(D, \frac{K}{2}, 2\sigma^2) = \int_0^D \frac{1}{\Gamma(K/2)(2\sigma^2)^K} x^{K-1} e^{-x/(2\sigma^2)} dx$ is the cumulative distribution function of a Gamma distribution with shape parameter $K/2$ and scale parameter $2\sigma^2$. This gives the detection threshold as

$$D = G^{-1}\left(1 - \beta, \frac{K}{2}, 2\sigma^2\right) \qquad (7.10)$$

where $G^{-1}(\cdot, \frac{K}{2}, 2\sigma^2)$ is the inverse function of $G(D, \frac{K}{2}, 2\sigma^2)$. Note that, the calculation of D in (7.10) does require knowledge of the noise power σ^2. Thus, it will suffer from noise uncertainty.

For (7.6), using a similar method, the detection threshold can be given by

$$D = \mu(R + \delta) + \frac{\sigma}{\sqrt{\mathbf{1}^T \times \mathbf{\Sigma}^{-1} \times \mathbf{1}}} Q^{-1}(\beta) \qquad (7.11)$$

where $Q(x) = \frac{1}{\sqrt{2\pi}} \int_x^\infty e^{-t^2/2} dt$ is the Gaussian Q function, $Q^{-1}(\cdot)$ is the inverse of the Gaussian Q function, and β is the predetermined probability of false alarm. In this case, the calculation of the detection threshold D requires the noise power too. Thus, it still suffers from noise uncertainty.

In the Neyman–Pearson rule, the most important performance measure is the receiver operating characteristics (ROC), where P_f is the x axis and P_d is the y axis. Figure 7.6 gives an example of ROC for the energy detector in (7.4). One sees that a larger value K or a larger value of α^2/σ^2 improves the detection performance and that α^2/σ^2 has a larger impact on the performance.

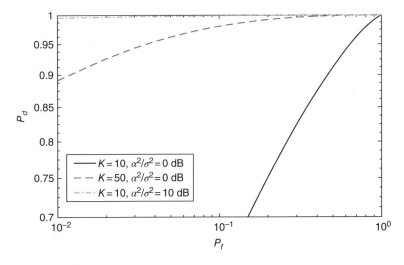

Figure 7.6 ROC of the energy detector in (7.4).

One sees from these expressions that the sensing accuracy depends on the number of samples K (or the number of users N in collaborative sensing) and the SNR $\frac{\alpha^2}{\sigma^2}$. Next, the feature-based detectors will be discussed.

7.2.2.2 Feature Detection

Energy is actually also a feature of the signal but feature detection uses other features of the signal, such as covariance and eigenvalues. In particular, the binary hypothesis testing problem is given by

$$H_0(\text{channel free}) : \mathbf{y} = \mathbf{n} \tag{7.12a}$$

$$H_1(\text{channel occupied}) : \mathbf{y} = \mathbf{s} + \mathbf{n} \tag{7.12b}$$

where $\mathbf{n}$ are independent and identically distributed Gaussian random variables with mean zero and variance σ^2, but the PU signals $\mathbf{s}$ are Gaussian random variables with mean zero and covariance matrix $\mathbf{\Sigma}$ and the (i,j)th element of $\mathbf{\Sigma}$ is $\rho^{|i-j|}$ with ρ being a constant $0 \leq \rho \leq 1$ and $|i-j|$ being the time difference between two samples. In this case, one has to assume correlated signals so that it can be distinguished from the uncorrelated noise samples. Otherwise, the feature detectors will fail. Most signals in practical systems are correlated, due to modulation, coding and other signal processing operations.

The sample covariance matrix of $\mathbf{y}$ is given by

$$R_y(K) = \begin{pmatrix} \vartheta(0) & \vartheta(1) & \cdots & \vartheta(L-1) \\ \vartheta(1) & \vartheta(0) & \cdots & \vartheta(L-2) \\ \vdots & \vdots & \vdots & \vdots \\ \vartheta(L-1) & \vartheta(L-2) & \cdots & \vartheta(0) \end{pmatrix}, \tag{7.13}$$

where $\vartheta(l) = \frac{1}{K}\sum_{k=1}^{K} y_k y_{k-l}$ is the sample covariance with a lag of l, and L is the smoothing factor. For pure noise or the null hypothesis, $R_y(K)$ will be very close to an identity matrix $\sigma^2 \mathbf{I}$, while for the signal-plus-noise case or the alternative hypothesis H_1, $R_y(K)$ will be close to $\mathbf{\Sigma} + \sigma^2 \mathbf{I}$. This allows us to detect the presence of the PU. Different properties of the sample covariance matrix can be used.

One can use the largest eigenvalue of $R_y(K)$. In this case, the sample covariance matrix is calculated from (7.13). The maximum eigenvalue of $R_y(K)$ is calculated from the sample covariance matrix. Then, it is compared with a detection threshold to be determined later. It can be derived that the probability of false alarm and the probability of detection are (Zeng et al. 2008)

$$P_f = 1 - TW_1\left(\frac{KD_{ME} - \gamma}{\epsilon}\right) \tag{7.14}$$

$$P_d = 1 - TW_1\left(\frac{KD_{ME} - \frac{K\rho_{max}}{\sigma^2} - \gamma}{\epsilon}\right) \tag{7.15}$$

where D_{ME} is the detection threshold of the maximum eigenvalue (ME) detector, $TW_1()$ represents the Tracy–Widom distribution of order 1 (Tracy and Widom 1996), $\epsilon = (\sqrt{K-1} + \sqrt{L})(\frac{1}{\sqrt{K-1}} + \frac{1}{\sqrt{L}})^{1/3}$, $\gamma = (\sqrt{K-1} + \sqrt{L})^2$, and ρ_{max} is the maximum eigenvalue of $R_y(K)$.

One can also use the ratio of the maximum eigenvalue to the minimum eigenvalue. In this case, first, the sample covariance matrix is calculated from (7.13). The maximum and minimum eigenvalues of $R_y(K)$ are calculated from the sample covariance matrix. Then, their ratio is calculated and compared with a detection threshold to be determined later. The probabilities of false alarm and detection can be derived as (Zeng and Liang 2009a)

$$P_f = 1 - TW_1 \left[\frac{D_{MME}(\sqrt{K} - \sqrt{L})^2 - \gamma}{\epsilon} \right] \tag{7.16}$$

$$P_d = 1 - TW_1 \left[\frac{KD_{MME} + \frac{K(D_{MME}\rho_{min} - \rho_{max})}{\sigma^2} - \gamma}{\epsilon} \right] \tag{7.17}$$

where D_{MME} is the detection threshold of the maximum-to-minimum eigenvalue (MME) detector, ρ_{min} is the minimum eigenvalues of $R_y(K)$, and other symbols are defined as before.

One can also use the ratio of the average energy to the minimum eigenvalue. In this case, the sample covariance matrix is calculated from (7.13). The minimum eigenvalue will be calculated from $R_y(K)$. Then, the average energy will be calculated as $\bar{E} = \frac{1}{K} \sum_{k=1}^{K} |y_k|^2$. After that, their ratio is calculated and compared with a detection threshold. The probabilities of false alarm and detection are derived as (Zeng and Liang 2009a)

$$P_f \approx Q \left[\frac{D_{EME}(\sqrt{K} - \sqrt{L})^2 - K}{\sqrt{2K}} \right] \tag{7.18}$$

$$P_d \approx Q \left\{ \frac{D_{EME}[\rho_{min} + \frac{\sigma^2}{\sqrt{K}}(\sqrt{K} - \sqrt{L})] - \frac{tr[R_y(K)]}{L} - \sigma^2}{\sqrt{\frac{2}{K}}\sigma^2} \right\} \tag{7.19}$$

where D_{EME} is the detection threshold of the average energy to minimum eigenvalue (EME) detector and $tr[R_y(K)]$ is the trace of $R_y(K)$.

Finally, one can use the covariance of the sample directly. In this case, the sample covariance matrix is calculated from (7.13). Denote $R_y(i,j)$ as the element in the ith row and jth column. Then, the average covariance and the average variance are calculated as $T_1 = \frac{1}{K} \sum_{i=1}^{K} \sum_{j=1}^{K} |R_y(i,j)|$ and $T_2 = \frac{1}{K} \sum_{i=1}^{K} |R_y(i,i)|$, respectively. Finally, the ratio of T_1/T_2 is calculated and compared with a detection threshold. The probabilities are given by (Zeng and Liang 2009b)

$$P_f \approx 1 - Q \left\{ \frac{\frac{1}{D_{cov}}[1 + (L-1)\sqrt{\frac{2}{K\pi}}] - 1}{\sqrt{2/K}} \right\} \tag{7.20}$$

$$P_d \approx 1 - Q \left[\frac{\frac{1}{D_{cov}} + \frac{r_L \sigma^2}{D_{cov}(\sigma_s^2 + \sigma^2)} - 1}{\sqrt{2/K}} \right] \tag{7.21}$$

where D_{COV} is the detection threshold, $\sigma_s^2 = E[s^2(n)]$ is the average energy of the PU signal and $r_L = \frac{2}{L}\sum_{i=1}^{L-1}(L-i)|E[s(n)s(n-i)]/E[s^2(n)]|$. All these equations are obtained from random matrix theories.

Using the Neyman–Pearson rule, the detection thresholds can be determined by letting $P_f = \beta$ and solving the equations for D_{ME}, D_{MME}, D_{EME}, and D_{COV} in (7.14), (7.16), (7.18), and (7.20), respectively. For example, for the ME detector

$$D_{ME} = \frac{\epsilon}{K} \cdot TW_1^{-1}(1-\beta) + \frac{\gamma}{K} \qquad (7.22)$$

from (7.14), where $TW_1^{-1}(\cdot)$ is the inverse function of the Tracy–Widom distribution of order 1.

One can see that (7.14), (7.16), (7.18), and (7.20) are only functions of the sample size K and the smoothing factor L. Consequently, the detection thresholds will be functions of K and L too. They will not depend on the noise power σ^2. Thus, they do not have the noise uncertainty problem.

On the other hand, in order to have an accurate estimate of the sample covariance, the sample size K often needs to be large. This incurs more computational complexity as well as longer sensing time. Thus, feature detectors are more complicated than the energy detector. Figure 7.7 compares different feature detectors in terms of their signal processing procedures.

Figure 7.8 compares the performances of different feature detectors, when $K = 100$ and $\sigma_s^2/\sigma^2 = 1$. The PU is assumed to leave or arrive during spectrum sensing with dynamic traffic (Chen et al. 2011b). The average arrival time and the average departure time are 4 s. One sees that the EME detector is the worst, while the ME detector is the best.

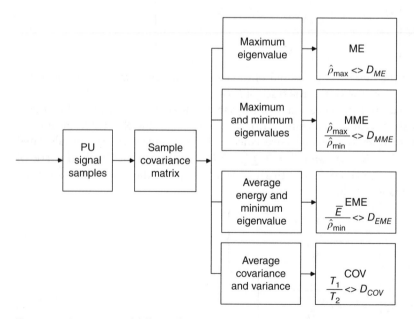

Figure 7.7 Comparison of different feature detectors.

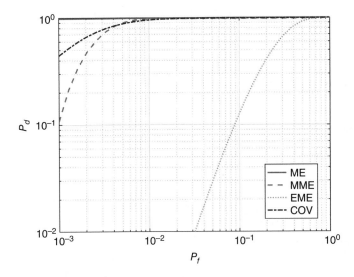

Figure 7.8 Performances of different feature detectors.

There are other feature detectors, such as cyclo-stationarity and spectral estimators. These are even more complicated. Also, feature detection can be combined with collaborative sensing to improve accuracies further. Most studies in energy harvesting CR adopt the energy detection due to its simplicity. In the following, without explicit explanation, spectrum sensing means energy detection.

7.3 Types of Energy Harvesting Cognitive Radio

With the background knowledge introduced in the previous section, from this section on, energy harvesting CR will be discussed. As mentioned before, the difference between energy harvesting CR and conventional CR is the extra energy harvesting capability at the CR. This not only adds a new function to the CR but also makes its energy supply random, leading to the energy causality problem.

7.3.1 Protocols

There are many different types of energy harvesting CR systems. They can be classified based on their protocols.

Some systems do not have an energy storage device at the CR so that the harvested energy must be used immediately. This is the "harvest-use" protocol (Kansal et al. 2007), similar to the variable-power transmission discussed in Chapter 4. In this protocol, the information rate is randomly changing, because the information rate is determined by the transmission power and the transmission power is determined by the random amount of energy harvested. Also, the instantaneous energy harvesting rate must always be larger than the energy consumption rate.

Other systems may have an energy storage device at the CR so that the harvested energy will be stored first before it can be used. The energy storage device serves as a

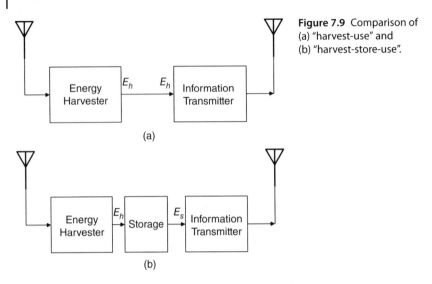

Figure 7.9 Comparison of (a) "harvest-use" and (b) "harvest-store-use".

buffer to level up the randomness in the amount of energy harvested. This is the "harvest-store-use" protocol (Yin et al. 2014), similar to the fixed-power transmission discussed in Chapter 4. In this case, it is possible to schedule energy harvesting and energy consumption, such as sensing and transmission, for the best resource allocation. Figure 7.9 compares these two protocols. The used energy E_s is not necessarily smaller than E_h because of previously stored energy. However, E_s is a fixed value, while E_h could be random.

The "harvest-use" protocol simplifies the energy harvesting CR design by removing energy storage, but this leads to variable rate. The "harvest-store-use" protocol requires energy storage, which increases the complexity of energy harvesting CR software and hardware, but it allows fixed-power and fixed-rate data performance with guaranteed QoS.

In the literature, the difference between these two protocols is reflected by different energy causality constraints. The optimization problems in the "harvest-use" protocol assume a battery with a size of 0 in the constraints. The optimization problems in the "harvest-store-use" protocol assume a battery with a finite size B in the constraints. As an extension of these studies, the optimization problems can also assume a battery with an infinite size in the constraints. These different constraints will lead to different optimal solutions. However, most studies in the literature have considered the "harvest-store-use" protocol with finite battery.

7.3.2 Energy Sources

The energy harvesting CR systems can also be classified based on the energy sources of the PU and the CR. Table 7.1 compares energy harvesting CR systems with different energy sources. Each reference in the table is only an example and the list is not exhaustive.

In some systems (Zhang et al. 2015), both PU and CR harvest energy from the same ambient source, for example, the Sun or the radio environment. In this case, the PU transmission is determined by the energy arrival process of the ambient source, which also determines the energy harvested by the CR. Essentially, sharing the same energy

Table 7.1 CR systems with different energy sources for PU and CR.

Reference	PU source	CR source
Zhang et al. (2015)	Ambient environment	Ambient environment
Azmat et al. (2018)	CR	PU
El Shafie et al. (2015)	Ambient environment	Fixed supply
Zhai et al. (2016a)	CR	Fixed supply
Zhang et al. (2016)	Fixed supply	Ambient environment
Lee and Zhang (2015)	Fixed supply	HAP
Zheng et al. (2014)	Fixed supply	PU
Usman and Koo (2014)	Fixed supply	PU and ambient

CR, cognitive radio; HAP, hybrid access point; PU, primary user.

source makes the PU and CR operations correlated and this correlation can be utilized to improve the sensing accuracy. However, this is only applicable to the case when the PU and CR use all the energy harvested, or the "harvest-use" protocol. In the cases when energy storage is used, this correlation will be less important or even negligible due to energy buffering.

In some systems (Azmat et al. 2018), PU and CR harvest energy from each other. When the PU transmits data, CR will harvest energy from this transmission. Similarly, when the CR transmits data over the spectrum holes, the PU harvests energy from the CR transmission too. In this case, since both PU and CR have three stages (data transmission, energy harvesting, and idle), more cases need to be considered for the sensing-throughput tradeoff at the CR and the PU to maximize the throughput and the harvested energy.

There are also systems where the CR does not harvest energy but the PU harvests energy from either the ambient source or the CR (El Shafie et al. 2015; Zhai et al. 2016a). They cannot be considered as energy harvesting CR systems, as the CR does not have the energy harvesting feature. Nevertheless, their designs will be related to energy harvesting CR designs, as the energy arrival affects the power supply of the PU. This in turn affects its data transmission due to the energy causality. The PU data transmission will then affect the spectrum sensing and the transmission opportunities at the CR.

In many studies on energy harvesting CR, such as Zhang et al. (2016), the PU has fixed energy supply, and only the CR harvests energy. In this case, as mentioned before, the energy arrival process determines the energy causality, while the PU traffic process determines the collision avoidance. Both need to be considered to optimize the resource allocation at the CR.

In Lee and Zhang (2015), the CR harvests energy from the secondary base station in the CR network. Thus, coordination between power transfer from the secondary base station and the data transmission from the CR users is required to make sure that there is enough energy for data transmission but also to make sure that the transmission power of the CR users will not cause noticeable interference to the PU.

In Zheng et al. (2014), the CR harvests energy from the PU. This is an incentive for the CR to forward its signal to the PR receiver. This idea is very similar to energy harvesting relaying to be discussed in the next chapter, except that the CR as a relay node is not in the same network as the PU users as source and destination nodes. The benefit for the CR user is that the CR user can combine its own signal with the PU signal during the relaying to use the licensed frequency band. This is an overlay system. In both Zheng et al. (2014) and Lee and Zhang (2015) the power transfer is intentional to make the designs easier by only considering the collision avoidance, unlike the ambient source in Zhang et al. (2016).

In Usman and Koo (2014), the CR harvests energy from both the PU and the ambient source. This makes the system designs more complicated, because the PU traffic will not only affect the collision avoidance but also the energy causality. In any case, such a system may not be used in practice, as it needs two sets of energy harvesters with increased complexity.

In the literature, the majority of the studies focus on energy harvesting CR where the PU has fixed energy supply and the CR harvests energy from either an ambient source, or the PU, or the secondary base station. If the CR harvests energy from the ambient source, the energy arrival process is often considered in the joint optimization of transmission time and transmission power for resource allocation. If the CR harvests energy from the PU, the sensing time, the detection threshold and the spectrum access are often jointly optimized subject to energy and collision constraints. In the systems where the CR harvests energy from the secondary base station, the energy harvesting CR is much simpler, as only the collision constraint needs to be considered.

The rest of this chapter will discuss these energy harvesting CR systems. We will first discuss the energy harvesting CR systems that harvest energy from the secondary base station. Then, the CR systems that harvest energy from the PU signal and the CR systems that harvest energy from ambient sources will be discussed.

7.4 From the Secondary Base Station

Figure 7.10 shows a diagram of these energy harvesting CR systems. In this case, the CR system is actually a wireless powered communications system, similar to the hybrid access point (HAP) system discussed in Chapter 6. The only difference is that now this CR system has to share the licensed frequency band with the PU and hence, its downlink power transfer and uplink information delivery may be affected by the PU. Specifically, if the PU is transmitting at the same time as the CR downlink power transfer from the base station to the CR users, the CR users can harvest more energy but the PU receiver will be interfered by the CR wireless power. If the PU is transmitting at the same time as the CR uplink information delivery from the CR users to the base station, the base station will be interfered by the PU signal, while the PU receiver will be interfered by the CR users.

Consider an energy harvesting CR network with one base station and K CR nodes. The CR network operates at the same channel as the PU system with one PU transmitter and one PU receiver. Similar to the HAP system discussed in Chapter 6, the CR nodes do not have fixed power supply. In this case, the base station broadcasts wireless power in the downlink for τT seconds to charge the CR nodes, where T is the total link time. The harvested energy is then used by the CR nodes to transmit data to the base station in the uplink for $(1 - \tau)T$ seconds. There is a slight difference between this model and the model of the HAP system in Chapter 6, as the nodes

Figure 7.10 CR system that has its secondary base station as the energy source.

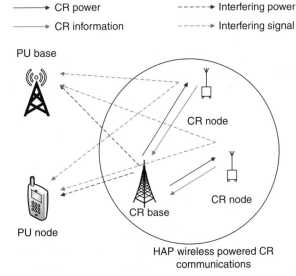

HAP wireless powered CR communications

in the HAP system in Chapter 6 transmit their data sequentially with allocated time intervals of $\tau_1, \tau_2, \cdots, \tau_K$, while here all K nodes adopt simultaneous transmission during $(1 - \tau)T$. Thus, the HAP system in Chapter 6 uses time-division-multiple-access (TDMA), while here the CR network can use code-division-multiple-access (CDMA) or orthogonal-frequency-division-multiple-access (OFDMA) that allow simultaneous transmission of multiple users. Non-orthogonal multiple access (NOMA) may also be used. The purpose of simultaneous transmission is to reduce the chance of interfering the PU as much as possible.

In the downlink, the received signals at the CR nodes can be expressed as

$$y_i = \sqrt{P_s} h_i s + \sqrt{P_p} g_i x + n_i \tag{7.23}$$

where $i = 1, 2, \cdots, K$ represents different CR nodes, P_s is the transmission power of the base station or the HAP, h_i is the channel gain from the base station to the ith CR node, s is the transmitted symbol of the base station, P_p is the transmission power of the PU transmitter, g_i is the channel gain from the PU transmitter to the ith CR node, x is the transmitted symbol of the PU, and n_i is the AWGN. In the following, assume that h_i and g_i are fixed complex values during each transmission but vary from transmission to transmission and that $E[|s|^2] = E[|x|^2] = 1$. Hence, the system operates in block fading channels with normalized symbols. Also, the noise is a complex Gaussian random variable with mean zero and variance $2\sigma^2$. An important assumption made in (7.23) is that the PU transmitter is always transmitting at a fixed transmission power. In some applications, the PU may have on–off traffic such that the PU term may not always be in (7.23). In this case, more complicated models for the PU term in (7.23), such as a Markov process, need to be used.

Using the signal in (7.23), the harvested energy at the ith CR node is given by

$$E_i = Q_i \tau = \eta_i (P_s |h_i|^2 + P_p |g_i|^2) \tau \tag{7.24}$$

where η_i is the conversion efficiency of the energy harvester at the ith CR node, Q_i is the harvested power, and $T = 1$ for convenience. One sees that the PU transmission leads to more harvested energy at the CR node, compared with the HAP system discussed in Chapter 6.

The harvested energy in (7.24) is then used for the following data transmission in the uplink. Assume that e_i of the harvested energy is used, where $e_i = c_i E_i$ and $0 \le c_i \le 1$ indicates the part of harvested energy used for data transmission. The other part $(1 - c_i) E_i$ is saved for future transmissions. Thus, the transmission power of the ith CR is given by

$$P_i = \frac{e_i}{(1 - \tau)} = \frac{\tau}{1 - \tau} c_i \eta_i (P_s |h_i|^2 + P_p |g_i|^2).$$

(7.25)

The received signal at the base station in the uplink can be expressed as

$$y_0 = \sum_{i=1}^{K} \sqrt{P_i} u_i s_i + \sqrt{P_p} g_0 x + n_0$$

(7.26)

where P_i is the transmission power of the ith CR node in (7.25), u_i is the channel gain from the ith CR node to the base station, s_i is the information transmitted by the ith CR node, g_0 is the channel gain from the PU transmitter to the base station, and n_0 is the AWGN with mean zero and variance $2\sigma^2$. Thus, the information rate in the uplink can be derived as

$$R = (1 - \tau) \log_2 \left(1 + \frac{\sum_{i=1}^{K} P_i |u_i|^2}{2\sigma^2 + P_p |g_0|^2} \right).$$

(7.27)

In the derivation of (7.27), the multi-user interference has been ignored so that the total signal power is the sum of each user's power. One sees that the PU transmitter degrades the CR performance in the uplink due to interference.

The above derivations apply to any wireless powered system with interference. For CR systems, since the CRs operate with the PU using the underlay principle, the transmission powers of the base station and the CR nodes must be limited. Specifically, assume an interference temperature determined by a peak transmission power limit Γ at the PU, one has

$$P_s = \min \left(\frac{\Gamma}{|h_0|^2}, P_{max} \right)$$

(7.28)

where h_0 is the channel gain from the base station to the PU receiver such that the interference caused by the base station power transfer $P_s |h_0|^2$ is smaller than or equal to the limit Γ, and P_{max} is the maximum power that the base station can physically use.

Similarly, to limit interference caused by the CR nodes, one has

$$\sum_{i=1}^{K} P_i |v_i|^2 \le \Gamma$$

(7.29)

where v_i is the channel gain from the ith CR node to the PU receiver. Several points can be made. First, in (7.28) and (7.29), knowledge of h_0 and v_i is available at the base station and the ith CR node, respectively, in order to satisfy the interference temperature requirement. Secondly, (7.28) and (7.29) use a peak transmission power limit. One may also limit the average transmission power or the power outage probability. In this case, only the average fading power or the distribution are needed. Finally, there have been

Table 7.2 List of channel gains.

Symbols	Channels
g_0	From PU transmitter to CR base station
g_i	From PU transmitter to ith CR
h_0	From CR base station to PU receiver
h_i	From CR base station to ith CR
u_i	From ith CR to CR base station
v_i	From ith CR to PU receiver

CR, cognitive radio; PU, primary user.

a lot of channel gains defined in the above. Table 7.2 lists all the channel gains used in the discussion for clarity.

Putting all these equations together, the optimization problem in this energy harvesting CR system can be derived as

$$\max_{\tau, e_i} (1 - \tau) \log_2 \left(1 + \frac{1}{1 - \tau} \sum_{i=1}^{K} \frac{|u_i|^2 e_i}{2\sigma^2 + P_p |g_0|^2} \right) \tag{7.30}$$

$$0 \le \tau \le 1 \tag{7.31}$$

$$0 \le e_i \le E_i \tag{7.32}$$

$$\frac{1}{1 - \tau} \sum_{i=1}^{K} |v_i|^2 e_i < \Gamma. \tag{7.33}$$

A global optimization is difficult. Hence, a suboptimal solution to (7.30) can be derived in two steps. First, one fixes τ to $\hat{\tau}$ to have

$$\max_{e_i} (1 - \hat{\tau}) \log_2 \left(1 + \frac{1}{1 - \hat{\tau}} \sum_{i=1}^{K} \frac{|u_i|^2 e_i}{2\sigma^2 + P_p |g_0|^2} \right) \tag{7.34}$$

$$0 \le e_i \le \hat{\tau} Q_i \tag{7.35}$$

$$\frac{1}{1 - \hat{\tau}} \sum_{i=1}^{K} |v_i|^2 e_i < \Gamma. \tag{7.36}$$

This optimization problem can be solved using the method in Lee and Zhang (2015). Specifically, it was shown in Lee and Zhang (2015) that, for $0 < \hat{\tau} < \frac{\Gamma}{\Gamma + \sum_{i=1}^{K} |v_i|^2 Q_i}$, the optimum solution of e_i is given by

$$\hat{e}_i = \hat{\tau} Q_i \tag{7.37}$$

and for $\dfrac{\Gamma}{\Gamma+\sum_{i=1}^{k}|v_{(i)}|^2 Q_{(i)}} < \hat{\tau} \le \dfrac{\Gamma}{\Gamma+\sum_{i=1}^{k-1}|v_{(i)}|^2 Q_{(i)}}$, $k = 1, 2, \cdots, K$, the optimum solution of e_i is

$$
\hat{e}_i = \begin{cases}
\hat{\tau} Q_{(i)}, & i = 1, 2, \cdots, k-1 \\[2mm]
\dfrac{1}{|v_{(i)}|^2}[(1 - \hat{\tau})\Gamma - \hat{\tau}\displaystyle\sum_{i=1}^{k-1}|v_{(i)}|^2 Q_{(i)}], & i = k \\[4mm]
0 & i = k+1, k+2, \cdots, K
\end{cases} \tag{7.38}
$$

where (i) indexes the ith largest value of $\dfrac{|u_i|^2}{(2\sigma^2 + P_p|g_0|^2)|v_i|^2}$ by ordering all these K values according to $\dfrac{|u_{(1)}|^2}{(2\sigma^2 + P_p|g_0|^2)|v_{(1)}|^2} > \cdots > \dfrac{|u_{(K)}|^2}{(2\sigma^2 + P_p|g_0|^2)|v_{(K)}|^2}$. Thus, for any value of $\hat{\tau}$ between 0 and 1, it has been divided into $K+1$ intervals, from 0 to $\dfrac{\Gamma}{\Gamma+\sum_{i=1}^{K}|v_i|^2 Q_i}$, from $\dfrac{\Gamma}{\Gamma+\sum_{i=1}^{K}|v_i|^2 Q_i}$ to $\dfrac{\Gamma}{\Gamma+\sum_{i=1}^{K-1}|v_{(i)}|^2 Q_{(i)}}$, until from $\dfrac{\Gamma}{\Gamma+\sum_{i=1}^{1}|v_{(i)}|^2 Q_{(i)}}$ to 1. For each interval, the optimum value of $\hat{e}_i$ can be found.

The first step fixes τ to optimize e_i. Using the optimized e_i, in the second step, the optimum values of $\hat{e}_i$ are put back into (7.30) to replace e_i. The objective function is then only dependent on τ. Specifically, for $0 < \hat{\tau} < \dfrac{\Gamma}{\Gamma+\sum_{i=1}^{K}|v_i|^2 Q_i}$, one has

$$
R(\hat{\tau}) = (1 - \hat{\tau})\log_2\left(1 + \frac{\hat{\tau}}{1-\hat{\tau}}\sum_{i=1}^{K}\frac{|u_i|^2 Q_i}{2\sigma^2 + P_p|g_0|^2}\right) \tag{7.39}
$$

and for $\dfrac{\Gamma}{\Gamma+\sum_{i=1}^{k}|v_{(i)}|^2 Q_{(i)}} < \hat{\tau} \le \dfrac{\Gamma}{\Gamma+\sum_{i=1}^{k-1}|v_{(i)}|^2 Q_{(i)}}$, $k = 1, 2, \cdots, K$, one has

$$
R(\hat{\tau}) = (1 - \hat{\tau})\log_2\left\{1 + \frac{\hat{\tau}}{1-\hat{\tau}}\sum_{i=1}^{k}\frac{|u_{(i)}|^2 Q_{(i)}}{2\sigma^2 + P_p|g_0|^2}\right.
$$
$$
\left. + \frac{1}{1-\hat{\tau}}\frac{|u_{(k)}|^2}{(2\sigma^2 + P_p|g_0|^2)|v_{(k)}|^2}[(1-\hat{\tau})\Gamma - \hat{\tau}\sum_{i=1}^{k-1}|v_{(i)}|^2 Q_{(i)}]\right\}. \tag{7.40}
$$

Finally, (7.39) and (7.40) can be used to find the optimum value of $\hat{\tau}$, which can then be used in (7.37) and (7.38) to find the optimum value of $\hat{e}_i$ in the second iteration. The iterative process continues until the updated objective function is within a threshold of the previous objective function.

There is no analytical solution due to the non-linearity of the functions in the expression but numerical methods can be used to search for the optimum values. Figure 7.11 gives an example of $R(\hat{\tau})$ for illustration purposes, where $K = 5$, $\eta_i = 1$, $P_s = 1$, $P_p = 0.01$, $h_i = g_i = u_i = v_i = g_0 = 1$, $\Gamma = 1$, and $2\sigma^2 = 1$. An optimum value of $\hat{\tau}$ can be seen from the figure. More details for the above results can be found in Lee and Zhang (2015).

Other optimizations can also be performed in (7.30) under different conditions. For example, if all the harvested energy is used for data transmission, $c_i = 1$. Then, one can optimize τ only. This optimization is very similar to the HAP system in Chapter 6, except that there is an additional constraint on the transmission power. One may also optimize P_s and τ jointly, or P_s, P_p and τ jointly. These will lead to a joint time and power allocation problem. Also, instead of fixing the downlink time τ and the uplink time $1 - \tau$, the two operation modes of the CR network can be adapted to the fading state so that uplink information delivery and downlink power transfer only operate when necessary (Ji et al. 2017).

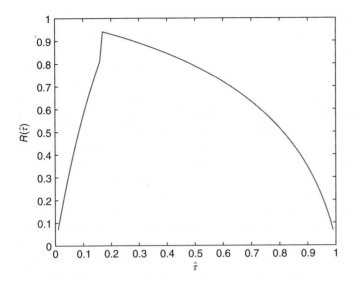

Figure 7.11 $R(\hat{\tau})$ versus $\hat{\tau}$.

The results for TDMA can be obtained in a similar way. In this case, each CR node transmits within the allocated time sequentially. This will make the optimization problem more complicated, as instead of τ one will have τ_1, τ_2, and τ_K for the K CR nodes, where $\tau_1 + \tau_2 + \cdots + \tau_K = 1 - \tau$. The optimal value still exists but the solution is difficult to obtain. For example, in Xu and Li (2017), K CR nodes were assumed to transmit during τ_1, τ_2 until τ_K, similar to the HAP wireless powered system discussed in Chapter 6. In this case, the ith CR node has a rate of

$$R_i = \tau_i \log_2 \left[1 + \frac{\eta_i(P_s|h_i|^2 + P_p|g_i|^2)|u_i|^2 \tau}{\tau_i(2\sigma^2 + P_p|g_0|^2)} \right] \tag{7.41}$$

and the total interference to the PU is

$$Q = \frac{1}{T} \left(P_s|h_0|^2 \tau + \sum_{i=1}^{K} P_i|v_i|^2 \tau_i \right). \tag{7.42}$$

The optimization problem becomes (Xu and Li 2017)

$$\max_{\tau, \tau_1, \cdots, \tau_K, P_s} \left\{ \sum_{i=1}^{K} R_i \right\} \tag{7.43}$$

$$Q \le Q_{max} \tag{7.44}$$

$$P_s \le P_{max}, P_s \tau \le P_{avg} \tag{7.45}$$

$$\tau + \sum_{i=1}^{K} \tau_i = 1 \tag{7.46}$$

where Q_{max} is the peak interference power allowed by the PU, P_{max} is the maximum transmission power for CR, and P_{avg} is the average transmission power for CR.

This is a complicated problem. In Cheng et al. (2017) the proportional fairness was also considered, where the sum rate becomes $\sum_{i=1}^{K} \omega_i \log(R_i)$ for optimization, where ω_i is the weighting factor for the ith CR node.

In the above discussion, the underlay principle is considered. One can easily extend the results to the interweave principle or the overlay principle. In fact, Lee and Zhang (2015) have also studied the overlay principle, where the CR system has more knowledge about the PU transmitter and the PU receiver. It was reported by Lee and Zhang (2015) that the overlay principle outperforms the underlay principle, as expected, as extra knowledge is available and extra coordination between CR and PU can be performed, but this also leads to higher complexity. For the interweave principle, spectrum sensing will be required, in addition to uplink data transmission and downlink power transfer. Thus, more cases need to be discussed.

In summary, the above discussion gives an example of how the energy harvesting CR system works by harvesting energy from the base station. The key is the interference temperature imposed by the PU that limits the transmission power of both the CR node and the base station, giving different optimum solutions compared with the wireless powered systems without spectrum sharing in Chapter 6. Next, the energy harvesting CR systems that harvest energy from the PU will be discussed.

7.5 From the Primary User

7.5.1 Conventional PU

Figure 7.12 shows the diagram of an energy harvesting CR system where the CR harvests energy from the signal transmitted by a conventional PU system without wireless power. In this case, the energy harvesting CR has three operation modes: transmission mode; harvesting mode; and idle mode. The PU transmitter is powered by either batteries or

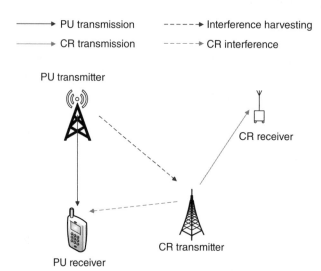

Figure 7.12 CR system that harvests energy from the PU.

mains connections and is randomly distributed in the space. The area around the PU transmitter is divided into three rings. The inner most ring is called the harvesting zone, where the PU signal is the strongest so that it is good for the CR to harvest the energy. The outer most ring is called the guard zone, where the PU signal is the weakest so that it is good for the CR to transmit the data by exploiting the spatial opportunities. The ring between the harvesting zone and the guard zone is neither good for harvesting nor good for transmission. The CR is also randomly distributed. If the CR finds itself in the harvesting zone and its battery is not full, it will switch to the harvesting mode to harvest energy from the PU signal. If the CR finds itself in the guard zone and it has sufficient energy, it will switch to the transmission mode to transmit data on the licensed channel. In other cases, the CR will switch to the idle mode. We are interested in finding out the transmission probability of the CR (Lee et al. 2013).

Also, this discussion assumes that the distance between the CR and the PU is known through some localization and feedback channel so that the CR switches its mode based on the distance. In this case, spectrum sensing is not needed. If the location information is not available, spectrum sensing will be required, as discussed in Section 7.2.

One can see from the above description that this is actually an underlay spectrum sharing system, where the CR utilizes the spectrum holes in the spatial domain (the guard zone) for opportunistic spectrum access. Also, this is a "harvest-store-use" protocol, as the CR only transmits when it has the required energy. Compared with a conventional CR system, the random location of the CR node makes its harvesting time random and the random signal from the PU also makes the amount of harvested energy random so that the energy supply at the CR is dynamic.

Consider a CR network, where the active PU transmitters (PUs that are transmitting data) follow a homogeneous Poisson point process with density λ_p and the CR transmitters follow another independent homogeneous Poisson point process with density λ_c. The PU transmitters transmit signals at a fixed power of P_p and the CR transmitters transmit signals at a fixed power of P_s. The distance between PU transmitter and receiver is d_p, and the distance between CR transmitter and receiver is d_c. Each PU transmitter has an associated guard zone, which is centered around itself with a radius of r_g. The probability that a CR transmitter is in a guard zone of any PU transmitter, or that there is no PU transmitter in the zone centered around the CR transmitter with radius r_g is given by (Lee et al. 2013)

$$p_g = e^{-\pi r_g^2 \lambda_p} \tag{7.47}$$

as the number of PU transmitters follows a Poisson distribution with mean $\pi r_g^2 \lambda_p$, where πr_g^2 is the area of the circle, and λ_p is the density of the PU transmitters.

Most energy harvesters have an activation level or sensitivity, below which the energy from the PU is too small to be harvested. As the signal strength decreases with the distance, if the CR wants to harvest energy from the PU signal, this sensitivity can be translated into a harvesting zone centered around the PU transmitter. Denote r_h as the distance from the PU transmitter beyond which no energy can be harvested. Thus, the disc around the PU transmitter with a radius of r_h is the harvesting zone, and the smallest power at the edge of this zone is $P_p r_h^{-\alpha}$, where P_p is the PU transmission power defined before, α is the path loss exponent, and $r_h^{-\alpha}$ is the path loss at a distance of r_h. The probability that a CR transmitter is in a harvesting zone of any PU transmitter, or that there is at least one PU transmitter in the zone centered around the CR transmitter with radius

r_h is given by (Lee et al. 2013)

$$p_h = 1 - e^{-\pi r_h^2 \lambda_p} \tag{7.48}$$

as the number of PU transmitters inside this disc follows a Poisson distribution with mean $\pi r_h^2 \lambda_p$, where πr_h^2 is the area of the disc. The minimum harvested power is then $\eta P_p r_h^{-\alpha}$ at the edge of the zone, where η is the conversion efficiency of the energy harvester.

In the above, several assumptions have been made. First, $\lambda_p \ll \lambda_s$ and hence there are far less PU transmitters than CR transmitters. Also, $d_p \ll r_g$, which means that the PU receiver is very close to the PU transmitter such that r_g is enough to protect both of them. Finally, $r_h \ll r_g$ so that the harvesting zone πr_h^2 is much smaller than the guard zone πr_g^2.

The CR transmitter only transmits data when it is in the guard zone and it has the required energy in the battery. Thus, the probability of transmission is given by

$$p_t = p_b p_g \tag{7.49}$$

where p_b is the probability that the battery is full. The probability p_b is discussed in the following. The minimum harvested power is $\eta P_p r_h^{-\alpha}$, and the required power for transmission at the CR is P_s. Thus, the CR transmitter needs at least $M = \lceil \frac{P_s}{\eta P_p r_h^{-\alpha}} \rceil$ times of charging to be ready for transmission. This means that the CR transmitter may need to be in the harvesting zone at most M times during M time slots, if it is too far away from the PU transmitter.

When $0 < P_s \leq \eta P_p r_h^{-\alpha}$, $M = 1$. Thus, the battery state is either 0 or P_s with at most one charge. The charging process can be modeled as a two-state Markov chain with state transition probability matrix

$$\begin{bmatrix} 1 - p_h & p_h \\ p_g & 1 - p_g \end{bmatrix} \tag{7.50}$$

where the probability from 0 to 0 is $1 - p_h$, the probability from 0 to P_s is p_h when the CR transmitter is in the harvesting zone, the probability from P_s to 0 is p_g when the CR transmitter is outside the guard zone and performs one data transmission, and the probability from P_s to P_s is $1 - p_g$ where the CR has full battery but is inside the guard zone. This model has been discussed in Chapter 3 as well. The average probability of full battery p_b is given by the steady-state probability of P_s and can be calculated as (Lee et al. 2013)

$$p_b = \frac{p_h}{p_h + p_g}. \tag{7.51}$$

When $\eta P_p r_h^{-\alpha} < P_s \leq 2\eta P_p r_h^{-\alpha}$, $M = 2$. In this case, the CR transmitter needs at most two charges to be ready for data transmission. If it is close to the PU transmitter, enough power can be harvested in one charge. If it is too far away from the PU transmitter, two charges are required. Thus, the battery state is either 0, or between $\frac{1}{2}P_s$ and P_s, or P_s. It has three states and thus, can be described by a three-state Markov chain with state transition probability matrix

$$\begin{bmatrix} 1 - p_h & p_2 & p_1 \\ 0 & 1 - p_h & p_h \\ p_g & 0 & 1 - p_g \end{bmatrix} \tag{7.52}$$

where $p_2 = p_h - p_1 = e^{-\pi h_1^2 \lambda_p} - e^{-\pi r_h^2 \lambda_p}$ is the probability from 0 to a value between $\frac{P_s}{2}$ and P_s when the CR transmitter is the harvesting zone but not close enough to the PU transmitter, $p_1 = 1 - e^{-\pi h_1^2 \lambda_p}$ is the probability from 0 to P_s when the CR transmitter is close enough to the PU transmitter such that only one charge is needed, $h_1 = (\frac{P_s}{\eta P_p})^{-\frac{1}{\alpha}}$ is the distance to the PU transmitter that determines if one charge is enough, and other probabilities are defined as before. The average probability of full battery p_b can again be obtained from the steady-state probability of P_s and is given by (Lee et al. 2013)

$$p_b = \frac{p_h}{p_h + p_g(1 + \frac{p_2}{p_h})}. \tag{7.53}$$

When $2\eta P_p r_h^{-\alpha} < P_s, M > 2$. This case is quite complicated. The probability of full battery can only be bounded as (Lee et al. 2013)

$$\frac{p_1 + p_2'}{(p_1 + p_2') + p_g(1 + \frac{p_2'}{p_1 + p_2'})} < p_b < \frac{p_h}{p_h + p_g(1 + \frac{p_3 + p_2'}{p_h})} \tag{7.54}$$

where $p_2' = e^{-\pi h_2^2 \lambda_p} - e^{-\pi h_1^2 \lambda_p}$, $p_3 = p_h - p_1 - p_2' = e^{-\pi h_1^2 \lambda_p} - e^{-\pi r_h^2 \lambda_p}$, and $h_2 = (\frac{P_s}{2\eta P_p})^{-\frac{1}{\alpha}}$ is the distance to the PU transmitter that determines if one charge to the CR transmitter is enough.

Next, the outage probability of the PU receiver and the CR receiver will be discussed. This will determine the information rate later. For the PU receiver, the outage probability can be defined as

$$P_{OP} = Pr\left\{\frac{|h|^2 P_p |d_p|^{-\alpha}}{I_p + I_c + 2\sigma^2} < \Gamma_p\right\} \tag{7.55}$$

where h is the channel gain between the PU transmitter and the PU receiver and is a complex Gaussian random variable with mean 0 and variance 0.5 so that $|h|^2$ is an exponential random variable with mean 1, I_p is the interference power caused by other active PU transmitters in the area, I_c is the interference power caused by other active CR transmitters nearby, $2\sigma^2$ is the noise power, and Γ_p is the interference temperature in terms of signal-to-interference-plus-noise ratio (SINR) that the PU can tolerate. It has been derived in Lee et al. (2013) that

$$P_{OP} = 1 - e^{-a_p} \tag{7.56}$$

where $a_p = [\lambda_p + p_t \lambda_c (P_s/P_p)^{2/\alpha}]\Gamma_p^{2/\alpha} d_p^2 m + \frac{\Gamma_p d_p^\alpha 2\sigma^2}{P_p}$, $m = \pi \frac{2}{\alpha}\Gamma(\frac{2}{\alpha})\Gamma(1 - \frac{2}{\alpha})$ and $\Gamma(\cdot)$ is the Gamma function.

Similarly, the outage probability at the CR receiver can be defined as

$$P_{OC} = Pr\left\{\frac{|g|^2 P_s d_c^{-\alpha}}{I_p + I_c + 2\sigma^2} < \Gamma_c\right\} \tag{7.57}$$

where g is the fading gain between the CR transmitter and the CR receiver, and $|g|^2$ is an exponential random variable with mean 1. This outage probability was also derived in Lee et al. (2013) as

$$P_{OC} \approx 1 - \frac{1}{p_g}e^{-a_c} \tag{7.58}$$

where $a_c = [\lambda_p(P_s/P_p)^{-2/\alpha} + p_t\lambda_c]\Gamma_c^{2/\alpha}d_c^2 m + \frac{\Gamma_c d_c^\alpha 2\sigma^2}{P_s}$. Details of the derivations of (7.56) and (7.58) can be found in Lee et al. (2013) and its relevant references.

Finally, the optimization problem is described as

$$\max_{P_s, \lambda_c}\{p_t\lambda_c\log_2(1+\Gamma_c)\} \tag{7.59}$$

$$P_{OP} \le \epsilon_p \tag{7.60}$$

$$P_{OC} \le \epsilon_c. \tag{7.61}$$

One sees that this problem aims to maximize the information rate with respect to the CR transmission power and CR density, subject to a constraint on the PU performance, which can be considered as the power outage constraint in underlay systems, and a constraint on the CR outage performance. In some applications, the CR does not have any QoS requirements and the system works in the best-effort manner. In this case, the constraint on the CR outage can be removed.

For interference-limited scenarios where the noise power $2\sigma^2$ can be ignored and $M \le 2$, the optimal solution to (7.59) can be derived analytically as (Lee et al. 2013)

$$R_{max} = \frac{b_c(b_p - m\Gamma_p^{2/\alpha}d_p^2\lambda_p)}{\Gamma_c^{2/\alpha}d_c^2 b_p m}\log_2(1+\Gamma_c) \tag{7.62}$$

where $b_c = -\ln[(1-\epsilon_c)p_g]$ and $b_p = -\ln(1-\epsilon_p)$. The optimal power and density are

$$P_s^{opt} = \frac{\Gamma_c}{\Gamma_p}\left(\frac{d_c}{d_p}\right)^\alpha\left(\frac{b_c}{b_p}\right)^{-\alpha/2}P_p \tag{7.63}$$

and

$$\lambda_c^{opt} = \frac{b_c(b_p - m\Gamma_p^{2/\alpha}d_p^2\lambda_p)}{p_t P_s^{opt}\Gamma_c^{2/\alpha}d_c^2 b_p m}, \tag{7.64}$$

respectively. Figure 7.13 show the maximum information rate in (7.62) with respect to λ_p. In this case, $d_p = d_c = 2$, $\alpha = 3$, $\Gamma_p = \Gamma_c = 7$, and $r_g = 2$. One sees that it increases with ϵ_p or ϵ_c but ϵ_c has a larger impact. This means that one can trade the CR outage performance for the CR rate performance. However, in general, the rate is very low. More details on the above results can be found in Lee et al. (2013).

In the case when the CR does not have any QoS requirement such that the constraint on the CR outage is dropped, one also has

$$\max_{P_s, \lambda_c}\{p_t\lambda_c\log_2(1+\Gamma_c)\} \tag{7.65}$$

$$P_{OP} \le \epsilon_p. \tag{7.66}$$

The maximum rate can be derived as

$$R_{max} = \log_2(1+\Gamma_c)\left(\frac{P_p}{P_s^{opt}}\right)^{2/\alpha}\left(\frac{b_p - \Gamma_p d_p^\alpha 2\sigma^2/P_p}{\Gamma_p^{2/\alpha}d_p^2 m} - \lambda_p\right) \tag{7.67}$$

and the optimum density is given by

$$\lambda_c^{opt} = \frac{1}{p_t(P_s^{opt})}\left(\frac{P_p}{P_s^{opt}}\right)^{2/\alpha}\left(\frac{b_p - \Gamma_p d_p^\alpha 2\sigma^2/P_p}{\Gamma_p^{2/\alpha}d_p^2 m} - \lambda_p\right) \tag{7.68}$$

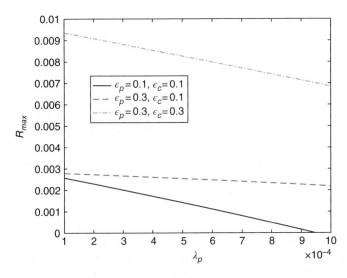

Figure 7.13 The optimum information rate in (7.62) versus λ_p.

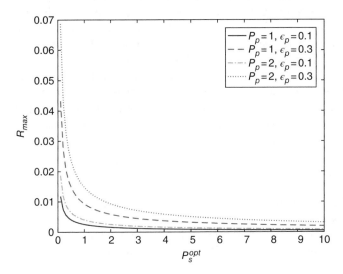

Figure 7.14 The optimum information rate in (7.67) versus P_s^{opt}.

where $p_t(P_s^{opt})$ is calculated by replacing P_s with P_s^{opt} in the expression of p_t. The optimum transmission power P_c^{opt} should be as small as possible, as expected, as there is no constraint on the CR outage (no minimum requirement on the CR performance) so that from the PU's point of view (the PU outage constraint), it should be as small as possible. Figure 7.14 shows the relationship between R_{max} and P_s^{opt} in (7.67). One sees that the transmission power of the CR can be traded for its rate. However, this tradeoff is only useful when the transmission power is small. For large transmission power, the decrease of the transmission power does not lead to noticeable increase of the rate.

Similarly, one could remove the constraint on the PU outage and conduct the optimization with the CR outage only. In this case, it can be shown that the optimum transmission power of the CR transmitter should be as large as possible, as expected. Details can be found in Lee et al. (2013).

Using (7.59), other optimization problems can also be formulated. For example, one may optimize the radius of the harvesting zone r_h and the radius of the guard zone r_g for fixed transmission power of P_s and fixed density of λ_c. This will be useful for CR applications with a fixed number of nodes and fixed transmission power due to hardware limitations.

The above design takes advantage of the spectrum holes in the spatial domain. One can also extend these results to the spectrum holes in the time domain. In this case, the PU and the CR are mixed in the space and the PU traffic is dynamic in time. Thus, the received power from the PU transmitter will be sampled at different time instants to decide when the licensed channel will be free and when the licensed channel will be occupied using spectrum sensing. If the channel is free and the battery of the CR is sufficient, the CR node will switch to the transmission mode. If the channel is occupied and the battery of the CR is not full, it switches to the harvesting mode. In other cases, it stays idle. This is very similar to the system discussed above where the channel status is determined by the distance instead. Such studies can be found in Pratibha et al. (2017).

One can also extend the above results to the interweave principle. In this case, the first energy detection method in Section 7.2.2.1 can be used to find the spectrum holes in the time domain, while the second energy detection method in Section 7.2.2.2 can be used to find the spectrum holes in the space domain. Then, the throughput and the harvested energy will be determined by the probabilities of false alarm and detection. The overlay principle can also be used. The result is very similar to energy harvesting relaying. Also, the above use the "harvest-store-use" protocol. The CR uses a fixed transmission power. One can also adopt the "harvest-use" protocol, where the CR uses a variable transmission power. In this case, $p_t = p_g$ will simplify the designs.

7.5.2 Wireless Powered PU

The previous subsection has considered the case when the PU is powered in a conventional way, such as batteries and mains connections. In this subsection, the PU also uses wireless power and thus is a HAP system as discussed in Chapter 6. Figure 7.15 shows a diagram of the system considered, where the PU network assumes a HAP structure using wireless power.

From Figure 7.15, we consider a wireless powered PU network with one access point (AP) and N PUs. These N PUs harvest energy from the AP for a time duration of τ_0. The harvested energy is used to transmit data from the PUs to the AP during τ_n, where τ_n represents the time duration allocated to the nth PU, $n = 1, 2, \cdots, N$, and $\tau_0 + \sum_{n=1}^{N} \tau_n = T$ is the total time. Meanwhile, CRs operate in the same area on the same channel. We consider one CR transmitter and one CR receiver. When the AP transfers energy to the PUs during τ_0, the cognitive transmitter transmits its own data to the cognitive receiver. When the PUs send data to the AP during $\tau_1, \tau_2, ..., \tau_N$, the cognitive transmitter starts to harvest energy from these transmissions. Finally, when the PU network is idle so that neither the AP is transferring energy nor the PUs are transmitting data, the cognitive

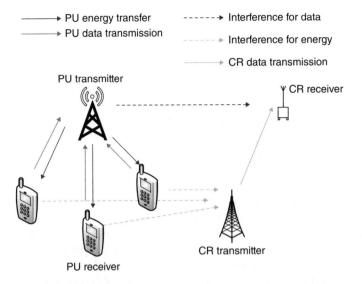

Figure 7.15 Diagram of energy harvesting CR where the PU uses wireless power.

transmitter uses this opportunity to transmit its data to the cognitive receiver. Denote this time as τ_{free}. Thus, the CR needs to detect the status of the PU network to decide its strategy. This detection could be erroneous so that several scenarios need to be discussed.

Assume that the transmission power of the AP for energy transfer to the PUs is much higher than the transmission power of the PUs for data transmission to the AP. This is normally the case, because the energy receiver has a much lower sensitivity than the information receiver. To this end, we define two decision thresholds as λ_1 and λ_2, where $\lambda_2 > \lambda_1$. Using energy detection, from Section 7.2.2, the detection variable is $T = \sum_{k=1}^{K} |y_k|^2$, where y_k is the received signal from the PUs or the AP at the CR. The PU network has three statuses based on energy detection:

$$H_0 : \text{Neither AP nor PUs are transmitting} \Rightarrow T < \lambda_1$$
$$H_1 : \text{PUs are transmitting data} \Rightarrow \lambda_1 < T < \lambda_2$$
$$H_2 : \text{AP is transferring energy} \Rightarrow T > \lambda_2$$

The probabilities of detection and false alarm for each status can be derived as in the following. For H_0,

$$P(H_2|H_0) = P(T > \lambda_2|H_0) = \frac{\Gamma(K, \frac{\lambda_2}{2\sigma^2})}{\Gamma(K)}$$

$$P(H_1|H_0) = P(\lambda_1 < T < \lambda_2|H_0) = \frac{\Gamma(K, \frac{\lambda_1}{2\sigma^2})}{\Gamma(K)} - \frac{\Gamma(K, \frac{\lambda_2}{2\sigma^2})}{\Gamma(K)}$$

$$P(H_0|H_0) = 1 - P(H_1|H_0) - P(H_2|H_0)$$

where $2\sigma^2$ is the noise power, $\Gamma(\cdot, \cdot)$ is the upper incomplete Gamma function, and $\Gamma(\cdot)$ is the Gamma function.

Similarly, for H_1, one has

$$P(H_2|H_1) = P(T > \lambda_2|H_1) = Q_K(\sqrt{2\gamma_1}, \sqrt{\lambda_2})$$
$$P(H_1|H_1) = P(\lambda_1 < T < \lambda_2|H_1) = Q_K(\sqrt{2\gamma_1}, \sqrt{\lambda_1}) - Q_K(\sqrt{2\gamma_1}, \sqrt{\lambda_2})$$
$$P(H_0|H_1) = 1 - P(H_1|H_1) - P(H_2|H_1)$$

and for H_2,

$$P(H_2|H_2) = P(T > \lambda_2|H_2) = Q_K(\sqrt{2\gamma_2}, \sqrt{\lambda_2})$$
$$P(H_1|H_2) = P(\lambda_1 < T(y) < \lambda_2) = Q_K(\sqrt{2\gamma_2}, \sqrt{\lambda_1}) - Q_K(\sqrt{2\gamma_2}, \sqrt{\lambda_2})$$
$$P(H_0|H_2) = 1 - P(H_2|H_2) - P(H_1|H_2)$$

where γ_1 is the primary SNR during data transmission, γ_2 is the primary SNR during energy transfer, and $Q_u(a, b)$ is the generalized Marcum Q-function.

In summary, the cognitive transmitter transmits data to the cognitive receiver when H_0 and H_2 are detected and harvests energy when H_1 is detected. The cognitive transmission during H_2 allows the PUs to harvest more energy, in addition to the energy from their AP. This is the purpose of the new scheme. Denote $H_p|H_q$ as one of the possible scenarios in the transmission, where H_p represents the PU status detected by the CR and H_q represents the actual PU status with $p, q = \{0, 1, 2\}$. Hence, when $p = q$, a correct detection is made. When the AP is actually transferring energy to the PUs, the received signal at the nth PU during their energy transfer is given by

$$y_n = \begin{cases} \left(h_n\sqrt{\dfrac{P_a}{L_{h_n}}}\right)x_a + \left(g_n\sqrt{\dfrac{P_s}{L_{s_n}}}\right)x_s + w_n, & H_0|H_2 \\[3ex] \left(h_n\sqrt{\dfrac{P_a}{L_{h_n}}}\right)x_a + w_n, & H_1|H_2 \\[3ex] \left(h_n\sqrt{\dfrac{P_a}{L_{h_n}}}\right)x_a + \left(g_n\sqrt{\dfrac{P_s}{L_{s_n}}}x_s\right) + w_n, & H_2|H_2 \end{cases} \tag{7.69}$$

where w_n represent the noise, P_a and P_s represent the transmitted power of the AP and the cognitive transmitter, respectively, h_n and g_n are the channel gains between AP and the nth PU and between the cognitive transmitter and the nth PU, respectively, and they are fixed during one transmission but change randomly from transmission to transmission, L_{h_n} and L_{s_n} represent the path loss between AP and the nth PU and between the cognitive transmitter and the nth PU, respectively, and x_a and x_s represent the transmitted signal by the AP and the cognitive transmitter, respectively. To explain this equation, the condition $H_0|H_2$ is the case when the AP is actually transferring energy while the CR detects it as free and hence transmits its own data to the cognitive receiver represented by the second term in the equation, $H_1|H_2$ is the case when the AP is actually transferring energy while the CR thinks that the PUs are transmitting data and hence the CR does not transmit any data, while $H_2|H_2$ is the case when the AP is actually transferring energy and this status is correctly detected by the CR so that it also sends its own data represented by the second term in the equation to the cognitive receiver. Using y_n, the average harvested energy at each PU is derived as (Azmat et al. 2018)

$$E[PE] = \eta \tau_0 \left(\frac{|h_n|^2 P_a}{L_{h_n}} + \frac{|g_n|^2 P_s}{L_{g_n}} \right) [1 - P(H_1|H_2)] + \eta \tau_0 \left(\frac{|h_n|^2 P_a}{L_{h_n}} \right) P(H_1|H_2) \quad (7.70)$$

where η represents the conversion efficiency of the energy harvester.

On the other hand, the cognitive receiver receives data from the cognitive transmitter only when H_0 or H_2 are detected. Thus, the received signal at the cognitive receiver can be expressed as

$$z = \begin{cases} \left(d\sqrt{\dfrac{P_s}{L_d}} \right) x_s + w_d, & H_0|H_0 \\[2ex] \left(d\sqrt{\dfrac{P_s}{L_d}} \right) x_s + w_d, & H_2|H_0 \\[2ex] \left(d\sqrt{\dfrac{P_s}{L_d}} \right) x_s + \left(u_n \sqrt{\dfrac{P_n}{L_{u_n}}} \right) x_n + w_d, & H_0|H_1 \\[2ex] \left(d\sqrt{\dfrac{P_s}{L_d}} x_s \right) + \left(u_n \sqrt{\dfrac{P_n}{L_{u_n}}} \right) x_n + w_d, & H_2|H_1 \\[2ex] \left(d\sqrt{\dfrac{P_s}{L_d}} \right) x_s + \left(v \sqrt{\dfrac{P_a}{L_v}} \right) x_a + w_d & H_0|H_2 \\[2ex] \left(d\sqrt{\dfrac{P_s}{L_d}} \right) x_s + \left(v \sqrt{\dfrac{P_a}{L_v}} \right) x_a + w_d, & H_2|H_2 \end{cases} \quad (7.71)$$

where w_d represents the noise at the cognitive receiver with noise power $2\sigma^2$, and the channel gains and the path losses are defined accordingly. Using z, the average throughput of the CR transmission can be derived as (Azmat et al. 2018)

$$E[R_{SUR}] = \tau_{free} \log_2 \left(1 + \frac{|d|^2 P_s}{2 L_d \sigma^2} \right) [1 - P(H_1|H_0)]$$

$$+ \tau_0 \log_2 \left(1 + \frac{\frac{|d|^2 P_s}{L_d}}{\frac{|v|^2 P_a}{L_v} + 2\sigma^2} \right) [1 - P(H_1|H_2)]$$

$$+ \tau_n \log_2 \left(1 + \frac{\frac{|d|^2 P_s}{L_d}}{\frac{|u_n|^2 P_n}{L_{u_n}} + 2\sigma^2} \right) [1 - P(H_1|H_1)]. \quad (7.72)$$

In the conventional energy harvesting CR, the CR operates with a different strategy. When the data transmission from the PUs to the AP is detected, the cognitive transmitter does not transmit any data. When the energy transfer from the AP to the PUs is detected, the CR harvests energy. When no data transmission or energy transfer is detected from the PU network, the cognitive transmitter sends its own data. This is based on the traditional interweave principle that the CR must not use the licensed channel when the PU activity is detected, whether or not the PU is transmitting data or transferring energy.

In this strategy, the received signal at the nth PU becomes

$$
y_n = \begin{cases}
\left(h_n \sqrt{\dfrac{P_a}{L_{h_n}}} \right) x_a + \left(g_n \sqrt{\dfrac{P_s}{L_{g_n}}} \right) x_s + w_n, & H_0|H_2 \\[3mm]
\left(h_n \sqrt{\dfrac{P_a}{L_{h_n}}} \right) x_a + w_n, & H_1|H_2 \\[3mm]
\left(h_n \sqrt{\dfrac{P_s}{L_{h_n}}} \right) x_a + w_n, & H_2|H_2
\end{cases}
\tag{7.73}
$$

which gives the average harvested energy at the PU as (Azmat et al. 2018)

$$
E[PE] = \eta \tau_0 \left(\frac{|h_n|^2 P_a}{L_{h_n}} + \frac{|g_n|^2 P_s}{L_{g_n}} \right) P(H_0|H_2) + \eta \tau_0 \left(\frac{|h_n|^2 P_a}{L_{h_n}} \right) [1 - P(H_0|H_2)].
\tag{7.74}
$$

Also, the received signal at the cognitive receiver is

$$
z = \begin{cases}
\left(d \sqrt{\dfrac{P_s}{L_d}} \right) x_s + w_d, & H_0|H_0 \\[3mm]
\left(d \sqrt{\dfrac{P_s}{L_d}} \right) x_s + \left(u_n \sqrt{\dfrac{P_n}{L_{u_n}}} \right) x_n + w_d, & H_0|H_1 \\[3mm]
\left(d \sqrt{\dfrac{P_s}{L_d}} \right) x_s + \left(v \sqrt{\dfrac{P_a}{L_v}} \right) x_a + w_d, & H_0|H_2
\end{cases}
\tag{7.75}
$$

which gives the average throughput for the CR transmission as (Azmat et al. 2018)

$$
\begin{aligned}
E[R_{SUR}] = {} & \tau_{\text{free}} \log_2 \left(1 + \frac{\frac{|d|^2 P_s}{L_d}}{2\sigma^2} \right) P(H_0|H_0) \\[3mm]
& + \tau_0 \log_2 \left(1 + \frac{\frac{|d|^2 P_s}{L_d}}{\frac{|v|^2 P_a}{L_v} + 2\sigma^2} \right) P(H_0|H_2) \\[3mm]
& + \tau_n \log_2 \left(1 + \frac{\frac{|d|^2 P_s}{L_d}}{\frac{|u_n|^2 P_n}{L_{u_i}} + 2\sigma^2} \right) P(H_0|H_1)
\end{aligned}
\tag{7.76}
$$

Next, we will compare these different strategies.

In the comparison, it is assumed that $P_s = 5.000$ dB and $\eta = 0.400$. Also, for the high channel gain case, $h_n = d = 3.000$ dB, $g_n = 2.000$ dB, $u_n = 1.300$ dB, and $v = 1.200$ dB are set, and for the low channel gain case, $h_n = 0.969$ dB, $d = 3.000$ dB, $g_n = 2.000$ dB, $u_n = 0.969$ dB, and $v = 1.200$ dB. It is also assumed that $L_{h_n} = L_d = -1.023$ dB, $L_{g_n} = -1.500$ dB, $L_{u_n} = 0.773$ dB, and $L_v = 0.569$ dB. Further, $\gamma_1 = 1.120$ dB and $\gamma_2 = 1.170$ dB

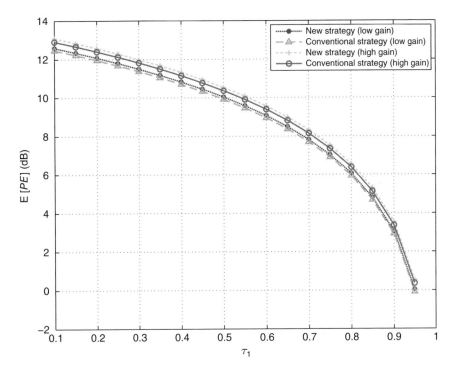

Figure 7.16 E[*PE*] versus τ_1.

are set, where it is assumed that SNR in H_1 is higher than that in H_2 for both strategies. Similarly, since λ_2 is larger than λ_1, we set $\lambda_1 = 5$ and $\lambda_2 = 8$.

Figures 7.16 and 7.17 show the average harvested energy at the PU and the average throughput of the CR transmission, respectively, for different values of τ_1, assuming one PU in the PU network and $T = 1$ for simplicity. One sees that the average harvested energy decreases when τ_1 increases. This is expected. When τ_1 increases, τ_0 decreases such that less energy will be harvested. The amount of harvested energy for different strategies in different cases is similar. This means the strategy and the channel gain do not change the average harvested energy much in the cases considered here. One also sees that the average throughput decreases slightly with τ_1 and then increases as τ_1 keeps increasing. When τ_1 increases, there are less chances for the CR to transmit its own data, if energy detection is correct, so that the throughput decreases. On the other hand, when τ_1 increases, τ_0 decreases so that the transmission power of the PU will be reduced due to less harvested energy. In this case, it is harder for the CR to detect its presence and hence more missed detection will occur to give the CR more transmission opportunities. From these two figures, in general, the new strategy has higher average throughput and higher average harvested energy than the conventional strategy.

In summary, the energy harvesting CR system in the previous section only harvests energy from its secondary base station so that the CR designs there only need to consider the interference temperature imposed by the PU and the wireless power is relatively stable. The energy harvesting CR system in this section harvests energy from the PU signal

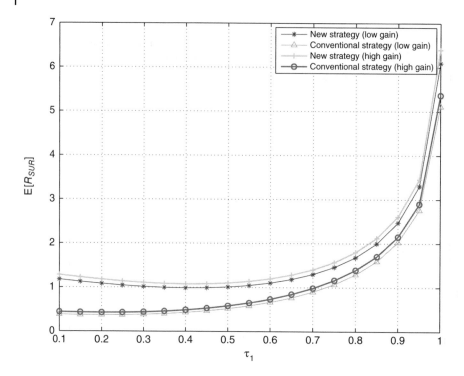

Figure 7.17 E[R_{SUR}] versus τ_1.

so that the CR designs need to consider both the interference temperature imposed by the PU or the unstable energy supply from the PU with random locations. Similarly, if the time-domain spectrum holes are utilized, the CR designs will need to schedule the transmission time to account for the unstable energy supply from the random traffic of the PU. This could be another line of research with many optimization problems to solve.

Next, energy harvesting CR systems that harvest energy from ambient sources will be discussed. In this case, the energy availability will be determined by the energy arrival process of the ambient source.

7.6 From the Ambient Environment

Figure 7.18 shows a diagram of the energy harvesting CR systems that harvest energy from the ambient environment. In this case, the interweave principle is considered so that spectrum sensing is necessary. The sensing accuracy will determine the collision between PU and CR, while the energy arrival rate of the ambient source will determine the energy causality. The two constraints are related to each other by the fact that, if the sensing time is long or the sensing threshold is large, a lot of energy will be consumed for spectrum sensing. This may violate the energy causality constrain due to insufficient harvested energy. However, a long sensing time or large sensing threshold may increase the sensing accuracy to reduce collision. The problem discussed in this section is to

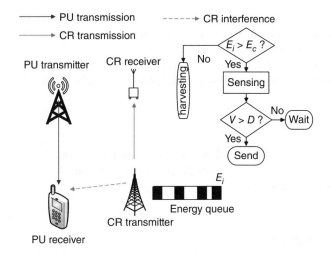

Figure 7.18 CR system that harvests energy from the ambient environment. *D*, detection threshold; E_c, consumed energy; E_i, incoming energy; and *V*, decision variable.

optimize the average throughput of the CR with respect to the sensing time and the sensing threshold, subject to the collision and energy causality constraints.

Consider the CR network where there is a pair of PU transmitter and PU receiver as well as a pair of CR transmitter and CR receiver, operating at the same location on the same frequency band. The CR network uses a slotted structure that is synchronous with the PU. The CR network assumes opportunistic spectrum access to the licensed channel, based on the sensing result and the energy availability. The licensed channel is idle with a probability of $P[H_0]$ and occupied with a probability of $P[H_1]$. The CR performs spectrum sensing in the first τ seconds and possible data transmission in the next $T - \tau$ seconds, within one time slot of T seconds. If the sensing result is correct, the CR transmission will have a throughput of $C_0 = \log_2(1 + \gamma_s)$, where γ_s is the SNR of the received CR signal. If the sensing result is incorrect, the CR transmission will suffer from the interference caused by the PU and thus have a throughput of $C_1 = \log_2(1 + \frac{\gamma_s}{1+\gamma_p})$, where γ_p is the SNR of the PU signal received at the CR receiver (Liang et al. 2008).

The energy arrival process is an important part of the system. Assume that the energy harvested during the ith time slot is E_i^h, which is an independent and identically distributed random variable with mean e_h. The actual distribution is not important, as only the mean will be used later. This is similar to the Bernoulli model discussed in Chapter 2. The battery capacity is assumed infinite so that all harvested energy can be stored and there is no energy overflow. On the energy consumption, assume that the spectrum sensing power is P_s and the data transmission power is P_b. Thus, the energy consumption of spectrum sensing is $P_s\tau$ and the energy consumption of data transmission is $P_b(1 - \tau)$. There are two decisions the CR transmitter needs to make.

First, it needs to decide whether there is enough energy to activate the sensing and transmission operations or remain idle to save energy. This is denoted by the mode indicator for the ith time slot as

$$a_i = \begin{cases} 1 & E_i \geq P_s\tau + P_b(1 - \tau) \\ 0 & E_i < P_s\tau + P_b(1 - \tau). \end{cases} \tag{7.77}$$

The decision is that the CR should not be activated and hence $a_i = 0$, if it is known in advance that there will not be enough energy for transmission, or vice versa. This is the energy causality constraint in the problem.

After the CR is activated, another decision that needs to be made is whether the CR should start data transmission. This depends on the spectrum sensing results for the ith time slot as

$$b_i = \begin{cases} 1 & V < D \\ 0 & V > D \end{cases} \tag{7.78}$$

where V is the decision variable and D is the detection threshold. Using the energy detector discussed in Section 7.2.2.1, one has $V = \mathbf{yy}^T$. Thus, based on the Neyman–Pearson rule, the probability of false alarm is given by

$$P_f = 1 - G\left(D, \frac{K}{2}, 2\sigma^2\right) \tag{7.79}$$

and the probability of detection is given by

$$P_d = 1 - G\left[D, \frac{K}{2}, 2\sigma^2(1 + \gamma_p)\right]. \tag{7.80}$$

If one assumes a sampling frequency of f_s during spectrum sensing, one has $K = \tau f_s$. Thus,

$$P_f(\tau, D) = 1 - G\left(D, \frac{\tau f_s}{2}, 2\sigma^2\right) \tag{7.81}$$

and

$$P_d(\tau, D) = 1 - G\left[D, \frac{\tau f_s}{2}, 2\sigma^2(1 + \gamma_p)\right]. \tag{7.82}$$

Based on the above, the overall energy consumption during the ith time slot is given by

$$E_i^c = a_i[P_s \tau + (1 - b_i)P_b(1 - \tau)] \tag{7.83}$$

and thus, the residual energy for the $(i + 1)$th time slot is

$$E_{i+1} = E_i + E_i^h - E_i^c. \tag{7.84}$$

The energy causality is determined by the relationship between the harvested energy E_i^h and the energy consumption E_i^c. It is difficult to consider the instantaneous energy causality, as different time slots are correlated through the residual energy in (7.84). To simplify this analysis, one could use the average energy. In this case, the energy causality constraint becomes the constraint where the average energy consumption must be smaller than the average harvested energy. One has

$$Pr[a_i = 1] \leq \min(1, \lambda(\tau, D, e_h)) \tag{7.85}$$

where

$$\lambda(\tau, D, e_h) = \frac{e_h}{P_s \tau + P_b(1 - \tau)[P[H_0](1 - P_f(\tau, D)) + P[H_1](1 - P_d(\tau, D))]}. \tag{7.86}$$

Detailed discussions can be found in Park and Hong (2013) and Chung et al. (2014). If the CR transmitter always activates whenever there is enough energy, the equality in (7.85) will be taken and the probability of activation can be approximated as

$$P_a(\tau, D, e_h) = \min(1, \lambda(\tau, D, e_h)). \tag{7.87}$$

The value of $\lambda(\tau, D, e_h)$ actually gives the ratio of the average harvested energy to the average energy consumption. Using (7.86), three system statuses can be defined.

If $\lambda(\tau, D, e_h) > 1$, this is the energy-surplus status. In this status, the average harvested energy is always enough for sensing and transmission. This status is actually the conventional CR systems without energy harvesting.

If $\lambda(\tau, D, e_h) < 1$, this is the energy-deficit status. In this status, there might be a lack of energy such that the CR transmitter will go to the idle mode. This is a unique status for energy harvesting CR.

If $\lambda(\tau, D, e_h) = 1$, this is the energy-equilibrium status.

Using the above results, the CR will start data transmission in two cases. In the first case when the licensed channel is free, $a_i = 1$ and $b_i = 0$, the throughput is given by C_0. This happens with a probability of $P[H_0]P_a(\tau, D, e_h)(1 - P_f(\tau, D))$. In the second case when the licensed channel is occupied, $a_i = 1$ and $b_i = 0$, the throughput is given by C_1. This happens with a probability of $P[H_1]P_a(\tau, D, e_h)(1 - P_d(\tau, D))$. Thus, the average throughput is

$$R(\tau, D, e_h) = \frac{T - \tau}{T}[C_0 P[H_0]P_a(\tau, D, e_h)(1 - P_f(\tau, D)) $$
$$+ C_1 P[H_1]P_a(\tau, D, e_h)(1 - P_d(\tau, D))] \tag{7.88}$$

where $T - \tau$ takes the penalty of sensing into account. This average throughput can be further approximated as

$$R(\tau, D, e_h) \approx \frac{T - \tau}{T}[P[H_0]P_a(\tau, D, e_h)(1 - P_f(\tau, D)) $$
$$+ P[H_1]P_a(\tau, D, e_h)(1 - P_d(\tau, D))]C_0 $$
$$= \min(1, \lambda(\tau, D, e_h))\hat{R}(\tau, D) \tag{7.89}$$

because when γ_p is large, $(1 - P_d(\tau, D)) \approx 0$ and when γ_p is small, $C_1 \approx C_0$. Finally, the optimization problem is

$$\max_{\tau, D}\{\min(1, \lambda(\tau, D, e_h))\hat{R}(\tau, D)\} \tag{7.90}$$

$$\min(1, \lambda(\tau, D, e_h))(1 - P_d(\tau, D)) \leq \bar{P}_c \tag{7.91}$$

where $\bar{P}_c$ is the maximum possibility of collision allowed by the PU. The solution to this problem depends on the system status.

If the system is in the energy-surplus status, one has $\lambda(\tau, D, e_h) > 1$. Thus, the optimization becomes

$$\max_{\tau, D}\{\hat{R}(\tau, D)\} \tag{7.92}$$

$$(1 - P_d(\tau, D)) \leq \bar{P}_c \tag{7.93}$$

and one has (Chung et al. 2014)

$$\tau^{opt} = \max_{\tau}\left\{\frac{T - \tau}{T}[P[H_0](1 - P_f(\tau, D)) + P[H_1]\bar{P}_c]C_0\right\} \tag{7.94}$$

$$D^{opt} = P_d^{-1}(\tau^{opt}, 1 - \bar{P}_c) \tag{7.95}$$

where $P_d^{-1}(\cdot)$ is the inverse function of $P_d(\cdot)$. This status only appears when $\lambda(\tau^{opt}, D^{opt}, e_h) > 1$ or $e_h > P_s\tau^{opt} + P_bT/C_0\hat{R}(\tau^{opt}, D^{opt})$. Thus, the optimum solutions in (7.94) need to be checked against the condition on e_h. In this case, the optimum sensing time τ^{opt} is actually the same as those in Liang et al. (2008) when there is no energy harvesting.

If the system is in the energy-deficit status, $\lambda(\tau, D, e_h) < 1$ and

$$\max_{\tau,D}\{\lambda(\tau, D, e_h)\hat{R}(\tau, D)\} \tag{7.96}$$

$$\lambda(\tau, D, e_h)(1 - P_d(\tau, D)) \le \bar{P}_c, \lambda(\tau, D, e_h) < 1. \tag{7.97}$$

There is no analytical solution but the optimum values of τ and D can be obtained from

$$\max_{\tau,D}\left\{\min\left(1, \frac{\bar{P}_c}{1 - P_d(\tau, D)}\right)\hat{R}(\tau, D)\right\} \tag{7.98}$$

using numerical methods.

If the system is in the energy-equilibrium status, there is only one pair of optimum values obtained from

$$D^{opt} = P_d^{-1}(\tau^{opt}, 1 - \bar{P}_c) \tag{7.99}$$

$$e_h = P_s\tau^{opt} + P_b(T - \tau^{opt})[P[H_0](1 - P_f(\tau^{opt}, D^{opt})) + \bar{P}_cP[H_1]]. \tag{7.100}$$

A full derivation of these equations can be found in Chung et al. (2014).

The above designs optimize the average throughput with respect to the sensing time and the sensing threshold jointly. One can also optimize the sensing threshold alone for a fixed sensing time. This was done in Park et al. (2013) and Park and Hong (2014), where the probability of accessing an idle channel was maximized subject to a constraint on the collision probability. Also, the above design simplifies the problem by considering a fixed slot structure. The CR always accesses the spectrum when it is activated and the channel is detected free. One can also optimize the access policy so that the optimum sensing strategy can be derived (Park and Hong 2013). A joint optimization is also possible (Yin et al. 2015). Essentially, one can derive different performance measures that are of interest for different applications and optimize them with respect to different parameters in the design. For all these designs, since the energy source is independent of the PU, the energy causality and the collision constraints are separate, which has simplified the problem greatly. If the energy source is the PU signal, then sensing interval, harvesting interval, and transmission interval have to be jointly optimized so that the above designs have many variants.

Also, the above designs used energy detection. Feature detectors can also be used. In this case, different solutions to the optimization problems can be obtained (Gao et al. 2016b). Also, the battery capacity is assumed infinite. In practice, they are of finite size such that there will be extra constraints on the energy process. These results can also be extended to the case of multiple CR nodes. In this case, collaboration is possible so that a joint optimization across different nodes can be conducted (Bae and Baek 2015).

Finally, instead of using the average energy causality, one can use the instantaneous energy. In this case, the problem will be more complicated, as each time slot will be considered to design an optimum spectrum access policy. At the beginning of each time slot, one needs to make the two decisions on activation and transmission. This involves the best resource allocation in the time domain as well as online energy management (Zhang et al. 2016).

All the above results, regardless of their energy sources, have focused on either the underlay system or the interweave system. Next, the overlay system will be discussed, where energy and information become two important resources for exchange between the CR and the PU.

7.7 Information Energy Cooperation

In this case, the CR cooperates with the PU when the PU link is in outage, in exchange for the transmission opportunities in the licensed channel. This follows the overlay principle. A diagram of this system is given in Figure 7.5. To achieve this, the CR and the PU must share more information about the system via dedicated control channels.

Consider a CR network with a pair of PU transmitter and PU receiver and a pair of CR transmitter and CR receiver. All nodes have a single antenna except the CR transmitter that has N antennas. This makes beamforming possible and therefore makes the CR transmission more efficient. The communication time T is evenly divided into two parts. In the first $\frac{T}{2}$, the PU transmitter broadcasts its data. This signal will be received at the PU receiver as

$$y_{SD} = \sqrt{P_p}hs_p + n_{SD} \tag{7.101}$$

where P_p is the transmission power of the PU, h is the complex Gaussian fading gain between the PU transmitter and the PU receiver and can be considered as fixed in block-fading channels, s_p is the symbol transmitted by the PU with $E\{|s_p|^2\} = 1$, and n_{SD} is the AWGN with mean zero and variance $2\sigma^2$. Also, the same signal will be received by the CR transmitter as

$$\mathbf{r}_{SR} = \sqrt{P_p}\mathbf{h}s_p + \mathbf{n}_{SR} \tag{7.102}$$

where $\mathbf{h}$ is the fading gain vector between the PU transmitter and different antennas at the CR transmitter and $\mathbf{n}_{SR}$ is the AWGN at the CR transmitter each element of which has mean zero and variance $2\sigma_a^2$. The signal in (7.101) does not provide enough rates for the PU link and hence help from the CR transmitter is needed.

The received signal at the CR transmitter is split into two parts: one for energy harvesting; and one for information forwarding. This is similar to the PS scheme discussed in Chapter 6 for SWIPT. The signal for forwarding is given by

$$\bar{\mathbf{r}}_{SR} = \sqrt{1-\rho}\mathbf{r}_{SR} + \bar{\mathbf{n}}_R \tag{7.103}$$

where ρ is the power splitting factor, as discussed in Chapter 6, and $\bar{\mathbf{n}}_R$ is the AWGN during the signal processing after the splitting with mean zero and variance $2\sigma_d^2$. One has $2\sigma^2 = 2\sigma_a^2 + 2\sigma_d^2$. The other part $\sqrt{\rho}\mathbf{r}_{SR}$ is used for energy harvesting to give the harvested energy

$$E_h = \frac{T}{2}\eta\rho P_p||\mathbf{h}||^2 \tag{7.104}$$

where η is the conversion efficiency and $||\mathbf{h}||^2$ is the total energy harvested from the PU. As before, the noise energy is ignored, as it is negligible compared with the signal energy. If the CR transmitter has an initial power of P_s, the total power or the maximum power available at the CR transmitter becomes $P_{max} = P_s + \eta \rho P_p ||\mathbf{h}||^2$, since the transmission time in the second half is also $\frac{T}{2}$.

The CR transmitter combines its own signal and the PU signal via superposition as

$$\mathbf{x} = \mathbf{w}_s s_s + \mathbf{w}_p \mathbf{h}^H \tilde{\mathbf{r}}_{SR} \tag{7.105}$$

where $\mathbf{w}_s$ is the beamforming vector for its own data s_s with $E\{|s_s|^2\} = 1$, $\mathbf{w}_p$ is the beamforming vector for the PU data, and $(\cdot)^H$ represents the conjugate transpose operation. This signal is transmitted in the second half of T. During this transmission, the PU transmitter remains silent. This is only possible when there is cooperation between CR and PU to avoid interference.

Thus, the received signal at the CR receiver is

$$r_{RR} = \mathbf{g}^H \mathbf{x} + n_{RR} \tag{7.106}$$

where $\mathbf{g}$ is the fading gain vector from different antennas at the CR transmitter to the CR receiver, and n_{RR} is the AWGN at the CR receiver with mean zero and variance $2\sigma^2$. The received signal at the PU receiver is

$$y_{RD} = \mathbf{f}^H \mathbf{x} + n_{RD} \tag{7.107}$$

where $\mathbf{f}$ is the fading gain vector from different antennas at the CR transmitter to the CR receiver, and n_{RD} is the AWGN at the PU receiver with mean zero and variance $2\sigma^2$. From (7.106), the SNR at the CR receiver is

$$\Gamma_s = \frac{|\mathbf{g}^H \mathbf{w}_s|^2}{|\mathbf{g}^H \mathbf{w}_p|^2 ||\mathbf{h}||^2[(1-\rho)P_p||\mathbf{h}||^2 + (1-\rho)2\sigma_a^2 + 2\sigma_d^2] + 2\sigma^2}. \tag{7.108}$$

Thus, the achievable rate for the CR link is

$$R_s = \log_2(1 + \Gamma_s). \tag{7.109}$$

Also, from (7.107), the SNR at the PU receiver is

$$\Gamma_p = \frac{\rho P_p ||\mathbf{h}||^4 |\mathbf{f}^H \mathbf{w}_p|^2}{|\mathbf{f}^H \mathbf{w}_s|^2 + |\mathbf{f}^H \mathbf{w}_p|^2 ||\mathbf{h}||^2[2\sigma_d^2 + (1-\rho)2\sigma_a^2] + 2\sigma^2}. \tag{7.110}$$

Thus, the achievable rate for the PU link is

$$R_p = \frac{1}{2}\log_2\left(1 + \frac{P_p|h|^2}{2\sigma^2} + \Gamma_p\right) \tag{7.111}$$

where maximum ratio combining is used to combine the signals from the CR transmitter and the PU transmitter. Thus, for this energy harvesting CR system, the optimization problem is

$$\max_{\mathbf{w}_s, \mathbf{w}_p, \rho} \{R_s\} \tag{7.112}$$

$$R_p \geq r_0 \tag{7.113}$$

$$E\{||\mathbf{x}||^2\} \leq P_{max} \tag{7.114}$$

$$0 \leq \rho \leq 1. \tag{7.115}$$

where

$$E\{||\mathbf{x}||^2\} = ||\mathbf{w}_s||^2 + ||\mathbf{w}_p||^2||\mathbf{h}||^2[P_p(1-\rho)||\mathbf{h}||^2 + (1-\rho)2\sigma_a^2 + 2\sigma_d^2] \qquad (7.116)$$

and r_0 is the minimum rate that is required by the PU. More details on the above equations can be found in Zheng et al. (2014).

This optimization problem does not have a closed-form solution in the general case. However, two special cases can be discussed. First, if zero-forcing is used for beamforming, one has $\mathbf{w}_p = \sqrt{\omega_p}\frac{(\mathbf{I}-\frac{\mathbf{g}\mathbf{g}^H}{||\mathbf{g}||^2})\mathbf{f}}{||(\mathbf{I}-\frac{\mathbf{g}\mathbf{g}^H}{||\mathbf{g}||^2})\mathbf{f}||}$ and $\mathbf{w}_s = \sqrt{\omega_s}\frac{(\mathbf{I}-\frac{\mathbf{f}\mathbf{f}^H}{||\mathbf{f}||^2})\mathbf{g}}{||(\mathbf{I}-\frac{\mathbf{f}\mathbf{f}^H}{||\mathbf{f}||^2})\mathbf{g}||}$. In this case, closed-form expressions for the optimal solutions are available. Secondly, if the CR transmitter only has one antenna, no beamforming is needed. In this case, the optimization will be much simpler. The above results use power splitting. One can also use time switching to obtain a similar optimization problem. Details can be found in Zheng et al. (2014).

Note that this energy harvesting CR system is very similar to the energy harvesting relaying to be discussed in Chapter 8 as it uses the overlay principle. The main difference is that, in Chapter 8, the relay only forwards the signal received from the source, while here the CR combines its own signal with the PU signal before forwarding. This superposition comes at a price, for example, increased peak-to-average-power ratio.

The above results also have many other variants. For example, in Zhai et al. (2016b), a similar cooperative strategy was considered between CR and PU for more complicated cases, such as the Alamouti space-time coding. In Hsieh et al. (2016), the CR becomes the node with fixed power supply, while the PU harvests energy from the CR transmission if it does not have enough energy. In this case, the precoding matrix at the CR and the energy harvesting parameter at the PU have been jointly optimized to maximize the CR information rate, while satisfying the minimum required rate and the harvested energy at the PU. In Zhai et al. (2016a), the PU uses wireless power, while the CR has a fixed power supply. The CR close to the PU power transmitter is chosen to relay the energy to the remote PU node. Finally, in Yin et al. (2014), the optimal cooperation strategy for the CR was studied. In this case, the CR can choose to cooperate with the PU if it has enough energy for its own data transmission but otherwise can wait until the PU finishes transmission.

In many of these studies, energy harvesting is performed by the PU, not the CR, and the PU trades the transmission opportunity for energy from the CR. In other studies, the CR helps the PU with its information delivery in exchange for the transmission opportunity.

7.8 Other Important Issues

The previous sections have mainly discussed the effect of different energy sources on the designs of energy harvesting CR systems. In these sections, optimization problems have been formulated that maximize the achievable rate of the CR subject to constraints from energy causality and/or collision with the PU. These optimizations focus on the choice of the transmission power for CR data transmission and the sensing interval and threshold for CR spectrum sensing. In addition to these designs, other important issues in energy harvesting CR have also been looked into.

For example, the secrecy performance is very important in wireless communications due to the broadcast nature of the wireless media. In Singh et al. (2016), the secrecy outage of the cognitive HAP wireless powered system was analyzed. The difference between this analysis and the conventional HAP system is that the transmission power of Alice as a CR node needs to be limited by the interference temperature imposed by the PU. In Jiang et al. (2016), the information rate of the CR node was optimized with respect to the beamforming vector for the CR node, the beamforming vector for the PU and the power splitting factor, with a constraint on the minimum secrecy rate required by the PU and another constraint on the minimum SINR required by the PU. In Ng et al. (2016), artificial noise was generated and optimized to achieve the best energy harvesting efficiency or transmission power.

Channel selection is another important issue in CR communications. It is often the case in practice that there will be more than one licensed channel available to the CRs. From the data transmission point of view, the CR should choose a channel with as little PU traffic as possible to make use of the transmission opportunities. However, from the energy harvesting point of view, the CR should choose a channel with as much PU traffic as possible so that it can harvest the maximum amount of energy from the PU. For a CR node that performs both data transmission and energy harvesting, a tradeoff has to be made. In Pratibha et al. (2016), this tradeoff was studied to maximize the average information rate of the CR, averaged over different PU channels, using the same system model as Lee et al. (2013). Depending on the density of the CR nodes, there could be a significant rate increase compared with the scheme that chooses channels uniformly or equally. This work assumes that the CR harvests energy from the PU. On the other hand, if the CR harvests from the ambient source, as in Pradha et al. (2014), the effect of the energy availability on channel selection can be analyzed and then the channel selection policy can be optimized based on the amount of available harvested energy.

There are other important issues in CRs, such as spectrum handover, spectrum aggregation, and spectrum allocation. They can be studied in a similar way by considering the dynamics of the energy supply for energy harvesting CR.

7.9 Summary

Energy harvesting can be combined with CRs to provide energy- and spectral-efficient wireless communications. This chapter has discussed different energy harvesting CR systems. The discussion has been based on the source of energy for the CR, as the source of energy makes a fundamental difference to the CR design. Several important energy harvesting CR systems have been investigated, where the CR system is a wireless powered communications system, the CR harvests energy from the PU signal, or the CR harvests energy from random ambient sources. Within these systems, different spectrum sharing principles have been examined. For interweave systems, spectrum sensing is required and the interference on the PU is determined by the sensing accuracy. Different sensing methods have different sensing accuracies. For underlay systems, an interference temperature will be in place. For overlay systems, it resembles the energy harvesting cooperative communications system to be discussed later.

The design problems in these systems often involve an optimization, where the objective function could be information rate, the amount of harvested energy, or other performance measures, while the constraints mainly come from the energy causality that the harvested energy must be larger than the energy to be consumed and the collision that the PU must be protected from. Then, either numerical methods or standard optimization methods can be applied to solve these problems. Following this idea, if a previous study is in the CR literature, one needs to add the energy causality constraint or consider the dynamics of the energy supply, while if a previous study is in the energy harvesting literature, one needs to add the collision constraint. In some cases, both constraints need to be considered.

8

Energy Harvesting Relaying

8.1 Introduction

8.1.1 Wireless Relaying

The relaying concept has been used in wired and wireless communications as a repeater or bridge for decades. The interest in relaying has been reignited in recent years due to the advances in electronics (Laneman et al. 2004). As such, relaying has been used in many applications recently, such as in the fourth generation (4G) standards (Hoymann et al. 2012).

A typical relaying system consists of three nodes: the source; the destination; and the relay. The source is the node that has information to deliver, the destination is the node that intends to receive this information, while the relay helps the delivery from source to destination. There are two main benefits of using a relay as an intermediate node: diversity gain; and coverage extension.

For the diversity gain, the direct link between the source and the destination still works but the relaying link provides an extra copy of the same signal at the destination. By properly combining these copies, diversity gain can be achieved to fight fading or shadowing. Thus, such a relaying system is sometimes called a distributed or virtual multiple-input-multiple-output (MIMO) system. In this case, the relay acts as a virtual antenna for the source. This increases the reliability of the system. Figure 8.1 shows a diagram of such a relaying system. There are $N + 1$ copies of the same signal available at the destination so that the destination can perform diversity combining.

For the coverage extension, the direct link between the source and the destination does not exist, due to obstacles or long transmission distances. Thus, the destination is out of the communications range of the source (Peng et al. 2011; Zhou et al. 2013a). In such a scenario, the source could increase its transmission power or deploy new infrastructure if the problem persists. However, the relay can help extend the transmission range of the source without any extra infrastructure by relaying the signal so that either the obstacle can be circumvented or the transmission distance can be reduced. This increases the capacity of the system. Figure 8.2 shows a diagram of such a relaying system. When the direct link is broken, other routes via relays can be used.

The relay could be an idle peer node in the same network. In this case, it is called peer relaying, user relaying, or mobile relaying. The relay could also be a dedicated relaying station in the network that only has the basic functions of a base station. In this case, it is called infrastructure relaying or fixed relaying. For fixed relaying, since the dedicated

Energy Harvesting Communications: Principles and Theories, First Edition. Yunfei Chen.
© 2019 John Wiley & Sons Ltd. Published 2019 by John Wiley & Sons Ltd.

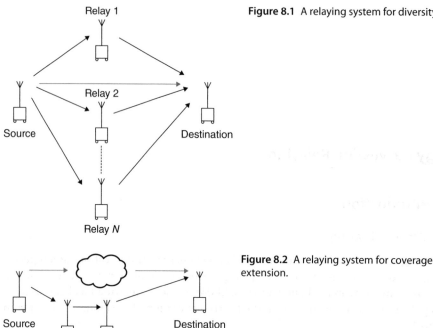

Figure 8.1 A relaying system for diversity gain.

Figure 8.2 A relaying system for coverage extension.

infrastructure is shared by all nodes, it can improve the network coverage and reliability at minimal extra cost but it is not flexible. For mobile relaying, reward mechanisms must be in place to encourage idle nodes to take part in relaying. In both structures, protocols are required to complete the relaying process.

8.1.2 Relaying Protocols

There are two main protocols for relaying: amplify-and-forward (AF); and decode-and-forward (DF) (Laneman et al. 2004). In AF, the source transmits the signal to the relay and the relay simply amplifies the received signal and forwards it to the destination without any further processing. In DF, after receiving the signal from the source, the relay first decodes the signal for information and then encodes the decoded information again before forwarding it to the destination.

In terms of complexity, DF has higher complexity than AF, due to the decoding and encoding operations. In terms of performance, DF is better than AF, as DF removes the noise at the relay by performing decoding and encoding, while AF amplifies the noise along with the signal so that the received signal at the destination will be more noisy. Thus, the choice of protocol depends on the specific application that either prefers performance or simplicity.

The above protocols are also called one-way relaying, as the information is sent from the source through the relay to the destination in one direction. In this case, the transmission happens in two phases. In the first phase, the source transmits the signal to the relay, or if the direct link works to the destination too. This is called the broadcast phase.

In the second phase, the source stops transmission to avoid interference, and the relay forwards its received signal to the destination. This is called the relaying phase. Thus, the source performs one transmission, the relay performs one reception and one transmission, while the destination performs one or two receptions, depending on the existence of the direct link from the source. Compared with the direct transmission, the relaying transmission doubles the transmission time so that its data rate is only half that of the direct transmission.

To remedy the data rate issue, two-way relaying can be used. In two-way relaying, the source and the destination transmit their signals to the relay at the same time in the broadcast phase. The relay will receive a combined copy of both signals and forward the combined copy to both source and destination in the relaying phase. Since simultaneous transmission is performed, the data rate is the same as that of the direct transmission. The self-interference in the received signals at both source and destination can be removed, as they know their own transmitted signals. There is also multi-way relaying, where more than two nodes transmit their signals at the same time. Both two-way relaying and multi-way relaying belong to a wider topic of network coding. Also, the above protocols assume that the same frequency band is used during different time slots to achieve orthogonal channels. One may also achieve the orthogonality in the frequency or code domains.

8.1.3 Energy Harvesting Relaying

From the above discussion, the relay node has to use its own power and spectrum resources to deliver the information for the source. This may be acceptable for infrastructure relaying but may discourage peer nodes from participating in mobile relaying, as they will have a shorter battery life by helping others. Thus, this issue needs to be solved.

Energy harvesting is a promising solution to this problem. In energy harvesting relaying, instead of asking the relay to use its own energy, the source node transfers a certain amount of energy to the relay node as wireless power first. The relay node then uses the harvested wireless power to forward the source signal to the destination. This minimizes the extra energy cost incurred at the relay and therefore this technique holds great potential.

In a wider sense, energy harvesting relaying not only refers to the system where the source transfers energy to the relay for signal forwarding but also refers to the relaying systems where any nodes perform energy harvesting. Similar to energy harvesting cognitive radio (CR), there are also different types of energy harvesting relaying systems. For example, the relay in the above scheme harvests energy from the source. In other schemes, the relay can harvest energy from the ambient energy source or from a dedicated power transmitter or even from its own transmission. The source node can also harvest energy from the ambient energy source, from a dedicated power transmitter or from the relay.

Similar to energy harvesting CRs, energy harvesting relaying needs to satisfy the energy causality constraint. Additionally, relaying needs to satisfy the signal causality constraint that the signal forwarded by the relay must be received from the source first. Next, we will first discuss conventional relaying.

8.2 Conventional Relaying

In this section, we will first discuss AF and DF in more detail. Then, several important performance metrics are examined to measure the reliability or the capacity of a relaying system. Finally, when multiple relays are available, relay selection will be investigated. Two-way relaying will also be discussed.

8.2.1 Amplify-and-Forward Relaying

Assume that all the nodes have a single antenna and operate in the half-duplex mode. The transmission is completed in two phases. In the first broadcast phase, the source sends signals to the relay. In the second relaying phase, the relay forwards the received signal to the destination. Assume that the total transmission is T seconds and that there is no direct link between the source and the destination. The case with multiple links, including a direct link, will be discussed in Section 8.2.4.

In the first $\frac{T}{2}$ seconds, the received signal at the relay is

$$y_r = \sqrt{P_s} hs + n_r \tag{8.1}$$

where P_s is the transmission power of the source, h is the channel gain from the source to the relay, s is the transmitted information symbol with unit power $E\{|s|^2\} = 1$, and n_r is the additive white Gaussian noise (AWGN) at the relay. The fading gain h is a complex circularly symmetric Gaussian random variable with mean zero and variance $2\alpha^2$, and the noise n_r is also Gaussian with mean zero and variance $2\sigma^2$.

In the second $\frac{T}{2}$ seconds, the signal in (8.1) is simply amplified and then forwarded to the destination so that the received signal at the destination is

$$y_d = \sqrt{P_r} agy_r + n_d \tag{8.2}$$

where P_r is the transmission power of the relay, a is the amplification factor that will be discussed later, g is the channel gain from the relay to the destination, y_r is the signal forwarded, and n_d is the AWGN at the destination. Similarly, g and n_d are complex Gaussian random variables with mean zero and variance $2\alpha^2$ and $2\sigma^2$, respectively. From (8.2), the transmission power consumed by the relay is $P_r a^2 E\{|y_r|^2\}$.

The choice of the amplification factor is very important, as it determines the amplification at the relay and affects the overall performance of relaying. In Laneman et al. (2004), the amplification factor is chosen as

$$a^2 = \frac{1}{E\{|y_r|^2\}} = \frac{1}{P_s|h|^2 + 2\sigma^2} \tag{8.3}$$

to normalize the power of the received signal at the relay. As one sees, if a satisfies (8.3), $a^2 E\{|y_r|^2\} = 1$ such that the transmission power consumed by the relay becomes P_r. Since the value of a depends on the channel gain h and h varies as a random variable, one must adjust the value of a for each relaying transmission. This is called variable-gain AF. In some studies, it is also called the channel-assisted AF (Amarasuriya et al. 2011). This is the most widely used amplification factor in the literature and will be the focus of our discussion.

For variable-gain AF, one must have the knowledge of h and $2\sigma^2$ in order to calculate the amplification factor for relaying. The noise variance does not change fast and can be estimated accurately. In channels with strong signals, $2\sigma^2$ may be negligible so that the amplification factor is given by

$$a^2 = \frac{1}{P_s|h|^2} \tag{8.4}$$

This is sometimes called ideal channel-assisted or inverting AF, as the relay always inverts the channel gain from the source to the relay (Hasna and Alouini 2003).

In both (8.3) and (8.4), the channel gain h has to be estimated for the relay. Thus, disintegrated channel estimation must be performed for variable-gain AF (Khan et al. 2012). This channel estimation incurs extra complexity at the relay. A solution to this problem is to use the average of the amplification gain in (8.3). In this case, one has

$$a^2 = E\left\{\frac{1}{P_s|h|^2 + 2\sigma^2}\right\}. \tag{8.5}$$

This average will only depend on the noise variance $2\sigma^2$ and the fading channel parameters. Thus, it is a fixed value for a homogeneous environment. This is called fixed-gain relaying.

8.2.2 Decode-and-Forward Relaying

Similar assumptions can be made for DF. In this case, the received signal at the relay is still given by (8.1) in the broadcast phase. Once the relay receives this signal, it tries to decode it. For example, if binary phase shift keying (BPSK) is used such that $s = 1$ or $s = -1$ with equal probabilities, the data decision is made as

$$\hat{s} = \begin{cases} 1, & Re\{y_r h^*\} > 0 \\ -1 & Re\{y_r h^*\} < 0 \end{cases} \tag{8.6}$$

where $Re\{y_r h^*\}$ is the decision device that multiplies the received signal y_r with the conjugate of the channel gain h and takes the real part of the product for decision. For other modulation schemes, similar decision devices can be applied to the received signal.

In the relaying phase, the decoded information $\hat{s}$ is encoded again and forwarded to the destination so that the received signal at the destination is

$$y_d = \sqrt{P_r}g\hat{s} + n_d \tag{8.7}$$

where all the symbols are defined as before. Since $\hat{s} = -1$ or $\hat{s} = 1$, one sees that the noise at the relay n_r will not appear in the received signal at the destination, unlike AF relaying. Thus, DF has a better performance than AF.

There are other variants of DF. For example, instead of decoding and encoding the signal from the source, the relay can first use the signal-to-noise ratio (SNR) in the first hop to decide if the received signal in (8.1) is of good quality. If it is, the relay will perform decoding, encoding, and then forwarding. If the SNR of the signal in (8.1) is below a threshold, an outage occurs such that the relay does not do anything. The quality check at the relay prevents the error in the first hop from propagating to the next hop. This is termed as incremental relaying in some references.

There are also other relaying protocols. For example, instead of forwarding the whole signal y_r, the relay can forward a quantized version of this signal. This is particularly useful in digital communications systems, where the signal will be quantized anyway. This is termed quantize-and-forward relaying. Also, if the received signal at the relay is sparse, it can be compressed before being forwarded. This is termed compress-and-forward relaying. These relaying protocols are useful in different applications but in general are not as widely used as AF and DF.

8.2.3 Performance Metrics

Next, we define several performance metrics that are commonly used in relaying systems to measure their performances.

8.2.3.1 Amplify-and-Forward

For AF, using (8.1) and (8.2), one has

$$y_d = \sqrt{P_r}ag(\sqrt{P_s}hs + n_r) + n_d = \sqrt{P_r P_s}aghs + \sqrt{P_r}agn_r + n_d. \tag{8.8}$$

The first term is the signal part, while the last two terms are the noise parts. One sees that the noise at the relay n_r has also been amplified to make y_d more noisy. This is the disadvantage of AF relaying. From (8.8), the end-to-end SNR or the overall SNR can be derived as

$$\gamma = \frac{P_r P_s a^2 |g|^2 |h|^2}{2\sigma^2 P_r a^2 |g|^2 + 2\sigma^2} \tag{8.9}$$

If the variable gain in (8.3) is used, the overall SNR can be rewritten as

$$\gamma = \frac{\gamma_r \gamma_d}{\gamma_r + \gamma_d + 1} \tag{8.10}$$

where $\gamma_r = \frac{P_s |h|^2}{2\sigma^2}$ is the SNR of the first hop from the source to the relay and $\gamma_d = \frac{P_r |g|^2}{2\sigma^2}$ is the SNR of the second hop from the relay to the destination. The overall SNR is determined by the weaker hop. For example, when γ_d is much smaller than γ_r, $\gamma \approx \gamma_d$, and when γ_r is much smaller than γ_d, $\gamma \approx \gamma_r$. The exact result in (8.10) is not mathematically tractable in many design problems. Hence, approximations can be used. One such approximation is

$$\gamma \approx \frac{\gamma_r \gamma_d}{\gamma_r + \gamma_d} \tag{8.11}$$

when the ideal channel-assisted AF is used or when the SNR is very large so that $\gamma_r + \gamma_d + 1 \approx \gamma_r + \gamma_d$. This is called the harmonic mean (Hasna and Alouini 2004). Other approximations are also possible, for example, by choosing the maximum or the minimum of γ_r and γ_d. They are not discussed here as they are less frequently used.

Using the end-to-end SNR, we can define the probability of power outage or the outage probability as

$$P_o(\gamma_0) = Pr\{\gamma < \gamma_0\} = F_\gamma(\gamma_0) \tag{8.12}$$

where γ_0 is the threshold SNR below which the receiver at the destination cannot detect any signal, and $F_\gamma(\cdot)$ is the cumulative distribution function (CDF) of γ. Thus, the outage probability is essentially the value of the CDF of γ at γ_0.

Also, using the end-to-end SNR, the error rates for different modulation schemes can be derived (Simon and Alouini 2005). For example, for M-ary square quadrature amplitude modulation, the symbol error rate is

$$P_e(\gamma) = 4\frac{\sqrt{M}-1}{\sqrt{M}}Q\left(\sqrt{\frac{3\gamma}{M-1}}\right) - 4\left(\frac{\sqrt{M}-1}{\sqrt{M}}\right)^2 Q^2\left(\sqrt{\frac{3\gamma}{M-1}}\right) \tag{8.13}$$

and for M-ary phase shift keying,

$$P_e(\gamma) = \frac{1}{\pi}\int_0^{(M-1)\pi/M} e^{-\gamma\frac{\sin^2(\pi/M)}{\sin^2\theta}}\, d\theta \tag{8.14}$$

where M is the constellation size, and $Q(\cdot)$ is the Gaussian Q function. In all cases, the error rate is a function of the end-to-end SNR. Since γ is random, the average error rate is calculated as

$$\bar{P}_e = \int_0^\infty P_e(x)f_\gamma(x)dx \tag{8.15}$$

where $f_\gamma(x)$ is the probability density function (PDF) of γ.

The above two measures, outage and error rate, determine the reliability of the relaying communications system. Next, we define the achievable rate or the capacity of the system. For delay-tolerant applications, the achievable rate can vary. Thus, from the Shannon's limit, one has

$$R(\gamma) = \frac{1}{2}\log_2(1+\gamma) \tag{8.16}$$

where $\frac{1}{2}$ takes the rate penalty of relaying into account. The average or ergodic capacity in a fading channel is given by

$$\bar{R} = \int_0^\infty R(x)f_\gamma(x)dx. \tag{8.17}$$

In other applications, delay is constrained so that the data rate must be larger than a certain value R_0. In this case, using (8.16), one has

$$R(\gamma) = \frac{1}{2}\log_2(1+\gamma) \geq R_0. \tag{8.18}$$

This gives $\gamma \geq 2^{2R_0} - 1$. Thus, the achievable rate in delay-constrained applications is given by

$$R = \frac{R_0}{2}[1 - P_o(2^{2R_0} - 1)] \tag{8.19}$$

where the outage probability P_o is given in (8.15).

8.2.3.2   Decode-and-Forward

For DF, we cannot define the end-to-end SNR in the same way as for AF but we can derive the equivalent end-to-end SNR. We start from the hop performances. From (8.1) and (8.7), the achievable rates of the source-to-relay link and the relay-to-destination link are (Nasir et al. 2014)

$$R_r = \log_2(1+\gamma_r) \tag{8.20}$$

and

$$R_d = \log_2(1+\gamma_d) \tag{8.21}$$

respectively. However, the rates on both sides of the relay must be balanced. If R_r is larger than R_d, a buffer will be required at the relay. Without a buffer, some data have to be discarded. Similarly, if R_d is larger than R_r, a rate outage occurs, as the relay does not have enough data to forward. Thus, the overall rate of the relaying link is the minimum of R_r and R_d, given by

$$R(\gamma) = \frac{1}{2}\min\{R_r, R_d\} = \frac{1}{2}\log_2(1 + \min\{\gamma_r, \gamma_d\}) = \frac{1}{2}\log_2(1 + \gamma) \tag{8.22}$$

where γ is the equivalent end-to-end SNR for DF in terms of rate given by

$$\gamma = \min\{\gamma_r, \gamma_d\} \tag{8.23}$$

and $\frac{1}{2}$ again takes the rate penalty of relaying into account.

For the error rate, the end-to-end SNR is not straightforward to obtain either. We only consider the case when BPSK is used. In this case, the error rate in the source-to-relay link is given by

$$P_e^r(\gamma_r) = Q(\sqrt{2\gamma_r}) \tag{8.24}$$

and the error rate in the relay-to-destination link is given by

$$P_e^d(\gamma_d) = Q(\sqrt{2\gamma_d}) \tag{8.25}$$

where $Q(\cdot)$ is the Gaussian Q function defined as before. The overall error rate of the relaying link from the source to the relay and then to the destination is given by

$$P_e(\gamma) = P_e^r(\gamma_r)[1 - P_e^d(\gamma_d)] + P_e^d(\gamma_d)[1 - P_e^r(\gamma_r)] \tag{8.26}$$

as the bit error only occurs when there is an error either in the source-to-relay link or in the relay-to-destination link, but not in both links. Thus, the equivalent end-to-end SNR γ can be found by solving the following equation

$$Q(\sqrt{2\gamma}) = Q(\sqrt{2\gamma_r})[1 - Q(\sqrt{2\gamma_d})] + Q(\sqrt{2\gamma_d})[1 - Q(\sqrt{2\gamma_r})] \tag{8.27}$$

for γ. This is a non-linear equation and is generally difficult to solve. It has been reported in Wang et al. (2007) that

$$\min\{\gamma_r, \gamma_d\} - 1.62 \leq \gamma \leq \min\{\gamma_r, \gamma_d\}. \tag{8.28}$$

When the hop SNRs are much larger than 1.62, roughly, the equivalent end-to-end SNR can be approximated as $\gamma \approx \min\{\gamma_r, \gamma_d\}$, which is the same as that from the achievable rate in (8.23). Figure 8.3 examines the accuracy of this approximation. One sees that when both γ_d and γ_r are small, there is some noticeable difference between the accurate bit error rate (BER) and the approximated BER. However, when γ_r is larger than 4 dB, this difference is negligible. For other modulation schemes, the equivalent end-to-end SNR can also be derived by solving $P_e(\gamma) = P_e^r(\gamma_r)[1 - P_e^d(\gamma_d)] + P_e^d(\gamma_d)[1 - P_e^r(\gamma_r)]$ but this is not a trivial problem.

In summary, the end-to-end SNR for DF is not as straightforward as that for AF but one may use (8.23) as either an exact result for achievable rate or an approximate result for error rate. Using this end-to-end SNR, the outage probability for DF can also be defined as

$$P_o(\gamma_0) = Pr\{\gamma < \gamma_0\} = F_\gamma(\gamma_0) \tag{8.29}$$

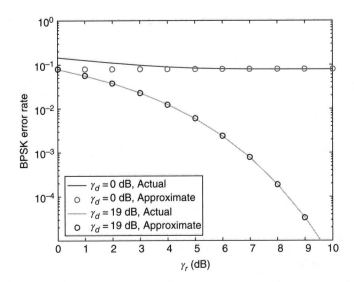

Figure 8.3 Accuracy of the equivalent end-to-end SNR for the BER of DF.

where $F_\gamma(\cdot)$ is the CDF of γ. Using (8.23), it can be derived that

$$F_\gamma(\gamma_0) = 1 - [1 - F_{\gamma_r}(\gamma_0)][1 - F_{\gamma_d}(\gamma_0)] \tag{8.30}$$

where $F_{\gamma_r}(\cdot)$ and $F_{\gamma_d}(\cdot)$ are the CDFs of γ_r and γ_d, respectively.

Also, for delay-constrained applications when the data rate must be larger than a threshold R_0, the achievable rate can be shown as

$$R = \frac{1}{2}R_0[1 - P_o(2^{2R_0} - 1)]. \tag{8.31}$$

Finally, γ in (8.23) is also a random variable. Its CDF is given by

$$F_\gamma(x) = 1 - [1 - F_{\gamma_r}(x)][1 - F_{\gamma_d}(x)]. \tag{8.32}$$

The average error rate is given by

$$\bar{P}_e = \int_0^\infty P_e(x)f_\gamma(x)dx \tag{8.33}$$

and the average achievable rate or ergodic capacity is given by

$$\bar{R} = \int_0^\infty R(x)f_\gamma(x)dx \tag{8.34}$$

where $f_\gamma(x) = \frac{dF_\gamma(x)}{dx}$ is the PDF of γ.

Figure 8.4 compares the ergodic capacities of AF and DF, where $\bar{\gamma}_r$ and $\bar{\gamma}_d$ are the average hop SNRs and Rayleigh fading is used. One sees that DF always has a higher capacity than AF. The performance metrics discussed above will be used in the following discussion.

8.2.4 Relay Selection

In the previous discussion, we have assumed that there is only one relaying node and that the direct link between the source and the destination is broken, in order to focus on

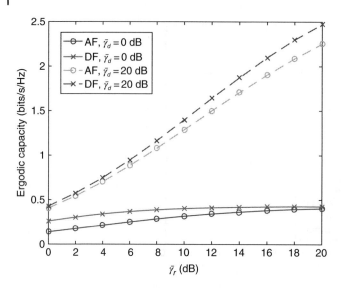

Figure 8.4 Comparison of ergodic capacities for AF and DF.

the relaying process from the source to the relay and then to the destination. In practical communications systems, the direct link may still work. Also, several peer nodes might be idle at the same time so that multiple relays are available. Consequently, the signal from the source can arrive at the destination through different routes. At the destination, all these copies can be coherently combined to achieve high diversity gain. However, the complexity of such a network increases with the number of relays. In some applications, such as wireless sensor networks, complexity is more of a concern than performance. In these applications, to reduce the network complexity, relay selection can be implemented that often chooses one out of all available links for the best tradeoff between complexity and performance.

It has been reported in Jing and Jafarkhani (2009) that, with properly designed schemes, the diversity order of relay selection is the same as the combination of all available links. Several relay selection schemes can be used. We use AF as an example. Consider an AF relaying system with one source, one destination, and N relays. Similarly, a two-phase transmission is assumed. In the first phase, the source broadcasts its signal to all the relays such that the received signal at the nth relay can be expressed as

$$y_{nr} = h_n \sqrt{P_s} s + w_n \qquad (8.35)$$

where $n = 1, 2, \cdots, N$ is the relay index, h_n is the complex fading gain in the channel between the source and the nth relay, P_s is the transmission power of the source, s is the transmitted symbol with unit power, and w_n is the AWGN at the nth relay with variance $2\sigma^2$. In the second phase, the received signals at the relays are amplified and forwarded such that the received signal from the nth relay at the destination is

$$y_{nd} = \sqrt{P_{nr}} a_n g_n y_{nr} + z_n \qquad (8.36)$$

where P_{nr} is the transmission power of the nth relay, g_n is the complex fading gain in the channel from the nth relay to the destination, $a_n = \sqrt{\dfrac{1}{P_s |h_n|^2 + 2\sigma^2}}$ is the amplification factor, and z_n is the complex Gaussian noise in the channel between the nth relay and the destination with the noise power $2\sigma^2$.

The end-to-end SNR of the nth relaying link can be shown as

$$\gamma_n = \frac{\gamma_{rn}\gamma_{dn}}{\gamma_{rn} + \gamma_{dn} + 1} \tag{8.37}$$

where $\gamma_{rn} = \frac{P_s|h_n|^2}{2\sigma^2}$ and $\gamma_{dn} = \frac{P_{nr}|g_n|^2}{2\sigma^2}$ are the instantaneous SNRs of the first hop and the second hop, respectively.

8.2.4.1 Full Selection

In the full selection scheme, the relay is selected according to

$$\hat{n} = \max_{n=1,2,\cdots,N} \{\gamma_n\} \tag{8.38}$$

that is, the relay that can provide the largest end-to-end SNR is selected. This is intuitively the best, because the achievable rate increases with the end-to-end SNR, while the error rate decreases with the end-to-end SNR. Thus, the largest end-to-end SNR will lead to the largest data rate or the smallest error rate. Denote this scheme as the $\max\{\gamma_n\}$ scheme.

For the full selection, the error rate can be derived as $P_e = \int_0^\infty P_e(x)dF_{\gamma_{\hat{n}}}(x)$, where $P_e(x)$ is the conditional error rate, conditioned on $\gamma_{\hat{n}}$, and $F_{\gamma_{\hat{n}}}(x) = F_{\gamma_n}^N(x)$ is the CDF of $\gamma_{\hat{n}}$ with $F_{\gamma_n}(x)$ given in Tsiftsis et al. (2006) for Nakagami-m fading channels.

8.2.4.2 Partial Selection

The performance of the full selection scheme should be the best, as it maximizes the end-to-end SNR. However, this means that, in order to make this selection, the source requires knowledge of the channel state information of both the source-to-relay link and the relay-to-destination link for all relays. This may incur a huge amount of overhead. To reduce this overhead, partial relay selection can be performed by using the hop SNR instead. For example, one can choose the link with the largest first hop SNR as

$$\hat{n} = \max_{n=1,2,\cdots,N} \{\gamma_{rn}\}. \tag{8.39}$$

In this case, only the channel state information of the first hops is required for selection. Denote this scheme as the $\max\{\gamma_{rn}\}$ scheme. This scheme reduces the complexity of relaying greatly. The error rate in this case is given by $P_e = \int_0^\infty P_e(x)dF_{\gamma_{\hat{n}}}(x)$, where $F_{\gamma_{\hat{n}}}(x)$ is the CDF of $\gamma_{\hat{n}}$ given by Chen et al. (2011a)

Another partial selection scheme is to choose the relaying link with the strongest second-hop SNR as

$$\hat{n} = \max_{n=1,2,\cdots,N} \{\gamma_{dn}\}. \tag{8.40}$$

Denote this scheme as the $\max\{\gamma_{dn}\}$ scheme. Its probability of error can be derived in a similar way and is given by Chen et al. (2011a).

The above two partial selection schemes use the hop SNRs for selection. This requires channel estimators for h_n or g_n or both, $n = 1, 2, \cdots, N$. To remove this complexity in the relaying system, it is also possible to use the received signal amplitude for selection directly. Thus, two other partial selection schemes are

$$\hat{n} = \max_{n=1,2,\cdots,N} \{|y_{nr}|\} \tag{8.41}$$

and

$$\hat{n} = \max_{n=1,2,\cdots,N}\{|y_{nd}|\}. \tag{8.42}$$

denoted as the max$\{|y_{nr}|\}$ scheme and the max$\{|y_{nd}|\}$ scheme, respectively.

The selection of the relaying link is made at the base station in a centralized network or at the group leader in a distributed network. The decision will be broadcast by the base station or the group leader to the source, the destination and all the relays before relaying. Thus, although there are N links, when the relaying starts, only one relaying link will be active and the relaying process is still the same as the two-phase transmission discussed before. On the other hand, if one wants to use all available relays, in order to avoid interference in the relaying phase, the relays have to forward their signals one by one, to achieve orthogonality in the time domain causing significant delay. This is another advantage of performing relay selection.

Next, we compare the performances of these schemes. In the comparison, BPSK is used. Also, consider Rayleigh fading channels with unit average fading power and unit noise power, while $UE = \frac{P_{rn}}{P_s}$, which is also the ratio of $E\{\gamma_{nr}\}$ to $E\{\gamma_{nd}\}$. Figure 8.5 compares different relay selection schemes when $N = 2$ and $UE = 0.1$. One sees that the max$\{\gamma_n\}$ scheme performs the best, as expected, as it uses the SNRs in both hops. Among the partial selection schemes, the max$\{\gamma_{nd}\}$ and max$\{|y_{nd}|\}$ schemes outperform the max$\{\gamma_{nr}\}$ and max$\{|y_{nr}|\}$ schemes. For example, at a BER of 10^{-2}, the max$\{\gamma_{nd}\}$ and max$\{|y_{nd}|\}$ schemes have performance gains of around 5 dB. Comparing the partial selection schemes, one sees that the max$\{|y_{nd}|\}$ scheme performs the best. Its performance is indistinguishable from the performance of the full selection max$\{\gamma_n\}$ scheme when the SNR is less than 20 dB. Figure 8.6 compares them when $N = 2$ and $UE = 10$. In this case, the max$\{\gamma_{nr}\}$ and max$\{|y_{nr}|\}$ schemes outperform the max$\{\gamma_{nd}\}$ and max$\{|y_{nd}|\}$ schemes. Among the partial selection schemes, the max$\{|y_{nr}|\}$ scheme performs the best. One concludes that one should choose the best idle node for

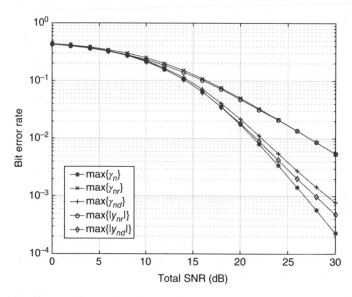

Figure 8.5 Different relay selection schemes when $N = 2$ and $UE = 0.1$.

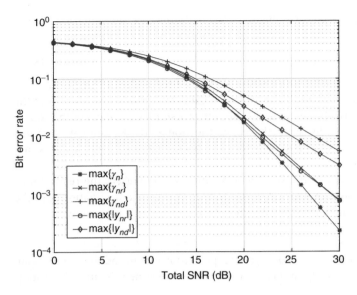

Figure 8.6 Different relay selection schemes when $N = 2$ and $UE = 10$.

the hop with smaller average SNR in order to achieve the best BER performance in partial selection.

8.2.5 Two-Way Relaying

Next, we discuss two-way relaying that was mentioned before. Both AF and DF in the previous discussion perform one-way relaying, where the source transmits data to the destination through the relay in one direction. In this scheme, the data rate is only half of that of the direct transmission, as the source has to stop for $\frac{T}{2}$ seconds to avoid interference. On the other hand, in two-way relaying, the source node and the destination node have data to exchange so that they transmit to the relay at the same time in the same channel to maintain the data rate. Figure 8.7 shows the frame structure and a diagram of two-way relaying.

In two-way relaying, there are also two phases. In the multiple-access phase, both the source node and the destination node send their data to the relaying node. In the broadcasting phase, the relaying node sends its received signal to both source and destination.

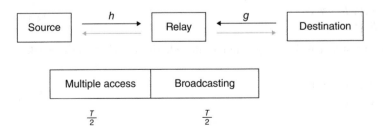

Figure 8.7 Diagram of two-way relaying.

Thus, in the multiple-access phase of the first $\frac{T}{2}$ seconds, the received signal at the relay can be given by

$$y_r = \sqrt{P_2}hs_1 + \sqrt{P_1}gs_2 + n_r \tag{8.43}$$

where P_1 and P_2 are the transmission power of the source node and the destination node, respectively, h and g are the fading channel gains from the source node and the destination node, respectively, and s_1 and s_2 are the transmitted symbols of the source node and the destination node, respectively, and they follow BPSK modulation in this case, while n_r is the AWGN defined as before. In this case, the symbols s_1 and s_2 are mixed in one received signal.

In the broadcasting phase of the second $\frac{T}{2}$ seconds, if AF is used, the relay simply amplifies the received signal and broadcasts it to both source and destination. Use the source node as an example. The received signal at the source node is given by

$$y_{d1} = \sqrt{P_r}ahy_r + n_d \tag{8.44}$$

where P_r, a, and h are transmission power, the amplification factor, and the channel gain from relay to source, respectively. In this equation, we have assumed channel reciprocity. Thus, one has

$$y_{d1} = \sqrt{P_r P_1}ah^2 s_1 + \sqrt{P_r P_2}aghs_2 + \sqrt{P_r}ahn_r + n_d. \tag{8.45}$$

In this equation, the first term is the self-interference caused by the transmission of the source's own signal s_1, the second term is the desired signal from the destination node that needs to be decoded, and the last two terms are the noise. Before decoding, the self-interference will be removed, as s_1 is known to the source node. Assuming perfect removal, that is, the first source knows a and h perfectly, one has

$$y'_{d1} = y_{d1} - \sqrt{P_r P_{s1}}ah^2 s_1 = \sqrt{P_r P_{s2}}aghs_2 + \sqrt{P_r}ahn_r + n_d \tag{8.46}$$

which is then used to decode s_2 from the destination node. Similar operations will be performed at the destination node to decode the information from the source node.

If DF is used, the relay tries to decode s_1 and s_2 using its received signal. This is usually difficult as s_1 and s_2 are mixed in the same received signal. However, when the two channels from the source and the destination are not balanced, or one is stronger than the other, decoding can still be performed. In this case, the received signal is used to decode the signal from the stronger node, while treating the signal from the other node as interference. After this decoding, the signal from the stronger node will be subtracted from the received signal to decode the signal from the weaker node. Assume that the signal from the source node is the stronger one. One has

$$\hat{s}_1 = \begin{cases} 1, & Re\{y_r h^*\} > 0 \\ -1 & Re\{y_r h^*\} < 0 \end{cases} \tag{8.47}$$

where y_r is the received signal at the relay. Then, to decode the signal from the weaker node, one has

$$\hat{s}_2 = \begin{cases} 1, & Re\{y'_r g^*\} > 0 \\ -1 & Re\{y'_r g^*\} < 0 \end{cases} \tag{8.48}$$

where

$$y'_r = y_r - \sqrt{P_1}h\hat{s}_1. \tag{8.49}$$

Once $\hat{s}_1$ and $\hat{s}_2$ are available, the relay forms a function of $x = \hat{s}_1 \oplus \hat{s}_2$ using network coding and then broadcasts it to both source and destination. The received signal at the source node is

$$y_{d1} = \sqrt{P_r} h x + n_d \tag{8.50}$$

This signal will be used to decode x and the effect of the self symbol s_1 can be easily removed by performing $\hat{x} \oplus s_1$, where $\hat{x}$ is the decoded symbol of x.

In one-way relaying, one symbol s is transmitted during T seconds, as the source has to stop transmission in the second $\frac{T}{2}$ seconds to avoid interference. On the other hand, in two-way relaying, two symbols s_1 and s_2 are transmitted during T seconds, using simultaneous transmission and network coding. Thus, two-way relaying has the same data rate as the direct transmission from source to destination.

The above two-way relaying is also called bi-directional relaying and it has to be combined with network coding for effective operation. As one can see from the above discussion, the success of this relaying protocol depends on accurate knowledge of all channel state information at all nodes, and the relay also consumes more power than that in one-way relaying. In the following, we only focus on one-way relaying.

8.3 Types of Energy Harvesting Relaying

From this section on, we focus on energy harvesting relaying. In general, energy harvesting relaying is different from conventional relaying in that at least one of the nodes in the relaying process harvests energy from either an ambient source, the other nodes, or a dedicated power transmitter. There are different types of energy harvesting relaying. For example, one can categorize them based on the relaying protocols: energy harvesting AF relaying; and energy harvesting DF relaying. One can also categorize them based on the energy harvesting methods: time switching energy harvesting relaying; and power splitting energy harvesting relaying.

We categorize different energy harvesting relaying systems based on the energy source, as the energy source has more fundamental impact than the relaying protocol or energy harvesting method on the relaying system designs. For example, if the relay harvests energy from the source, the source has to either increase its transmission power or increase its frame length for power transfer. If the relay harvests energy from the ambient source, the energy causality constraint applies such that the relay must accumulate enough energy before forwarding the signal. In all cases, either the energy causality constraint or the signal causality constraint or both need to be considered in the optimization of one of the performance metrics discussed in Section 8.2.3.

In some energy harvesting relaying systems, both the source and the relay harvest energy from the ambient sources, such as in Huang et al. (2013), Minasian et al. (2014), and Orhan and Erkip (2015). This is the most general case. In this case, optimal scheduling or power allocation algorithms can be formulated, as the energy arrival rate varies and hence the transmission time and power should be adapted to it. For example, in Minasian et al. (2014), assuming that both source and relay harvest energies from the ambient source, the optimal transmission policy was designed to maximize the achievable rate. The energy causality was considered in the optimization, along with the constraint on a finite battery. In Orhan and Erkip (2015) the optimal transmission policy was also obtained by maximizing the achievable rates of four different setups with respect to the transmission power of the source and the transmission power of the relay. In this optimization, both energy and signal causality constraints were accounted for. The battery has an infinite size but the buffer at the relay is finite so that overflow must be dealt with. In Huang et al. (2013) a similar optimization problem that maximizes the achievable rate over the transmission power of the source and the relay was formulated but the energy arrival follows a deterministic model with known arrival time and amount. Also, in Moradian and Ashtiani (2015), the relay has a fixed power supply as well as harvesting energy from the ambient source, while in Kashef and Ephremides (2016) the data traffic or the data arrival were studied along with the energy arrival in the optimization.

In some energy harvesting relaying systems, the source harvests energy from the ambient source, while the relay does not harvest (Luo et al. 2013). In this case, the relay could be an infrastructure relay and the randomness in the energy availability comes from the source. For example, in Luo et al. (2013) the throughput and the transmission time were optimized with respect to the transmission power of the source and the relay. They are adaptive to the energy harvested by the source. The problem is similar to that in Minasian et al. (2014) and Orhan and Erkip (2015) and other studies due to the randomness of the energy at the source, but it is simpler due to the fixed power at the relay.

In some energy harvesting systems, unlike Luo et al. (2013), the source does not harvest, while the relay harvests energy from the ambient source (Medepally and Mehta 2010; Qian et al. 2016). Then, the randomness in the energy availability comes from the relay, not from the source. One can also optimize the transmission policy in terms of transmission time and transmission power for the relay. Again, the energy causality, the battery size and the buffer size at the relay will affect the design problem.

In some energy harvesting relaying systems, both the source and the relay harvest energy from the hybrid access point (HAP) during its power transmission (Chen et al. 2015). The difference between this system and the HAP wireless powered system discussed in Chapter 6 is that, in this system, the uplink information delivery is completed via a relaying process, while in Chapter 6, it is a direct transmission to the HAP. A similar system was also considered in Liang et al. (2017). In another energy harvesting system, both the source and the relay harvest energy from the power beacon (Zhong et al. 2015b). These systems are similar to the wireless powered systems discussed in Chapter 6, except that the direct transmission there has been replaced by a relaying transmission. In another variant, the relay harvests energy from the HAP and uses this energy to forward signals from the user equipment to the HAP or from the HAP to the user equipment (Hwang et al. 2017; Ramezani and Jamalipour 2017). The source does not harvest in this case. In these problems, energy causality is often not a problem as there is dedicated wireless power supply.

In another interesting study, the relay first harvests energy from the source for relay transmission but during the relay transmission, the source also harvests energy from the relay for extra energy (Chen et al. 2017c). This is possible if the source and the relay are peer nodes of similar transmission power. This is part of a wider topic on energy self-recycling for the full-duplex node (Hwang et al. 2017) or for the whole network (Xie et al. 2017).

In most energy harvesting relaying systems, however, the relay node harvests energy from the source node and the source node has a fixed power supply (Nasir et al. 2013) as

Table 8.1 Energy harvesting relaying systems with different energy sources.

Reference	Source energy	Relay energy
Minasian et al. (2014)	Ambient environment	Ambient environment
Luo et al. (2013)	Ambient environment	Fixed supply
Qian et al. (2016)	Fixed supply	Ambient environment
Chen et al. (2015)	HAP	HAP
Zhong et al. (2015b)	PB	PB
Hwang et al. (2017)	Fixed supply	HAP
Chen et al. (2017c)	Relay node	Source node
Nasir et al. (2013)	Fixed supply	Source node

HAP, hybrid access point; PB, power beacon.

this removes the relay's concern on its battery life and therefore encourages idle nodes to participate in relaying.

Table 8.1 summarizes the different energy harvesting relaying systems based on their energy sources. These references are meant to be examples, not an exhaustive list of all the relevant studies. Also, in the literature, there are other terms, such as wireless powered relaying, relaying with wireless energy harvesting or radio frequency (RF) energy harvesting, and so on. In the following, we will mainly use some examples to discuss the system where both source and relay harvest energy from the ambient environment, the system where both source and relay harvest energy from a dedicated power transmitter, and the system where the relay harvests energy from the source only.

8.4 From the Ambient Environment

Figure 8.8 shows a diagram of the energy harvesting system where both source and relay harvest energy from the ambient environment. In this case, the energy uncertainty appears at both the source and the relay. Thus, energy causality must be considered for both the source and the relay. We aim to find the optimum transmission power that maximizes the achievable rate.

Consider AF. The instantaneous achievable rate for one relaying transmission with a duration of T is given by (8.16). Over a time horizon of N relaying transmissions, the total achievable rate is

$$R = \frac{1}{2} \sum_{i=1}^{N} \log_2 \left[1 + \frac{\gamma_r(i)\gamma_d(i)}{\gamma_r(i) + \gamma_d(i) + 1} \right] \tag{8.51}$$

where

$$\gamma_r(i) = \frac{P_s(i)|h(i)|^2}{2\sigma^2} \tag{8.52}$$

$$\gamma_d(i) = \frac{P_r(i)|g(i)|^2}{2\sigma^2}. \tag{8.53}$$

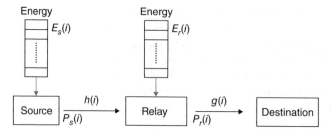

Figure 8.8 Both source and relay harvest energy from the ambient environment.

In these equations, $P_s(i)$, $h(i)$, $P_r(i)$, and $g(i)$ represent the transmission power of the source, the channel gain from the source to the relay, the transmission power of the relay, and the channel gain from the relay to the destination, respectively, during the ith transmission. This assumes a block fading channel.

We are going to optimize $P_s(i)$ and $P_r(i)$ for $i = 1, 2, \cdots, N$ to maximize R. This means that the data rate during each transmission varies due to the changes in the transmission power. There is no direct link between the source and the destination. It is quite easy to extend the results to the case with direct link by replacing the end-to-end SNR in (8.51) with the overall SNR using either relay selection or combination. For example, if relay selection is used and the direct link has a channel gain of $f(i)$, $\frac{\gamma_r(i)\gamma_d(i)}{\gamma_r(i)+\gamma_d(i)+1}$ can be replaced by $\max\left\{\frac{P_s(i)|f(i)|^2}{2\sigma^2}, \frac{\gamma_r(i)\gamma_d(i)}{\gamma_r(i)+\gamma_d(i)+1}\right\}$. If relay combination is used, $\frac{\gamma_r(i)\gamma_d(i)}{\gamma_r(i)+\gamma_d(i)+1}$ can be replaced by $\frac{P_s(i)|f(i)|^2}{2\sigma^2} + \frac{\gamma_r(i)\gamma_d(i)}{\gamma_r(i)+\gamma_d(i)+1}$. Other derivations are similar.

The source harvests $E_s(i)$ energy from the ambient environment at the beginning of the ith transmission. The relay harvests $E_r(i)$ energy from the ambient environment at the beginning of the ith transmission. Also, at the beginning of the ith transmission, the source has $S_s(i)$ energy in the battery and the relay has $S_r(i)$ energy in the battery. The battery has a finite size of S_{max} at both the source and the relay. The buffer size at the relay is infinite. This implies that the relay can always take as much data from the source as needed without any overflow. Putting all the conditions together, one has the optimization problem as

$$\max_{\mathbf{P_s}, \mathbf{P_r}} R \tag{8.54}$$

$$T \sum_{i=1}^{k} P_s(i) \leq \sum_{i=1}^{k} E_s(i), \ k = 1, 2, \cdots, N, \tag{8.55}$$

$$T \sum_{i=1}^{k} P_r(i) \leq \sum_{i=1}^{k} E_r(i), \ k = 1, 2, \cdots, N, \tag{8.56}$$

$$\sum_{i=1}^{k+1} E_s(i) - T \sum_{i=1}^{k} P_s(i) \leq S_{max}, \ k = 1, 2, \cdots, N-1, \tag{8.57}$$

$$\sum_{i=1}^{k+1} E_r(i) - T \sum_{i=1}^{k} P_r(i) \leq S_{max}, \ k = 1, 2, \cdots, N-1, \tag{8.58}$$

$$P_s(i) \geq 0, P_r(i) \geq 0, \ i = 1, 2, \cdots, N \tag{8.59}$$

where $\mathbf{P_s} = [P_s(1)P_s(2)\cdots P_s(N)]$ and $\mathbf{P_r} = [P_r(1)P_r(2)\cdots P_r(N)]$. The first two constraints in (8.55) and (8.56) come from the energy causality that the consumed energy must be smaller than the harvested energy. If the source or the relay do not harvest, one of these two constraints can be removed to have another optimization problem. The second two constraints in (8.57) and (8.58) come from the limited capacity of the battery. However, they are not necessary, as the battery overflow will not affect the transmission.

One sees from (8.54) that all of $E_s(i)$, $E_r(i)$, $h(i)$, and $g(i)$, $i = 1, 2, \cdots, N$, need to be known in order to calculate the optimum values. This knowledge is not causal, as the values for $i = k+1, k+2, \cdots, N$ are not available in the kth transmission. Thus, the optimization must be performed offline. For online optimization, one must remove or replace values of $E_s(i)$, $E_r(i)$, $h(i)$, and $g(i)$ for $i > k$ during the ith transmission.

To solve the optimization in (8.54), a two-step iterative procedure can be used. In the first step during the first iteration, $\mathbf{P_r} = \left[\frac{E_r(1)}{T} \ \frac{E_r(2)}{T} \ \cdots \ \frac{E_r(N)}{T} \right]$ can be assigned as initial values to optimize $\mathbf{P_s}$ only as

$$\max_{\mathbf{P_s}} R \tag{8.60}$$

$$T \sum_{i=1}^{k} P_s(i) \leq \sum_{i=1}^{k} E_s(i), \ k = 1, 2, \cdots, N, \tag{8.61}$$

$$\sum_{i=1}^{k+1} E_s(i) - T \sum_{i=1}^{k} P_s(i) \leq S_{max}, \ k = 1, 2, \cdots, N-1, \tag{8.62}$$

$$P_s(i) \geq 0, \ i = 1, 2, \cdots, N. \tag{8.63}$$

The optimized $\mathbf{P_s}$ can then be used in the second step to optimize $\mathbf{P_r}$ only as

$$\max_{\mathbf{P_r}} R \tag{8.64}$$

$$T \sum_{i=1}^{k} P_r(i) \leq \sum_{i=1}^{k} E_r(i), \ k = 1, 2, \cdots, N, \tag{8.65}$$

$$\sum_{i=1}^{k+1} E_r(i) - T \sum_{i=1}^{k} P_r(i) \leq S_{max}, \ k = 1, 2, \cdots, N-1, \tag{8.66}$$

$$P_r(i) \geq 0, \ i = 1, 2, \cdots, N. \tag{8.67}$$

Then another iteration starts where the first step uses the value of $\mathbf{P_r}$ from (8.64). The iteration stops when the achievable rate using the optimized values of $\mathbf{P_s}$ and $\mathbf{P_r}$ does not change above an accuracy threshold. This process converges, but sometimes not to a global optima. It was shown in Minasian et al. (2014) that performance gains can be achieved by using the proposed optimization. More details can be found in Minasian et al. (2014).

The above optimization problem does not consider the signal causality. When there is not enough data buffer at the relay, this constraint needs to be considered. In this case, the optimization problem in (8.54) can be modified as

$$\max_{\mathbf{P_s}, \mathbf{P_r}} R \tag{8.68}$$

$$T \sum_{i=1}^{k} P_s(i) \leq \sum_{i=1}^{k} E_s(i), \ k = 1, 2, \cdots, N, \tag{8.69}$$

$$T \sum_{i=1}^{k} P_r(i) \leq \sum_{i=1}^{k} E_r(i), \ k = 1, 2, \cdots, N, \tag{8.70}$$

$$\sum_{i=1}^{k+1} E_s(i) - T \sum_{i=1}^{k} P_s(i) \le S_{max}, \ k = 1, 2, \cdots, N-1, \tag{8.71}$$

$$\sum_{i=1}^{k+1} E_r(i) - T \sum_{i=1}^{k} P_r(i) \le S_{max}, \ k = 1, 2, \cdots, N-1, \tag{8.72}$$

$$P_s(i) \ge 0, P_r(i) \ge 0, \ i = 1, 2, \cdots, N \tag{8.73}$$

$$\sum_{i=1}^{k} \log_2[1 + \gamma_r(i)] \le \sum_{i=1}^{k} \log_2[1 + \gamma_d(i)], \ k = 1, 2, \cdots, N, \tag{8.74}$$

$$\sum_{i=1}^{k} \log_2[1 + \gamma_r(i)] - \sum_{i=1}^{k} \log_2[1 + \gamma_d(i)] \le B_{max}, \ k = 1, 2, \cdots, N \tag{8.75}$$

where (8.74) is the new constraint on the signal causality, that is, the amount of data forwarded by the relay to the destination cannot be larger than the amount of data received by the relay from the source. Also, (8.75) is a constraint on the buffer size at the relay, that is, the extra data from the source cannot be larger than the capacity of the buffer B_{max}.

The optimization in (8.68) is also an offline optimization. It can be solved by using a similar two-step iterative procedure as in Minasian et al. (2014). Several special cases can be obtained. If the source node does not harvest, the constraints on the source node (8.69) and (8.71) can be dropped. This corresponds to the case discussed in Medepally and Mehta (2010) and Qian et al. (2016). If the relay node does not harvest, the constraints on the relay node (8.70) and (8.72) can be dropped. This corresponds to the case discussed in Luo et al. (2013). In all these cases, if the battery and the buffer are large enough, the optimization can be simplified as

$$\max_{P_s, P_r} R \tag{8.76}$$

$$T \sum_{i=1}^{k} P_s(i) \le \sum_{i=1}^{k} E_s(i), \ k = 1, 2, \cdots, N, \tag{8.77}$$

$$T \sum_{i=1}^{k} P_r(i) \le \sum_{i=1}^{k} E_r(i), \ k = 1, 2, \cdots, N, \tag{8.78}$$

$$P_s(i) \ge 0, P_r(i) \ge 0, \ i = 1, 2, \cdots, N \tag{8.79}$$

$$\sum_{i=1}^{k} \log_2[1 + \gamma_r(i)] \le \sum_{i=1}^{k} \log_2[1 + \gamma_d(i)], \ k = 1, 2, \cdots, N \tag{8.80}$$

where only the energy causality and the signal causality need to be considered.

One can also consider the DF case by replacing R with (8.22). It is not discussed here. In Orhan and Erkip (2015), the transmission time has also been optimized. Instead of using a fixed time duration of T, they proposed to adjust the time duration as t_i for the ith transmission and optimize them. This makes the optimization more complicated. In all these optimizations, the amount of harvested energy $E_s(i)$ and $E_r(i)$ needs to be known at both the source and the relay. This is the deterministic model discussed in Chapter 3. However, they can also be random values that follow the stochastic models discussed in Chapter 3. In this case, since the optimized transmission power is a function of the harvested energy, the optimized transmission power is also random and needs to be adjusted from transmission to transmission. To overcome this problem, one may use the average constraints or the outage constraints in the optimization.

8.5 From the Power Transmitter

In this case, both the source and the relay harvest energy from a dedicated power transmitter. Since the energy is supplied as wireless power by the dedicated power transmitter, the uncertainty in the energy supply has been greatly reduced. In this case, the time allocation between power transfer and information delivery is more important, similar to the discussion in Chapter 6.

8.5.1 One User and Single Antenna

We start from the simplest case with one user and a single antenna. Figure 8.9 shows a diagram of the considered system. In this case, the signal is sent from the source to the relay and then to the HAP. Assume that the direct link is broken due to obstacles. In this case, there is a HAP, a user terminal and a relay node in the network. All nodes work in the half-duplex mode. In the downlink, the HAP transmits power for τT seconds so that the relay and the user terminal can harvest this signal for energy. The user terminal and the relay rely completely on the harvested energy without any other energy source. After harvesting the energy, in the next $\frac{1-\tau}{2}T$ seconds, the user terminal uses the harvested energy to send the signal to the HAP and the relay in the uplink. In this case, the user terminal is the source, and the HAP is the destination. In the last $\frac{1-\tau}{2}T$ seconds, the relay uses the harvested energy to forward the signal to the HAP using AF.

One sees that this system is similar to the HAP wireless powered system discussed in Chapter 6, except that the signal is now relayed via two hops. Following similar procedures to those in Section 6.4, the harvested energy at the relay and at the source can be given by

$$E_r = \eta_r P_a |h_r|^2 \tau \tag{8.81}$$

and

$$E_s = \eta_s P_a |h_s|^2 \tau \tag{8.82}$$

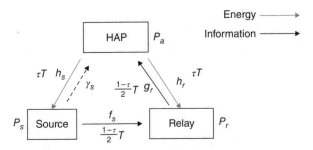

Figure 8.9 Both source and relay harvest from a dedicated power transmitter with one user.

where η_r and η_s are the conversion efficiencies of the harvesters at the relay and the source, respectively, P_a is the transmission power of the HAP, h_r is the channel gain from the HAP to the relay, h_s is the channel gain from the HAP to the source, and $T = 1$ has been assumed for convenience. Using them, the transmission power at the relay and the source can be derived as

$$P_r = \frac{E_r}{(1-\tau)/2} = \frac{2\eta_r\tau}{1-\tau}P_a|h_r|^2 \qquad (8.83)$$

and

$$P_s = \frac{E_s}{(1-\tau)/2} = \frac{2\eta_s\tau}{1-\tau}P_a|h_s|^2 \qquad (8.84)$$

respectively. Thus, from Section 8.2.3, one has the hop SNRs as

$$\gamma_r = \frac{P_s|f_s|^2}{2\sigma^2} = \frac{2\eta_s\tau}{(1-\tau)2\sigma^2}|f_s|^2 P_a|h_s|^2 \qquad (8.85)$$

and

$$\gamma_d = \frac{P_r|g_r|^2}{2\sigma^2} = \frac{2\eta_r\tau}{(1-\tau)2\sigma^2}P_a|h_r|^2|g_r|^2 \qquad (8.86)$$

where f_s is the channel gain from the source to the relay and g_r is the channel gain from the relay to the HAP. Thus, if only the relaying link is used, the achievable rate at the HAP in the uplink is given by

$$
\begin{aligned}
R &= \frac{1-\tau}{2}\log_2\left(1 + \frac{\gamma_r\gamma_d}{\gamma_r+\gamma_d+1}\right) \\
&= \frac{1-\tau}{2}\log_2\left[1 + \frac{\frac{2\eta_s\tau}{(1-\tau)2\sigma^2}|f_s|^2 P_a|h_s|^2 \frac{2\eta_r\tau}{(1-\tau)2\sigma^2}P_a|h_r|^2|g_r|^2}{\frac{2\eta_s\tau}{(1-\tau)2\sigma^2}|f_s|^2 P_a|h_s|^2 + \frac{2\eta_r\tau}{(1-\tau)2\sigma^2}P_a|h_r|^2|g_r|^2 + 1}\right].
\end{aligned}
\qquad (8.87)
$$

Thus, the optimization problem is

$$\max_{\tau}\{R\} \qquad (8.88)$$

$$0 \le \tau \le 1. \qquad (8.89)$$

Figure 8.10 shows R versus τ. For simplicity, we set $\eta_r = \eta_s = 0.5$, and $f_s = h_s = h_r = g_r = 2\sigma^2 = 1$. As expected, the rate first increases and then decreases when τ increases. An optimum value of τ exists that maximizes the rate. The rate also increases with P_a. A single-variable equation can also be obtained by taking the first-order derivative of R with respect to τ and then solving the equation for the optimum τ.

The above result assumes that only the relaying link is used. If the direct link from the source to the HAP exists with a SNR of γ_s and is also used, one can either perform relay selection or combination. In the case when relay selection is used, the performance was obtained and optimized in Chen et al. (2015).

8.5.2 Multiple Users and Single Antenna

Next, we consider the more practical case when multiple user terminals exist. Figure 8.11 shows a diagram of the considered system. In this case, relays and sources form pairs.

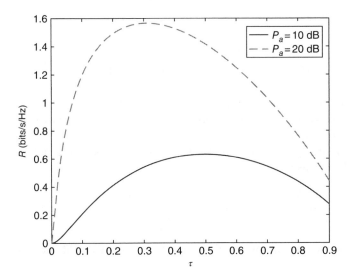

Figure 8.10 *R* versus τ.

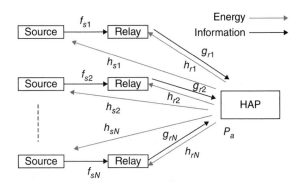

Figure 8.11 Both source and relay harvest from a dedicated power transmitter with multiple users and multiple relays.

Assume that there are N pairs of user terminals and relays. Each user terminal is helped by the nearest relay. In the first $\tau_0 T$ seconds, the HAP broadcasts signals in the downlink for harvesting. In the next $\frac{\tau_1}{2} T$ seconds, user terminal 1 uses the harvested energy to transmit the signal to relay 1, and in the next $\frac{\tau_1}{2} T$ seconds, the relay uses the harvested energy to forward the signal to the HAP. Then, user terminal 2 uses the harvested energy to transmit to relay 2 for $\frac{\tau_2}{2} T$ seconds, and relay 2 forwards the signal to the HAP for the next $\frac{\tau_2}{2} T$ seconds, and so on.

Following the similar procedures, the achievable rate from the kth user terminal to the HAP is

$$
R_k = \frac{\tau_k}{2} \log_2 \left(1 + \frac{\frac{2\eta_s \tau_0}{\tau_k 2\sigma^2} |f_{sk}|^2 P_a |h_{sk}|^2 \frac{2\eta_r \tau_0}{\tau_k 2\sigma^2} P_a |h_{rk}|^2 |g_{rk}|^2}{\frac{2\eta_s \tau_0}{\tau_k 2\sigma^2} |f_{sk}|^2 P_a |h_{sk}|^2 + \frac{2\eta_r \tau_0}{\tau_k 2\sigma^2} P_a |h_{rk}|^2 |g_{rk}|^2 + 1} \right)
\tag{8.90}
$$

where $k = 1, 2, \cdots, N$, and f_{sk}, h_{sk}, h_{rk}, and g_{rk} are the channel gains from the kth source to the kth relay, from the HAP to the kth source, from the HAP to the kth relay, and from the kth relay to the HAP, respectively. Thus, one has the sum rate

$$R_t = \sum_{k=1}^{N} \frac{\tau_k}{2} \log_2 \left(1 + \frac{\frac{2\eta_s \tau_0}{\tau_k 2\sigma^2} |f_{sk}|^2 P_a |h_{sk}|^2 \frac{2\eta_r \tau_0}{\tau_k 2\sigma^2} P_a |h_{rk}|^2 |g_{rk}|^2}{\frac{2\eta_s \tau_0}{\tau_k 2\sigma^2} |f_{sk}|^2 P_a |h_{sk}|^2 + \frac{2\eta_r \tau_0}{\tau_k 2\sigma^2} P_a |h_{rk}|^2 |g_{rk}|^2 + 1} \right) \tag{8.91}$$

which can be optimized with respect to τ_0 and τ_k, $k = 1, 2, \cdots, N$, subject to $\sum_{k=0}^{N} \tau_k = 1$ and $\tau_k \geq 0$. This is very similar to the optimization problem in Chapter 6, except that we use the end-to-end SNR of the kth link now.

In Ramezani and Jamalipour (2017) the system was simplified by considering energy harvesting at the relay only. In this case,

$$R_t = \sum_{k=1}^{N} \frac{\tau_k}{2} \log_2 \left(1 + \frac{\frac{P_s}{\sigma^2} |f_{sk}|^2 \frac{2\eta_r \tau_0}{\tau_k 2\sigma^2} P_a |h_{rk}|^2 |g_{rk}|^2}{\frac{P_s}{2\sigma^2} |f_{sk}|^2 + \frac{2\eta_r \tau_0}{\tau_k 2\sigma^2} P_a |h_{rk}|^2 |g_{rk}|^2 + 1} \right). \tag{8.92}$$

The closed-form expressions for the optimum values of τ_k were obtained in Ramezani and Jamalipour (2017) as

$$\tau_0^{opt} = \frac{1}{1 + \sum_{i=1}^{N} \frac{1}{a_i}} \tag{8.93}$$

$$\tau_k^{opt} = \frac{1}{a_k \left(1 + \sum_{i=1}^{N} \frac{1}{a_i} \right)}, \quad k = 1, 2, \cdots, N \tag{8.94}$$

where a_k is the solution to

$$\log \left(1 + \frac{A_k a_k}{B_k a_k + C_k} \right) - \frac{A_k C_k a_k}{(B_k a_k + C_k)(A_k a_k + B_k a_k + C_k)}$$
$$= \sum_{i=1}^{N} \frac{A_i C_i}{(B_i a_i + C_i)(A_i a_i + B_i a_i + C_i)} \tag{8.95}$$

and $A_k = \frac{P_s}{\sigma^2} |f_{sk}|^2 \frac{2\eta_r}{2\sigma^2} P_a |h_{rk}|^2 |g_{rk}|^2$, $B_k = \frac{2\eta_r}{2\sigma^2} P_a |h_{rk}|^2 |g_{rk}|^2$, and $C_k = \frac{P_s}{2\sigma^2} |f_{sk}|^2 + 1$, for $k = 1, 2, \cdots, N$. The optimization in (8.91) may be solved in a similar way by using the Lagrange multiplier. It can lead to an equation similar to (8.95) but will be more complicated. Figure 8.12 compares (8.91) and (8.92) for the case of two users, when $P_a = P_s = 10$ dB and all the channel gains have been set to 1 with $\eta_r = \eta_s = 0.5$. From this figure, one can clearly see a global maximum of R_t for certain values of τ_1 and τ_2. Also, (8.92) has a slightly higher rate than (8.91).

Other optimization problems can also be formulated. For example, instead of having a one-to-one pair for the relays and the sources, one source can use all relays to have diversity gain. Also, direct links can be added to further complicate the end-to-end SNR. Finally, one can assume that the source receives energy from the HAP while the relay does not.

8.5.3 One User and Multiple Antennas

Next, consider the case when multiple antennas are used. In this case, the HAP is equipped with K antennas, while the user terminal and the relay has a single antenna.

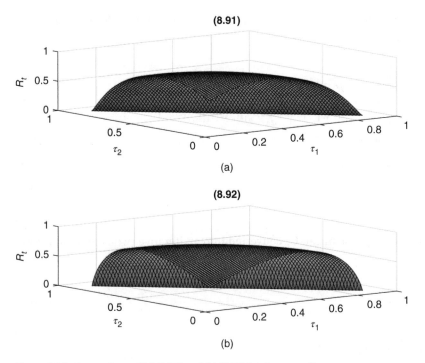

(8.91)

(a)

(8.92)

(b)

Figure 8.12 Comparison of (a) (8.91) and (b) (8.92) for the case of two users.

The system diagram is similar to Figure 8.9, except that there are multiple channels between the HAP and other nodes.

Similarly, the energy harvested at the relay and the energy harvested at the source can be given by

$$E_r = \eta_r P_a |\mathbf{h}_\mathbf{r}^T \mathbf{w}|^2 \tau \tag{8.96}$$

and

$$E_s = \eta_s P_a |\mathbf{h}_\mathbf{s}^T \mathbf{w}|^2 \tau \tag{8.97}$$

where $\mathbf{h}_\mathbf{r}$ is the $K \times 1$ channel vector from the HAP to the relay, $\mathbf{h}_\mathbf{s}$ is the $K \times 1$ channel vector from the HAP to the source, $\mathbf{w}$ is the energy beamforming vector at the HAP, and other symbols are defined as before. Using the harvested energy, the transmission power at the relay and the source can be derived as

$$P_r = \frac{E_r}{(1-\tau)/2} = \frac{2\eta_r \tau}{1-\tau} P_a |\mathbf{h}_\mathbf{r}^T \mathbf{w}|^2 \tag{8.98}$$

and

$$P_s = \frac{E_s}{(1-\tau)/2} = \frac{2\eta_s \tau}{1-\tau} P_a |\mathbf{h}_\mathbf{s}^T \mathbf{w}|^2 \tag{8.99}$$

respectively, where the source transmits the signal for the first $\frac{(1-\tau)}{2}$ seconds and the relay forwards the signal for the next $\frac{(1-\tau)}{2}$ seconds.

Assume that the channel gain from the source to the relay is still f_s but the $K \times 1$ channel vector from the relay to the HAP becomes $\mathbf{g}_\mathbf{r}$. Using these assumptions and following a similar procedure to Section 8.2.3, the end-to-end SNR and then the achievable rate can be derived as

$$\gamma = \frac{\gamma_r \gamma_d}{\gamma_r + \gamma_d + 1}$$

$$= \frac{\frac{2\eta_s \tau}{(1-\tau)2\sigma^2}|f_s|^2 P_a |\mathbf{h_s}^T \mathbf{w}|^2 \frac{2\eta_r \tau}{(1-\tau)2\sigma^2} P_a |\mathbf{h_r}^T \mathbf{w}|^2 ||\mathbf{g_r}||^2}{\frac{2\eta_s \tau}{(1-\tau)2\sigma^2}|f_s|^2 P_a |\mathbf{h_s}^T \mathbf{w}|^2 + \frac{2\eta_r \tau}{(1-\tau)2\sigma^2} P_a |\mathbf{h_r}^T \mathbf{w}|^2 ||\mathbf{g_r}||^2 + 1}$$

(8.100)

$$R = \frac{1-\tau}{2}\log_2(1+\gamma)$$

(8.101)

respectively. Hence, the optimization problem becomes

$$\max_{\tau,\mathbf{w}}\{R\}$$

(8.102)

$$0 \leq \tau \leq 1, ||\mathbf{w}||^2 = 1.$$

(8.103)

This optimization problem can be converted into an iterative two-step optimization as

$$\max_{\mathbf{w}}\{\gamma\}$$

$$||\mathbf{w}||^2 = 1$$

(8.104)

and then

$$\max_{\tau}\left\{\frac{1-\tau}{2}\log_2(1+\gamma)\right\}$$

$$0 \leq \tau \leq 1.$$

(8.105)

The optimization problem can also be simplified by assuming that the source or the relay does not harvest energy. Instead of AF, one can also use DF. In this case, the end-to-end SNR becomes

$$\gamma = \min\left\{\frac{2\eta_s \tau}{(1-\tau)2\sigma^2}|f_s|^2 P_a |\mathbf{h_s}^T \mathbf{w}|^2, \frac{2\eta_r \tau}{(1-\tau)2\sigma^2} P_a |\mathbf{h_r}^T \mathbf{w}|^2 ||\mathbf{g_r}||^2\right\}$$

(8.106)

and the achievable rate to be optimized becomes

$$R = \frac{1-\tau}{2}\log_2\left[1 + \frac{2\tau}{(1-\tau)2\sigma^2}\min\{\eta_s |f_s|^2 P_a |\mathbf{h_s}^T \mathbf{w}|^2, \eta_r P_a |\mathbf{h_r}^T \mathbf{w}|^2 ||\mathbf{g_r}||^2\}\right].$$

(8.107)

The optimization of (8.107) with respect to τ and $\mathbf{w}$ has closed-form expressions for the optimum values of τ and $\mathbf{w}$. The derivation for DF can be found in Zhong et al. (2015a).

One can also extend the above results to the case when multiple user terminals and multiple antennas are used. The derivation is similar. In summary, when the source and the relay harvest energy from a dedicated power transmitter, the optimization is performed for the achievable rate with respect to the time allocation and other relevant parameters. The energy causality and the signal causality in this case are not the main concerns. Next, we consider another important energy harvesting relaying system where the relay harvests energy from the source for forwarding.

8.6 From the Source

Figure 8.13 shows a diagram and the time structure of the energy harvesting system where the relay harvests energy from the source and the source has fixed power supply.

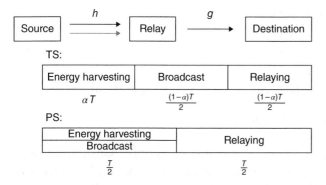

Figure 8.13 The relay harvests energy from the source.

In this system, both the source and the destination have fixed power supply but the relay either does not have enough energy or is not willing to use its own energy. For such a system, the source must provide the wireless power required for relaying and the relay can harvest this wireless power for forwarding. Next, we discuss AF relaying, followed by DF relaying.

8.6.1 Amplify-and-Forward Relaying

There are two energy harvesting methods, time switching (TS) and power splitting (PS), as discussed in Chapter 6 for SWIPT. If TS is used for AF relaying, the source first transmits the signal for αT seconds and the relay harvests energy during this period of time, where $0 \leq \alpha \leq 1$. Then, the source transmits the information for $\frac{1-\alpha}{2} T$ seconds to the relay, and the relay uses all the harvested energy to forward this signal to the destination for $\frac{1-\alpha}{2} T$ seconds. All nodes have a single antenna and operate in the half-duplex mode. One can see that if α increases, more energy can be harvested by the relay and hence, the transmission power of the relay increases for a larger achievable rate. On the other hand, since T is fixed, when α increases, the time used for information transmission will decrease to reduce the achievable rate. Thus, an optimum value of α exists.

Similar to (8.81), the amount of harvested energy is given by

$$E_r = \eta P_s |h|^2 \alpha \tag{8.108}$$

where η is the conversion efficiency of the harvester, P_s is the transmission power of the source, h is the channel gain between the source and the relay, and α is the harvesting time with $T = 1$ for simplicity. If this harvested energy is used for signal forwarding, the transmission power of the relay is given by

$$P_r = \frac{E_r}{(1 - \alpha)/2} = \frac{2\eta\alpha}{1 - \alpha} P_s |h|^2. \tag{8.109}$$

Thus, one has

$$\gamma_r = \frac{P_s |h|^2}{2\sigma^2} \tag{8.110}$$

and

$$\gamma_d = \frac{P_r |g|^2}{2\sigma^2} = \frac{2\eta\alpha}{2\sigma^2(1 - \alpha)} P_s |h|^2 |g|^2 \tag{8.111}$$

where g is the channel gain from the relay to the destination.

For AF, the end-to-end SNR is then derived as

$$\gamma = \frac{\frac{P_s|h|^2}{2\sigma^2} \frac{2\eta\alpha}{2\sigma^2(1-\alpha)} P_s|h|^2|g|^2}{\frac{2\eta\alpha}{2\sigma^2(1-\alpha)} P_s|h|^2|g|^2 + \frac{P_s|h|^2}{2\sigma^2} + 1}. \tag{8.112}$$

Using (8.112), the outage probability can be derived as (Nasir et al. 2013)

$$P_o(\gamma_0) = 1 - \frac{1}{e_h} \int_{d\gamma_0/c}^{\infty} e^{-\frac{x}{e_h} - \frac{a\gamma_0 x + b\gamma_0}{(cx^2 - d\gamma_0 x)e_g}} dx \tag{8.113}$$

where Rayleigh fading channels are assumed so that $|h|^2$ and $|g|^2$ follow exponential distributions with parameters e_h and e_g, respectively, $a = 2\sigma^2 P_s(1-\alpha)$, $b = 4\sigma^4(1-\alpha)$, $c = 2\eta P_s^2\alpha$, and $d = 4\eta P_s \sigma^2\alpha$. Thus, in delay-constrained applications, the information rate is

$$R = \frac{1-\alpha}{2} R_0[1 - P_o(2^{2R_0} - 1)] \tag{8.114}$$

where R_0 is the minimum rate that needs to be achieved and γ_0 has been replaced by $2^{2R_0} - 1$.

Also, using (8.112), the ergodic capacity can be derived for delay-tolerant applications as (Nasir et al. 2013)

$$C = \frac{1-\alpha}{2} \int_0^{\infty} \int_{dx/c}^{\infty} \frac{(axy + bx)cy^2}{(cy^2 - dxy)^2 e_g e_h x} e^{-\frac{y}{e_h} - \frac{axy+bx}{(cy^2-dxy)e_g}} \log_2(1+x) dy dx \tag{8.115}$$

by averaging the instantaneous rate over the random variables $|h|^2$ and $|g|^2$. Figure 8.14 shows R and C versus α for TS AF, where $P_s = 10$ dB, $e_h = e_g = 2\sigma^2 = 1$, and $R_0 = 2$. One sees that there is an optimum value of α for both R and C. These optimum values are different for R and C, showing that the choice of performance measure is relevant in optimum designs. Details of the derivations regarding TS and AF can be found in Section 6.3.2.1 and Section 8.2, respectively.

If PS is used, the source transmits the signal to the relay during the first $\frac{T}{2}$ seconds, and the relay forwards the signal to the destination in the second $\frac{T}{2}$ seconds. However, before the signal forwarding, the relay splits the received signal from the source into two parts: ρ part of the power is used for energy harvesting; and $(1 - \rho)$ part of the power is used for information forwarding.

In this case, the amount of harvested energy is

$$E_r = \frac{1}{2}\eta\rho P_s|h|^2 \tag{8.116}$$

where ρ is the PS factor and $0 \le \rho \le 1$. Since the relaying time is $\frac{1}{2}$ (again $T = 1$ has been used for simplicity.), the transmission power of the relay is

$$P_r = \frac{E_r}{1/2} = \eta\rho P_s|h|^2. \tag{8.117}$$

It turns out that the hop SNRs are given by

$$\gamma_r = \frac{(1-\rho)P_s|h|^2}{(1-\rho)2\sigma_a^2 + 2\sigma_d^2} \tag{8.118}$$

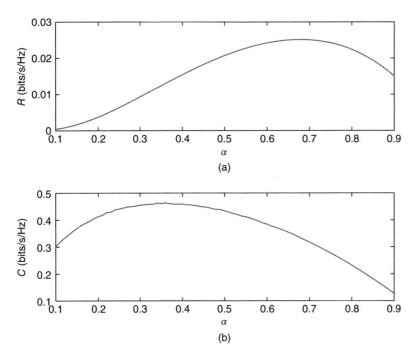

Figure 8.14 (a) R and (b) C versus α for TS AF.

and

$$\gamma_d = \frac{P_r|g|^2}{2\sigma^2} = \frac{\eta\rho}{2\sigma^2}P_s|h|^2|g|^2 \tag{8.119}$$

where g is the channel gain from the relay to the destination, $2\sigma_a^2$ is the noise power at the RF part, $2\sigma_d^2$ is the noise power at the baseband, and $2\sigma^2 = 2\sigma_a^2 + 2\sigma_d^2$. Section 6.3.2.2 explains how PS works and the associated noise power.

It can be shown that PS AF has the same outage probability as (8.113), except that $a = 2\sigma^2 P_s(1-\rho)$, $b = 2\sigma^2[(1-\rho)2\sigma_a^2 + 2\sigma_d^2]$, $c = \eta P_s^2\rho(1-\rho)$, and $d = [(1-\rho)2\sigma_a^2 + 2\sigma_d^2]\eta P_s\rho$. The rate and the ergodic capacity are given by

$$R = \frac{R_0}{2}[1 - P_o(2^{2R_0} - 1)] \tag{8.120}$$

and

$$C = \frac{1}{2}\int_0^\infty \int_{dx/c}^\infty \frac{(axy + bx)cy^2}{(cy^2 - dxy)^2 e_g e_h x} e^{-\frac{y}{e_h} - \frac{axy+bx}{(cy^2-dxy)e_g}} \log_2(1+x)\,dy\,dx \tag{8.121}$$

respectively. Figure 8.15 shows R and C versus ρ for PS AF, where $P_s = 10\,\text{dB}$, $e_h = e_g = 1$, $2\sigma_a^2 = 2\sigma_d^2 = 0.5$, and $R_0 = 2$. Again, there exists an optimum ρ for both R and C, and these optimum values are different. Compared with TS AF, the optimum ρ for R for PS is smaller than the optimum α for R, while the optimum ρ for C for PS is larger than the optimum α for C.

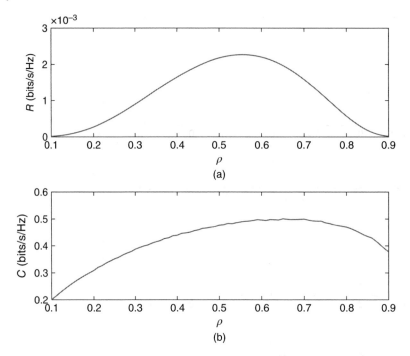

Figure 8.15 (a) R and (b) C versus ρ for PS AF.

8.6.2 Decode-and-Forward Relaying

In this section, we will discuss a DF relaying system where the relay node harvests energy from the source node to decode its received information before forwarding it to the destination node. In the discussion, the exact BER and throughput performances of DF relaying using PS energy harvesting are analyzed. Three different scenarios are considered: instantaneous transmission where the channel state information is fixed and known; delay- or error-constrained transmission where the source transmission rate or the error rate is restricted with a minimum requirement; and delay- or error-tolerant transmission where throughput or error rate is averaged over the channel state.

Again, a typical three-node system is considered. In this case, the source node transmits a signal to the relay node. The relay node splits the received signal into two parts. One part is used for energy harvesting and the other part is used for information decoding. The decoded information is then encoded again and forwarded to the destination by using the harvested energy. Assume a total relaying time of T, where $\frac{T}{2}$ is used in the broadcasting phase and $\frac{T}{2}$ is used in the relaying phase.

Using the above assumptions, the energy harvested at the relay node is

$$E_h = \eta \rho P_s |h|^2 \frac{T}{2} \tag{8.122}$$

and the transmission power of the relay is therefore

$$P_r = \frac{E_h}{T/2} = \eta \rho P_s |h|^2. \tag{8.123}$$

The symbols are defined the same as before. Also, for the later derivation, one has the PDF of the channel power as

$$f_{|h|^2}(x) = \frac{1}{\Omega_h} e^{-\frac{x}{\Omega_h}}, x > 0. \tag{8.124}$$

Thus, Rayleigh fading is assumed. The hop SNR for the source-to-relay link and the hop SNR for the relay-to-destination link can be derived as

$$\gamma_r = \frac{(1-\rho)P_s|h|^2}{2(1-\rho)\sigma_a^2 + 2\sigma_d^2} \tag{8.125}$$

$$\gamma_d = \frac{\eta\rho P_s|h|^2|g|^2}{2\sigma_a^2 + 2\sigma_d^2} \tag{8.126}$$

as before, where the fading power of the relay-to-destination link follows

$$f_{|g|^2}(x) = \frac{1}{\Omega_g} e^{-\frac{x}{\Omega_g}}, x > 0. \tag{8.127}$$

For later use, the CDFs of $|h|^2$ and $|g|^2$ are given by

$$F_{|h|^2}(y) = 1 - e^{\frac{y}{\Omega_h}}, y > 0 \tag{8.128}$$

and

$$F_{|g|^2}(y) = 1 - e^{\frac{y}{\Omega_g}}, y > 0, \tag{8.129}$$

respectively. Using the hop SNRs, the BERs for BPSK signals at the relay and at the destination, respectively, can be derived as

$$BER_r = \frac{1}{2}\text{erfc}(\sqrt{\gamma_r}) \tag{8.130}$$

and

$$BER_d = \frac{1}{2}\text{erfc}(\sqrt{\gamma_d}) \tag{8.131}$$

where $\text{erfc}(\cdot)$ is the complementary error function [Gradshteyn and Ryzhik 2000, eq. (8.250.4)]. Similarly, the throughput of the source-to-relay link and the relay-to-destination link can be derived as

$$C_r = \ln(1 + \gamma_r) \tag{8.132}$$

and

$$C_d = \ln(1 + \gamma_d), \tag{8.133}$$

respectively.

8.6.2.1 Instantaneous Transmission

In this scenario, the channel state information is known via channel estimation. The end-to-end BER of the whole relaying link from source to destination can be calculated as

$$BER = BER_r(1 - BER_d) + BER_d(1 - BER_r). \tag{8.134}$$

An error occurs only when either the source-to-relay link or the relay-to-destination link are erroneous but not both. Using (8.130) and (8.131), one has for BPSK

$$BER = \frac{1}{2}\text{erfc}\left(\sqrt{\frac{(1-\rho)P_s|h|^2}{2(1-\rho)\sigma_a^2 + 2\sigma_d^2}}\right) + \frac{1}{2}\text{erfc}\left(\sqrt{\frac{\eta\rho P_s|h|^2|g|^2}{2\sigma_a^2 + 2\sigma_d^2}}\right)$$

$$- \frac{1}{2}\text{erfc}\left(\sqrt{\frac{(1-\rho)P_s|h|^2}{2(1-\rho)\sigma_a^2 + 2\sigma_d^2}}\right)\text{erfc}\left(\sqrt{\frac{\eta\rho P_s|h|^2|g|^2}{2\sigma_a^2 + 2\sigma_d^2}}\right). \tag{8.135}$$

When the hop SNRs in the two links are reasonably large, the third term in (8.135) is very small. In this case, the first two terms are dominant. One sees that, when the value of ρ increases, the first term in (8.135) increases while the second term in (8.135) decreases. Thus, there exists an optimum value of ρ that minimizes the BER.

By taking the differentiation of (8.135) with respect to ρ and setting the derivative to zero, after some mathematical manipulations, the optimum value of ρ satisfies

$$e^{-\frac{\eta\rho_{opt}^B P_s|h|^2|g|^2}{2\sigma_a^2 + 2\sigma_d^2}} \frac{1}{\sqrt{\rho_{opt}^B}}\sqrt{\frac{\eta P_s|h|^2|g|^2}{2\sigma_a^2 + 2\sigma_d^2}}\left[1 - \text{erfc}\left[\sqrt{\frac{(1-\rho_{opt}^B)P_s|h|^2}{2(1-\rho_{opt}^B)\sigma_a^2 + 2\sigma_d^2}}\right]\right]$$

$$= e^{-\frac{(1-\rho_{opt}^B)P_s|h|^2}{2(1-\rho_{opt}^B)\sigma_a^2 + 2\sigma_d^2}} \frac{1}{\sqrt{1-\rho_{opt}^B}} \frac{\sqrt{P_s|h|^2}2\sigma_d^2}{[2(1-\rho_{opt}^B)\sigma_a^2 + 2\sigma_d^2]^{1.5}}$$

$$\times \left[1 - \text{erfc}\left(\sqrt{\frac{\eta\rho_{opt}^B P_s|h|^2|g|^2}{2\sigma_a^2 + 2\sigma_d^2}}\right)\right]. \tag{8.136}$$

This equation could be further simplified when the SNRs in the two links are large such that the two complementary error functions can be approximated as zero. However, due to the non-linear exponential functions in the equation, there is still no closed-form expression for the optimum value of ρ. Nevertheless, the single-variable equation can be numerically solved using standard mathematical software.

On the other hand, the end-to-end throughput of the DF relaying system can be derived as

$$C = \min\{C_r, C_d\} = \ln\left(1 + \min\left\{\frac{(1-\rho)P_s|h|^2}{2(1-\rho)\sigma_a^2 + 2\sigma_d^2}, \frac{\eta\rho P_s|h|^2|g|^2}{2\sigma_a^2 + 2\sigma_d^2}\right\}\right). \tag{8.137}$$

There also exists an optimum value of ρ. The optimum value of ρ can be derived as (Gao et al. 2017)

$$\rho_{opt}^C = \frac{(\sigma_a^2 + \sigma_d^2 + \eta|g|^2\sigma_d^2 + \eta|g|^2\sigma_a^2) - \sqrt{\Delta}}{2\eta|g|^2\sigma_a^2} \tag{8.138}$$

where $\Delta = [\eta|g|^2\sigma_a^2]^2 + (\eta|g|^2\sigma_a^2)(2\sigma_a^2 + 2\sigma_d^2 + 2\eta|g|^2\sigma_a^2) + (\sigma_a^2 + \sigma_d^2 - \eta|g|^2\sigma_a^2)^2$. One sees from (8.138) that the optimum value of ρ does not depend on $|h|^2$, the channel power of the source-to-relay link. Also, when $\frac{|g|^2}{2\sigma_a^2 + 2\sigma_d^2}$ is large and goes to infinity, ρ_{opt}^C approaches zero.

8.6.2.2 Delay- or Error-Constrained Transmission

In this scenario, the transmission rate of the source or the BER are restricted by a minimum requirement. Specifically, for the delay-constrained case, one has $C = R_0$ such that the throughput is constrained by

$$\hat{C} = \frac{R_0}{2}(1 - p_{out}^C) \tag{8.139}$$

where $\frac{1}{2}$ takes the throughput penalty of relaying into account and the outage probability is

$$p_{out}^C = Pr\{\gamma < \gamma_{th}^C\} \tag{8.140}$$

with

$$\gamma = \min\{\gamma_r, \gamma_d\} = \min\left\{\frac{(1 - \rho)P_s|h|^2}{2(1 - \rho)\sigma_a^2 + 2\sigma_d^2}, \frac{\eta\rho P_s|h|^2|g|^2}{2\sigma_a^2 + 2\sigma_d^2}\right\} \tag{8.141}$$

being the end-to-end SNR of the DF relaying system and $\gamma_{th}^C = 2^{R_0} - 1$ is the threshold of γ.

Similarly, for the error-constrained case, one has $BER = BER_r(1 - BER_d) + BER_d(1 - BER_r) = BER_0$. However, this form is not very convenient to use. We can define an equivalent end-to-end SNR as

$$BER = \frac{1}{2}\text{erfc}(\sqrt{\gamma_{eq}}) \tag{8.142}$$

where γ_{eq} is bounded as (Wang et al. 2007)

$$\gamma - 1.62 < \gamma_{eq} \leq \gamma \tag{8.143}$$

where γ is given by (8.141). When γ is large, the upper and lower bounds will converge such that $\gamma_{eq} \approx \gamma$. Thus, one has $\frac{1}{2}\text{erfc}(\sqrt{\gamma_{eq}}) < BER_0$ such that $\gamma_{eq} > \gamma_{th}^B$, where $\gamma_{th}^B = [\text{erfc}^{-1}(2BER_0)]^2$ and $\text{erfc}^{-1}(\cdot)$ is the inverse function of the complementary error function. We use the bit correct rate (BCR) to be constrained as

$$\hat{BCR} = (1 - BER_0)(1 - p_{out}^B) \tag{8.144}$$

where the outage probability in this case is given by

$$p_{out}^B = Pr\{\gamma < \gamma_{th}^B\} \tag{8.145}$$

when the upper bound of γ_{eq} is used or

$$p_{out}^B = Pr\{\gamma < \gamma_{th}^B + 1.62\} \tag{8.146}$$

when the lower bound of γ_{eq} is used.

Thus, the derivations of the constrained throughput and BCR boil down to the calculation of the CDF of γ. This CDF can be derived as (Gao et al. 2017)

$$F_\gamma(y) = F_{|g|^2}\left(\frac{\sigma_a^2 + \sigma_d^2}{(1 - \rho)\sigma_a^2 + \sigma_d^2}\frac{1 - \rho}{\eta\rho}\right) + F_{|h|^2}\left(\frac{2(1 - \rho)\sigma_a^2 + 2\sigma_d^2}{(1 - \rho)P_s}y\right)$$
$$-F_{|g|^2}\left(\frac{\sigma_a^2 + \sigma_d^2}{(1 - \rho)\sigma_a^2 + \sigma_d^2}\frac{1 - \rho}{\eta\rho}\right)F_{|h|^2}\left(\frac{2(1 - \rho)\sigma_a^2 + 2\sigma_d^2}{(1 - \rho)P_s}y\right)$$

$$-\frac{1}{\Omega_g}\int_0^{\frac{\sigma_a^2+\sigma_d^2}{(1-\rho)\sigma_a^2+\sigma_d^2}\frac{1-\rho}{\eta\rho}} e^{-\frac{1}{\Omega_2}t-\frac{(2\sigma_a^2+2\sigma_d^2)y}{\Omega_h\eta\rho P_s t}}\,dt \tag{8.147}$$

where $F_{|h|^2}(\cdot)$ and $F_{|g|^2}(\cdot)$ are the CDFs of $|h|^2$ and $|g|^2$ given before. The one-dimensional integral in (8.147) cannot be solved and has to be evaluated numerically. Using (8.147), one has

$$\hat{C} = \frac{R_0}{2}[1 - F_\gamma(\gamma_{th}^C)] \tag{8.148}$$

and

$$\begin{aligned}
B\hat{C}R &= (1 - BER_0)[1 - F_\gamma(\gamma_{th}^B)],\\
B\hat{C}R &= (1 - BER_0)[1 - F_\gamma(\gamma_{th}^B + 1.62)].
\end{aligned} \tag{8.149}$$

It is difficult to find an explicit expression for the optimum value of ρ from (8.148) and (8.149). They will be found numerically.

8.6.2.3 Delay- or Error-Tolerant Transmission

In this scenario, one has to calculate the ergodic capacity and the average BER by averaging them over the channel gains such that only the channel statistics are needed.

In the case of delay-tolerant transmission, the ergodic capacity can be calculated as

$$\bar{C} = \int_0^\infty \int_0^\infty \ln\left(1 + \min\left\{\frac{(1-\rho)P_s x}{2(1-\rho)\sigma_a^2 + 2\sigma_d^2}, \frac{\eta\rho P_s xy}{2\sigma_a^2 + 2\sigma_d^2}\right\}\right) f_{|h|^2}(x) f_{|g|^2}(y)\,dx\,dy\,z$$

by averaging the throughput over the fading gains. This integration can be derived as (Gao et al. 2017)

$$\begin{aligned}
\bar{C} = \frac{1}{\Omega_h}&\left[1 - F_{|g|^2}\left(\frac{\sigma_a^2 + \sigma_d^2}{(1-\rho)\sigma_a^2 + \sigma_d^2}\frac{1-\rho}{\eta\rho}\right)\right]W\left(\frac{(1-\rho)P_s}{2(1-\rho)\sigma_a^2 + 2\sigma_d^2}, \frac{1}{\Omega_h}\right)\\
&+\frac{1}{\Omega_h\Omega_g}\int_0^{\frac{\sigma_a^2+\sigma_d^2}{(1-\rho)\sigma_a^2+\sigma_d^2}\frac{1-\rho}{\eta\rho}} W\left(\frac{\eta\rho P_s y}{2\sigma_a^2 + 2\sigma_d^2}, \frac{1}{\Omega_h}\right)e^{-\frac{y}{\Omega_g}}\,dy
\end{aligned} \tag{8.150}$$

where $W(a, b) = \int_0^\infty \ln(1 + ax)e^{-bx}\,dx$.

In the case of error-tolerant transmission, the average BER can be calculated as

$$\begin{aligned}
B\bar{E}R = \int_0^\infty \int_0^\infty \Bigg[&\frac{1}{2}\mathrm{erfc}\left(\sqrt{\frac{(1-\rho)P_s x}{2(1-\rho)\sigma_a^2 + 2\sigma_d^2}}\right) + \frac{1}{2}\mathrm{erfc}\left(\sqrt{\frac{\eta\rho P_s xy}{2\sigma_a^2 + 2\sigma_d^2}}\right)\\
&-\frac{1}{2}\mathrm{erfc}\left(\sqrt{\frac{(1-\rho)P_s x}{2(1-\rho)\sigma_a^2 + 2\sigma_d^2}}\right)\mathrm{erfc}\left(\sqrt{\frac{\eta\rho P_s xy}{2\sigma_a^2 + 2\sigma_d^2}}\right)\Bigg] f_{|h|^2}(x) f_{|g|^2}(y)\,dx\,dy
\end{aligned}$$

This can be solved as (Gao et al. 2017)

$$\begin{aligned}
B\bar{E}R = \frac{1}{2\Omega_h}&U\left(\frac{(1-\rho)P_s}{2(1-\rho)\sigma_a^2 + 2\sigma_d^2}, \frac{1}{\Omega_h}\right) + \frac{1}{2\Omega_h\Omega_g}V\left(\frac{\eta\rho P_s}{2\sigma_a^2 + 2\sigma_d^2}, \frac{1}{\Omega_h}, 1, \frac{1}{\Omega_g}\right)\\
&-\frac{1}{2\Omega_h\Omega_g}\int_0^\infty \mathrm{erfc}\left(\sqrt{\frac{(1-\rho)P_s x}{2(1-\rho)\sigma_a^2 + 2\sigma_d^2}}\right)e^{-\frac{1}{\Omega_h}x}U\left(\frac{\eta\rho P_s x}{2\sigma_a^2 + 2\sigma_d^2}, \frac{1}{\Omega_g}\right)\,dx
\end{aligned}$$

where

$$U(a,b) = \frac{1}{b} - \frac{2\sqrt{a}\Gamma(1.5)}{\sqrt{\pi}(\sqrt{b})^3} {}_2F_1(0.5, 1.5; 1.5; -\frac{a}{b}) \tag{8.151}$$

and

$$V(a,b,c) = \frac{1}{bc} - \frac{2\sqrt{a/\pi}}{b^{1.5}c^{1.5}} \frac{E(0.5, 1.5; 1.5, 1.5; \frac{bc}{a})}{\Gamma(0.5)} \tag{8.152}$$

with ${}_2F_1(\cdot, \cdot; \cdot; \cdot)$ being the hypergeometric function and $E(\cdot, \cdot; \cdot, \cdot; \cdot)$ is the MacRobert's E function (Gradshteyn and Ryzhik 2000).

8.6.2.4 Numerical Examples

We show some numerical examples for the expressions derived. Without loss of generality, in the examples, $P_s = 1$, $2\sigma_a^2 = 2\sigma_d^2 = 1$, while $|h|^2$ and $|g|^2$ in the instantaneous transmission change with $\beta_h = \frac{|h|^2}{2\sigma_a^2 + 2\sigma_d^2}$ and $\beta_g = \frac{|g|^2}{2\sigma_a^2 + 2\sigma_d^2}$ and Ω_h and Ω_g in the delay- and error-tolerant or -constrained transmissions change with $\beta_h = \frac{\Omega_h}{2\sigma_a^2 + 2\sigma_d^2}$ and $\beta_g = \frac{\Omega_g}{2\sigma_a^2 + 2\sigma_d^2}$. The value of β_h and β_g can be considered as the quality indicators of the source-to-relay and relay-to-destination links and for fixed noise variances are proportional to the fading power or the average fading power of the links.

Figures 8.16–8.19 show the performances of instantaneous transmission using DF relaying with PS. In particular, Figure 8.16 shows the throughput versus ρ for different values of β_g and η. Several observations can be made. First, there does exist an optimum

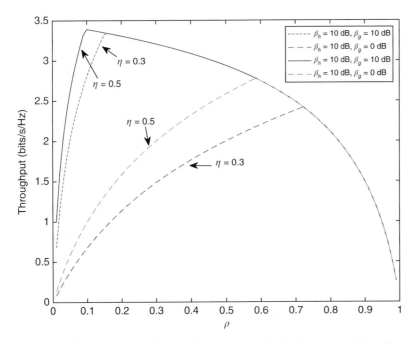

Figure 8.16 Throughput versus ρ for different values of η, β_h and β_g using instantaneous transmission.

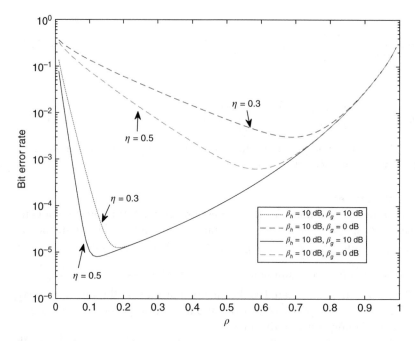

Figure 8.17 BER versus ρ for different values of η, β_h and β_g using instantaneous transmission.

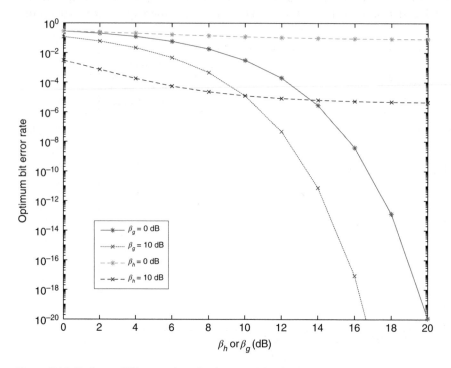

Figure 8.18 Optimum BER versus β_h or β_g when $\eta = 0.3$ using instantaneous transmission.

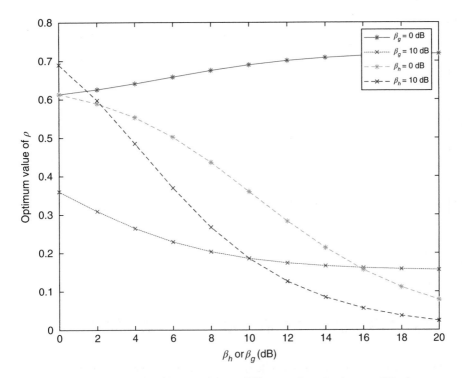

Figure 8.19 The optimum ρ achieving minimum BER versus β_h or β_g when $\eta = 0.3$ using instantaneous transmission.

value of ρ that maximizes the throughput, in all the cases considered. The optimum value of ρ lies at the only discontinuity of the throughput. This optimum value can also be calculated using (8.138). Secondly, when the value of η increases, the throughput increases and the optimum value of ρ decreases, but a large part of the throughput curves overlap with each other on the right side of the peak, indicating that the throughput performance is not sensitive to the value of η in this part. Thirdly, when β_g increases, the throughput increases. Specifically, when β_g increases from 0 to 10 dB, the throughput increases from around 2.4 to 3.3 bits/s/Hz. Figure 8.17 shows the BER versus ρ for different values of β_g and η. In this case, there exists an optimum value of ρ that minimizes the BER as well. The BER decreases when η or β_g increases, as expected. Again, on the right side of the peak, there is a large part of the BER curves that overlap with each other for different η, showing that the BER is not sensitive to η in this part either.

Figure 8.18 shows the optimum minimum BER using the optimized ρ for different values of β_h and β_g. When β_h is fixed in the legend, the x axis corresponds to β_g, and when β_g is fixed in the legend, the x axis corresponds to β_h. In this case, one sees that the optimized BER decreases significantly with β_h, when β_g is fixed to 0 and 10 dB, while the optimized BER remains almost the same when β_g increases and β_h is fixed to 0 and 10 dB. Thus, the BER is more sensitive to β_h. Figure 8.19 shows the optimized ρ used to calculate the minimum BER in Figure 8.18. When β_g is fixed to 0 or 10 dB, the optimum ρ only changes slightly with β_h. When β_h is fixed to 0 or 10 dB, the optimum ρ decreases with an increase of β_g, as less power needs to be harvested when the channel condition of the relay-to-destination link improves, under the same other conditions.

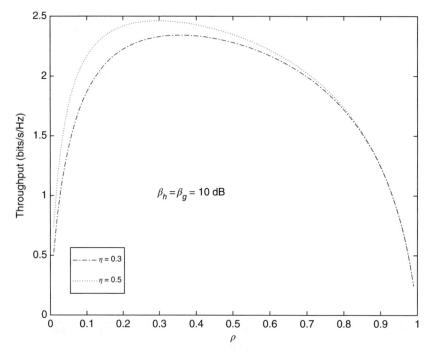

Figure 8.20 Throughput versus ρ for different values of η using delay-tolerant transmission.

Figures 8.20 and 8.21 show the performances of delay- and error-tolerant transmissions where the ergodic capacity or average BER are considered, respectively. Again, an optimum value of ρ exists that either maximizes the throughput or minimizes the BER. The value of η changes the performance slightly. In this case, the BER is less sensitive to η than the throughput.

Figures 8.22 and 8.23 show the performances of delay- or error-constrained transmissions. One sees that there exists an optimum value of ρ that maximizes the throughput or the BCR. The throughput increases quickly with ρ, when ρ is small, and decreases slowly with ρ, when ρ keeps increasing. The right sides of the peak of the throughput curves largely overlap with each other for different values of η. This shows that the throughput is not sensitive to η in delay-constrained transmission. Figure 8.23 shows the BCR versus ρ for different values of η. In general, the lower bound leads to higher BCR but the optimum value of ρ is approximately the same for both bounds.

In Ju et al. (2015), an initial amount of energy was added at the relay to derive the performances. The direct link was also considered. In Gao et al. (2017), the results were extended to Nakagami-m fading channels. One can obtain the above results by setting $m = 1$ in Gao et al. (2017) or by removing the direct link and setting the initial energy to zero in Ju et al. (2015). Recently, light energy harvesting was also studied, where the first hop uses visible light and the relay harvests the direct current from the light for forwarding signal in the second hop using RF (Rakia et al. 2016). In summary, these studies focus on the optimization of achievable rate with respect to the TS coefficient or the PS factor. In some cases, power allocation can be jointly considered. Since the energy is supplied by the source, the energy causality at the relay is not a constraint any more.

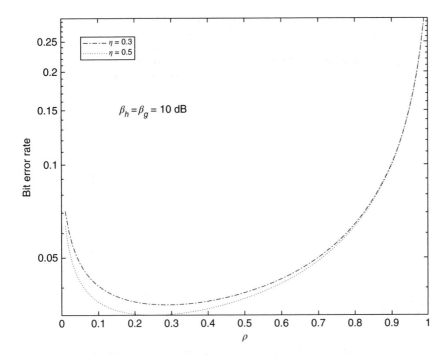

Figure 8.21 BER versus ρ for different values of η using error-tolerant transmission.

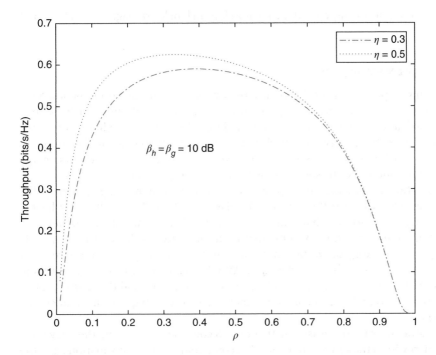

Figure 8.22 Throughput versus ρ for different values of η using delay-constrained transmission.

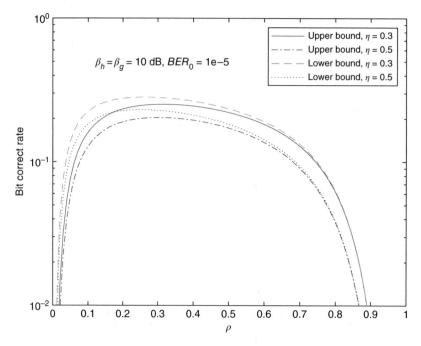

Figure 8.23 BCR versus ρ for different values of η using error-constrained transmission.

Also, since these studies only consider transmissions in blocks, the signal causality is not a concern either.

8.6.3 Energy Harvesting Source

An interesting improvement of the energy relaying system discussed in Section 8.6.1 and Section 8.6.2 is that, when the relay forwards the signal to the destination, the source node can also harvest energy from this transmission, as the wireless medium is of broadcast nature. The extra energy harvested by the source node may allow the source to perform more or better transmissions. In this system, the relay node harvests energy from the source node, and the source node harvests energy from the relay node.

This subsection studies such an improved energy harvesting relaying system. In the study, we consider a typical three-node AF relaying system. Again, all nodes have a single antenna and work in a half-duplex mode, and there is no direct link between the source and the destination. Assume that both the source node and the relay node are equipped with energy harvesters and thus, can harvest energies from the relevant radio-frequency (RF) signals. Consider static AWGN channels. Also, assume that there are E_t joules of total energy initially available at the source node and that one complete relaying transmission takes T seconds that includes broadcasting, relaying and energy harvesting phases.

If the TS strategy is used, we assume that the TS coefficient is α. Then, the relay node harvests energy from the source node for αT seconds and receives information from the source node for $\frac{1-\alpha}{2} T$ seconds in the broadcasting phase. The received signal at the relay can be given by

$$y_r = \sqrt{P_s} \frac{h}{\sqrt{d_{sr}^v}} s + n_{ra} + n_{rc} \tag{8.153}$$

where P_s is the transmission power of the source, h is the fixed channel gain of the source-to-relay link, d_{sr} is the distance between source and relay, v is the path loss exponent, $s = \pm 1$ is the transmitted symbol, n_{ra} is the AWGN incurred at the RF front as a Gaussian random variable with mean zero and variance $2\sigma_a^2$, and n_{rc} is the AWGN incurred in the RF-to-baseband conversion as a Gaussian random variable with mean zero and variance $2\sigma_d^2$. Thus, the harvested energy at the relay can be shown as

$$E_{hr} = \eta P_s \frac{h^2}{d_{sr}^v} \alpha T \tag{8.154}$$

where η is the conversion efficiency of the energy harvester and all other symbols are defined as before.

In the relaying phase, the relay node transmits y_r to the destination for $\frac{1-\alpha}{2} T$ seconds such that the received signal at the destination is

$$y_d = \frac{g}{\sqrt{d_{rd}^v}} \sqrt{P_r} by_r + n_{da} + n_{dc} \tag{8.155}$$

where g is the fixed channel gain of the relay-to-destination link, d_{rd} is the distance between relay and destination, $P_r = \frac{E_{hr}}{\frac{1-\alpha}{2} T} = \frac{2\alpha \eta}{1-\alpha} P_s \frac{h^2}{d_{sr}^v}$ is the transmission power of the relay, by_r is the normalized transmitted signal, normalized with respect to the average power of y_r, $b = \frac{1}{\sqrt{P_s \frac{h^2}{d_{sr}^v} + 2\sigma_a^2 + 2\sigma_d^2}}$ is the amplification factor, n_{da} is the AWGN at the destination incurred by the RF front as a Gaussian random variable with mean zero and variance $2\sigma_a^2$, and n_{dc} is the AWGN at the destination from RF-to-baseband conversion with mean zero and variance $2\sigma_d^2$.

Unlike the conventional relaying protocol, in the new protocol, the source node also harvests energy from the transmitted signal of the relay node during the broadcasting phase so that the received signal at the source node during relaying is

$$y_s = \frac{h}{\sqrt{d_{sr}^v}} \sqrt{P_r} by_r + n_{sa} \tag{8.156}$$

where the channel gain h and the distance d_{sr} are used due to channel reciprocity and n_{sa} is the AWGN at the source. Further, the received signal at the source node can be expanded as

$$y_s = \frac{\sqrt{P_r P_s} bh^2}{d_{sr}^v} s + \frac{\sqrt{P_r} bh}{\sqrt{d_{sr}^v}} (n_{ra} + n_{rc}) + n_{sa}. \tag{8.157}$$

In this case, the harvested energy at the source is

$$E_{hs} = \eta \frac{P_r P_s b^2 h^4}{d_{sr}^{2v}} \cdot \frac{(1 - \alpha) T}{2}. \tag{8.158}$$

Using $b = \frac{1}{\sqrt{P_s \frac{h^2}{d_{sr}^v} + 2\sigma_a^2 + 2\sigma_d^2}}$ and $P_r = \frac{E_{hr}}{\frac{1-\alpha}{2}T} = \frac{2\alpha\eta}{1-\alpha}P_s\frac{h^2}{d_{sr}^v}$ one has

$$E_{hs} = \eta^2 \alpha \frac{P_s^2 (h^2/d_{sr}^v)^3 T}{P_s h^2/d_{sr}^v + 2\sigma_a^2 + 2\sigma_d^2}. \tag{8.159}$$

For the conventional relaying protocol using TS, without energy harvesting at the source node, the source node transmits the signal for a duration of $\alpha T + \frac{1-\alpha}{2}T$ with a transmission power of P_s, where the first part is the energy transfer time and the second part is the information delivery time from the source to the relay. Thus, each transmission costs the source node an energy of

$$E_i = \left(\alpha T + \frac{1-\alpha}{2}T\right)P_s. \tag{8.160}$$

Using the total energy, the total number of transmissions the source can make in the conventional protocol using TS is

$$K_{TS}^{Con} = \frac{E_t}{E_i} = \frac{E_t}{P_s T}\frac{2}{1+\alpha}. \tag{8.161}$$

In the case of TS, the end-to-end SNR at the destination can be derived from its received signal in (8.155) as

$$\gamma_{TS} = \frac{P_s \gamma_d \gamma_r b^2}{\gamma_d b^2 + \frac{(1-\alpha)d_{sr}^v}{2\alpha\eta P_s h^2 (2\sigma_a^2 + 2\sigma_d^2)}} \tag{8.162}$$

where $\gamma_d = \frac{g^2}{d_{rd}^v(2\sigma_a^2 + 2\sigma_d^2)}$ and $\gamma_r = \frac{h^2}{d_{sr}^v(2\sigma_a^2 + 2\sigma_d^2)}$. Thus, the overall throughput in all transmissions using an initial energy of E_t at the source node is derived as

$$C_{TS}^{Con} = K_{TS}^{Cov} \cdot \log_2(1 + \gamma_{TS})\frac{1-\alpha}{2}. \tag{8.163}$$

For the new protocol, each relaying transmission generates an additional energy of E_{hs} at the source, which could be used for later transmissions. Two strategies are considered. In the first strategy, all the harvested energies at the source node will be stored until the K_{TS}^{Cov} transmissions are finished. Then, they will be used to make more transmissions. In this case, one has the new total number of relay transmissions as

$$K_{TS}^{New} = \left[\frac{E_t/(P_s T)}{\frac{1+\alpha}{2} - \eta^2\alpha\frac{P_s^2(h^2/d_{sr}^v)^3 T}{P_s h^2/d_{sr}^v + 2\sigma_a^2 + 2\sigma_d^2}}\right] \tag{8.164}$$

where $[\cdot]$ is the rounding function and $\frac{1+\alpha}{2} > \eta^2\alpha\frac{P_s^2(h^2/d_{sr}^v)^3 T}{P_s h^2/d_{sr}^v + 2\sigma_a^2 + 2\sigma_d^2}$. Thus, one has the total throughput in the first strategy as (Chen et al. 2017c)

$$C_{TS}^{New1} = K_{TS}^{New} \cdot \log_2(1 + \gamma_{TS})\frac{1-\alpha}{2}. \tag{8.165}$$

One sees that $C_{TS}^{New1} > C_{TS}^{Con}$, as $K_{TS}^{New} > K_{TS}^{Cov}$. Thus, the new protocol has throughput gain over the conventional protocol and this gain comes from using the extra energy harvested at the source node to make more transmissions. More specifically, the gain can be calculated as

$$Gain_1 = \frac{C_{TS}^{New1} - C_{TS}^{Con}}{C_{TS}^{Con}} = \frac{1}{1 - \eta^2 \frac{2\alpha}{1-\alpha} P_s \left(\frac{h^2}{d_{sr}^v}\right)^3 b^2} - 1 \tag{8.166}$$

which depends on P_s, η, $\frac{h^2}{d_{sr}^v}$, α, and b.

In the second strategy, instead of storing all harvested energy until the K_{TS}^{Cov} transmissions are finished, the harvested energy will be used immediately in the next transmission to increase its transmission power. This strategy has the advantages of requiring smaller energy storage at the source node as well as improving the quality of each relay transmission. In particular, one has the following iterative relationships (Chen et al. 2017c)

$$P_s^{(i+1)} = \frac{E_t/K_{TS}^{Con} + E_{hs}^{(i)}}{T} \frac{2}{1+\alpha}$$

$$\gamma_{TS}^{i+1} = \frac{P_s^{(i+1)} \gamma_d \gamma_r}{\gamma_d + \frac{(1-\alpha)d_{sr}^v (P_s^{(i+1)} \frac{h^2}{d_{sr}^v} + 2\sigma_a^2 + 2\sigma_d^2)}{2\alpha\eta P_s^{(i+1)} h^2 (2\sigma_a^2 + 2\sigma_d^2)}}$$

$$E_{hs}^{(i+1)} = \eta^2 \alpha \frac{(P_s^{(i)})^2 (h^2/d_{sr}^v)^3 T}{P_s^{(i)} h^2/d_{sr}^v + 2\sigma_a^2 + 2\sigma_d^2}. \tag{8.167}$$

where $i = 1, 2, \cdots, K_{TS}^{Con}$, $E_{hs}^1 = 0$, and $P_s^1 = P_s$. In this case, the total throughput is derived as (Chen et al. 2017c)

$$C_{TS}^{New2} = \sum_{i=1}^{K_{TS}^{Con}} \log_2(1 + \gamma_{TS}^{(i)}) \frac{1-\alpha}{2}. \tag{8.168}$$

There is no direction calculation of the gain in this strategy.

If the PS strategy is used, we assume that ρ is the PS factor. In this case, the source transmits the signal to the relay for $\frac{T}{2}$ seconds. Part of this signal is received at the relay for information delivery as

$$y_r = \sqrt{(1-\rho)P_s} \frac{h}{\sqrt{d_{sr}^v}} s + \sqrt{1-\rho} n_{ra} + n_{rc} \tag{8.169}$$

and part of this signal is harvested by the relay as

$$E_{hr} = \eta\rho P_s \frac{h^2}{d_{sr}^v} \frac{T}{2}. \tag{8.170}$$

In the relaying phase, the relay uses the harvested energy to forward the signal to the destination such that the received signal at the destination is

$$y_d = \frac{g}{\sqrt{d_{rd}^v}} \sqrt{P_r} by_r + n_{da} + n_{dc} \tag{8.171}$$

where $P_r = \frac{E_{hr}}{T/2} = \eta\rho P_s \frac{h^2}{d_{sr}^v}$ in this case and all other symbols are defined as before.

Also, unlike the conventional relaying protocol, in the relaying phase of the new system, the source node also harvests energy from the relay transmission with a received signal of

$$y_s = \frac{h}{\sqrt{d_{sr}^v}} \sqrt{P_r} by_r + n_{sa} \tag{8.172}$$

and the harvested energy is obtained as

$$E_{hs} = \eta^2 \rho (1 - \rho) \frac{P_s^2 (h^2/d_{sr}^v)^3 T/2}{P_s h^2/d_{sr}^v + 2\sigma_a^2 + 2\sigma_d^2}. \tag{8.173}$$

For PS, in the conventional protocol, each relay transmission costs the source node an energy of

$$E_i = \frac{T}{2} P_s \tag{8.174}$$

and the total number of transmissions is then

$$K_{PS}^{Con} = \frac{2E_t}{P_s T}. \tag{8.175}$$

Also, using PS, the end-to-end SNR at the destination is

$$\gamma_{PS} = \frac{P_s \gamma_d \gamma_p b^2}{\gamma_d b^2 + \frac{d_{sr}^v}{(1-\rho)[2\sigma_a^2 + 2\sigma_d^2/(1-\rho)]\eta\rho P_s h^2}} \tag{8.176}$$

where $\gamma_p = \frac{h^2}{d_{sr}^v [2\sigma_a^2 + 2\sigma_d^2/(1-\rho)]}$ and other symbols are defined as before. This gives the total throughput of the conventional protocol using PS as (Chen et al. 2017c)

$$C_{PS}^{Con} = \frac{K_{PS}^{Con}}{2} \log_2(1 + \gamma_{PS}). \tag{8.177}$$

In the new protocol where the source node harvests energy during the relaying phase, using the first strategy, the number of total transmissions is calculated as

$$K_{PS}^{New} = \left[\frac{2E_t/(P_s T)}{1 - \eta^2 \rho (1 - \rho) \frac{P_s^2 (h^2/d_{sr}^v)^3 T/2}{P_s h^2/d_{sr}^v + 2\sigma_a^2 + 2\sigma_d^2}} \right]. \tag{8.178}$$

where $1 > \eta^2 \rho (1 - \rho) \frac{P_s^2 (h^2/d_{sr}^v)^3 T/2}{P_s h^2/d_{sr}^v + 2\sigma_a^2 + 2\sigma_d^2}$ and the total throughput is (Chen et al. 2017c)

$$C_{PS}^{New1} = \frac{K_{PS}^{New}}{2} \log_2(1 + \gamma_{PS}). \tag{8.179}$$

The performance gain in this case is given by

$$Gain_2 = \frac{1}{1 - \eta^2 \rho (1 - \rho) P_s \left(\frac{h^2}{d_{sr}^v}\right)^3 b^2} - 1. \tag{8.180}$$

On the other hand, using the second strategy for the new protocol, one has

$$P_s^{(i+1)} = \frac{2(E_t/K_{PS}^{Con} + E_{hs}^{(i)})}{T}$$

$$\gamma_{PS}^{i+1} = \frac{P_s^{(i+1)}\gamma_d\gamma_p}{\gamma_d + \frac{d_{sr}^v(P_s^{(i+1)}\frac{h^2}{d_{sr}^v}+2\sigma_a^2+2\sigma_d^2)}{[2(1-\rho)\sigma_a^2+2\sigma_d^2]\eta\rho P_s^{(i+1)}h^2}}$$

$$E_{hs}^{(i+1)} = \eta^2\rho(1-\rho)\frac{(P_s^{(i)})^2(h^2/d_{sr}^v)^3 T/2}{P_s^{(i)}h^2/d_{sr}^v + 2\sigma_a^2 + 2\sigma_d^2}. \tag{8.181}$$

and the total throughput is (Chen et al. 2017c)

$$C_{PS}^{New2} = \sum_{i=1}^{K_{PS}^{Con}} \log_2(1 + \gamma_{PS}^{(i)})\frac{1}{2}. \tag{8.182}$$

Next, we use some numerical examples to show the gain of using the new energy harvesting relaying system where the source node and the relay node harvest energy from each other. The parameters of α and ρ are first calculated by finding their optimum values that maximize the throughput of a single transmission $\log_2(1 + \gamma_{TS})\frac{1-\alpha}{2}$ and $\log_2(1 + \gamma_{PS})\frac{1}{2}$, respectively, as is often the case in practice. We examine the effects of P_s, η, d_{sr}, and $2\sigma_a^2 + 2\sigma_d^2$ on the performance gain. Other parameters are set as $E_t = 100$ J, $T = 1$ s, $2\sigma_a^2 = 2\sigma_d^2 = \sigma^2$, $v = 2.7$, and $h = g = 1$. The path loss exponent $v = 2.7$ corresponds to an urban cellular environment. The channel gains $h = g = 1$ is chosen such that the operating SNR will be from 10 to 20 dB without any path loss, when σ^2 is from 0.01 to 0.1 as examined in this study. These are all realistic setups. The choices of distances are for illustration purposes only. Other choices of parameters can also be made, depending on the applications considered. The gain examined in the figures is calculated as the difference between the throughput of the new system and the throughput of the conventional system normalized by that of the conventional system, as given in (8.166) and (8.180).

Figure 8.24 shows the performance gain versus P_s. Several observations can be made. First, since the gain is always positive, the new scheme with energy harvesting source outperforms the conventional scheme without energy harvesting source, as expected, as the source node harvests extra energy in the relaying phase. Secondly, the new scheme using the first strategy has a larger performance gain than that using the second strategy at the cost of requiring a larger capacity for energy storage. Thirdly, the TS energy harvesting has a larger gain than the PS energy harvesting, as PS normally harvests less energy than TS.

Figure 8.25 shows the gain versus η. One sees that the performance gain increases when η increases. Figure 8.26 shows the gain versus d_{sr}. In this case, the performance gain decreases when d_{sr} increases. Also, the rate of change in Figure 8.26 is much higher than that in Figure 8.25. Again, the first strategy using TS has the largest gain. Figure 8.27 shows the gain versus σ^2. In this case, the gain increases when σ^2 increases, except when the first strategy is used with PS.

From these figures, for variable-gain AF relaying, one concludes that the distance d_{sr} has the largest effect on the performance gain, followed by the conversion efficiency η. To increase the performance gain of the new protocol, one needs to choose a large η or a small d_{sr}. Also, TS is preferred to PS, as it produces larger gains.

Figures 8.28–8.31 show the corresponding results for fixed-gain AF relaying, where $b = 1$ is used instead of $b = \dfrac{1}{\sqrt{P_s\frac{h^2}{d_{sr}^v}+2\sigma_a^2+2\sigma_d^2}}$ in (8.155). Again, the performance gain

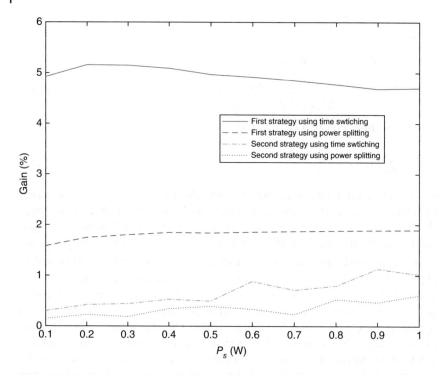

Figure 8.24 Performance gain versus P_s when $d_{sr} = 1.2$ m, $d_{rd} = 1.2$ m, $\eta = 0.5$ and $\sigma^2 = 0.01$ for variable-gain AF.

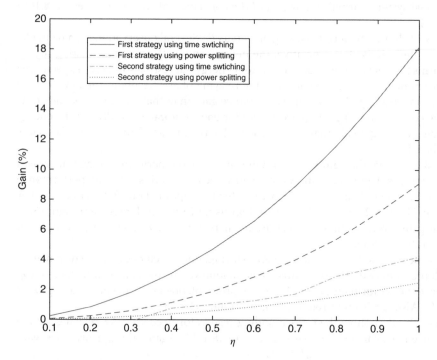

Figure 8.25 Performance gain versus η when $P_s = 1$ W, $d_{sr} = 1.2$ m, $d_{rd} = 1.2$ m, and $\sigma^2 = 0.01$ for variable-gain AF.

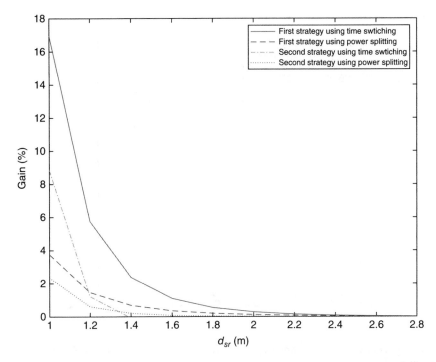

Figure 8.26 Performance gain versus d_{sr} when $P_s = 1$ W, $d_{rd} = 3 - d_{sr}$, $\eta = 0.5$, and $\sigma^2 = 0.01$ for variable-gain AF.

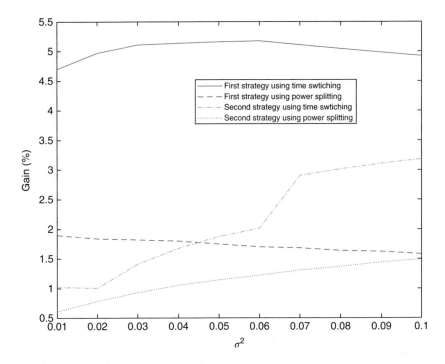

Figure 8.27 Performance gain versus σ^2 when $P_s = 1$ W, $d_{sr} = 1.2$ m, $d_{rd} = 1.2$ m, and $\eta = 0.5$ for variable-gain AF.

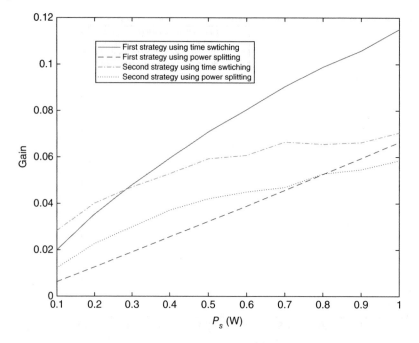

Figure 8.28 Performance gain versus P_s when d_{sr} = 1.2 m, d_{rd} = 1.2 m, η = 0.5, and σ^2 = 0.01 for fixed-gain AF, where b = 1.

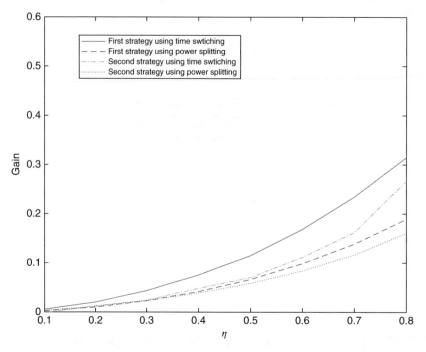

Figure 8.29 Performance gain versus η when P_s = 1 W, d_{sr} = 1.2 m, d_{rd} = 1.2 m, and σ^2 = 0.01 for fixed-gain AF, where b = 1.

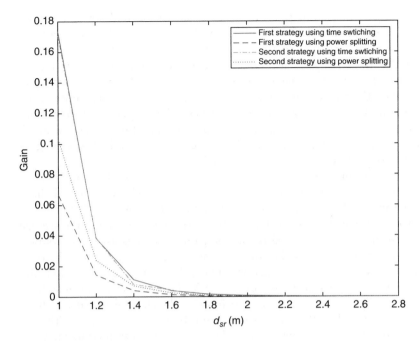

Figure 8.30 Performance gain versus d_{sr} when $P_s = 1$ W, $d_{rd} = 3 - d_{sr}$, $\eta = 0.5$, and $\sigma^2 = 0.01$ for fixed-gain AF, where $b = 1$.

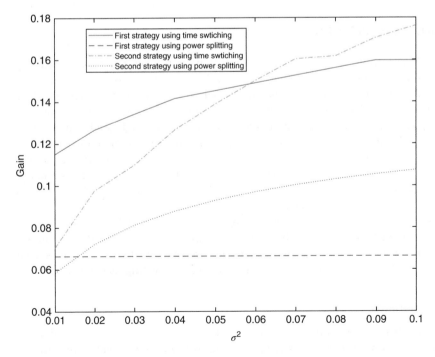

Figure 8.31 Performance gain versus σ^2 when $P_s = 1$ W, $d_{sr} = 1.2$ m, $d_{rd} = 1.2$ m, and $\eta = 0.5$ for fixed-gain AF, where $b = 1$.

increases with P_s. The first strategy is still better than the second strategy but PS outperforms TS when the value of P_s is small. The performance gain also increases with η, as expected, as more harvested energy always leads to better system performance. In Figure 8.29, the first strategy is always better than the second strategy and TS is always better than PS. On the other hand, from Figure 8.30, the performance gain drops quickly when the distance d_{sr} increases. There is little difference between the first strategy and the second strategy for TS but TS is still better than PS. Finally, from Figure 8.31, for fixed-gain AF relaying, the performance gain also increases with σ^2, except for the first strategy using PS. Again, TS is better than PS but the first strategy is better than the second strategy only when σ^2 is small.

Note that the above result considers a static AWGN channel for simplicity. In this case, channel estimation can be performed for each transmission, and the estimated channel information can then be used at the source node for rate adaptation. Alternatively, one can also average the instantaneous throughput derived above and use the average throughput at the source node. Note also that energy harvesting is used to encourage the nodes to be involved in relaying but this does not necessarily mean that the relay transmits very weak signals to the destination. In fact, the relay needs to transmit signals comparable with what is transmitted by the source in order for the destination to decode them correctly. Also, due to channel reciprocity, the channel attenuation from the source to the relay will be the same as that from the relay to the source. Thus, it is practically meaningful for the source to recycle the energy transmitted by the relay. More results on energy harvesting source node can be found in Chen et al. (2017c).

8.7 Other Important Issues

In the previous section, we have discussed several different energy harvesting relaying systems based on their sources of energy. As can be seen, the source of energy is one of the most important factors that has a fundamental impact on the designs of these systems. For systems harvesting energy from the ambient source, the energy causality and the signal causality must be accounted for. For systems with dedicated source or systems where only the relay harvests from the source, the time allocation or the harvesting parameter is more important. In all these problems, fixed resources are assumed so that one must find the best tradeoff between energy and information. This is very similar to the problems in Chapter 6, where the same signal or link is used for both energy and information and hence its use needs to be optimized.

Next, we will discuss several important issues in energy harvesting relaying. These issues occur in all types of energy harvesting relaying systems. In the discussion, we may assume only one specific type of those systems to make the discussion concise, but readers can easily extend them to other types of energy harvesting relaying systems.

8.7.1 Interference

As mentioned in Chapter 6, interference has been an important issue in almost all wireless communications systems. This is mainly due to the limited spectral resource in wireless systems such that most frequency bands are shared by all users in the communications system. However, as discussed in Chapter 6, for energy harvesting wireless communications, interference could actually be beneficial, as it provides an extra source of energy. The overall effect of interference depends on whether it provides more energy or reduces the signal-to-interference-plus-noise ratio (SINR) more.

Next, we analyze the effect of co-channel interference on energy harvesting relaying. In the analysis, consider a three-node AF relaying system where the source transmits signal and energy to the relay and the relay uses the harvested energy to forward the signal to the destination. All nodes have a single antenna and operate in half-duplex mode.

8.7.1.1 Time Switching

Using TS, the source transmits energy to the relay for αT seconds, followed by $\frac{1-\alpha}{2} T$ seconds during which the source transmits the information to the relay, and another $\frac{1-\alpha}{2} T$ seconds during which the relay uses the harvested energy to forward the information to the destination. The model is very similar to that in Section 8.6, except that there is interference so that the received signal at the relay is given by

$$y_r = \sqrt{P_s} h s + \sum_{i=1}^{N} \sqrt{P_i} h_i s_i + n_{ra} + n_{rd} \tag{8.183}$$

where P_s, h, and s are the same as defined before, P_i is the transmission power of the ith interferer, h_i is the channel gain from the ith interferer to the relay, s_i is the symbol transmitted by the ith interferer, n_{ra} is the antenna noise with mean zero and variance $2\sigma_a^2$, and n_{rd} is the baseband noise with mean zero and variance $2\sigma_d^2$. Together one has $n_{ra} + n_{rd} = n$ with mean zero and variance $2\sigma^2 = 2\sigma_a^2 + 2\sigma_d^2$. One sees that the only difference between (8.183) and the previous model is the second term that represents the interference.

From (8.183), the harvested energy is given by

$$E_r = \eta \left(P_s |h|^2 + \sum_{i=1}^{N} P_i |h_i|^2 \right) \alpha \tag{8.184}$$

where $T = 1$ and $E\{|s|^2\} = E\{|s_i|^2\} = 1$ have been used for simplicity. The extra energy from the interference is evident in (8.184). The transmission power of the relay is given by

$$P_r = \frac{E_r}{(1-\alpha)/2} = \frac{2\alpha\eta}{1-\alpha} \left(P_s |h|^2 + \sum_{i=1}^{N} P_i |h_i|^2 \right). \tag{8.185}$$

Using this energy for relaying, the received signal at the destination is

$$y_d = \sqrt{P_r} a g y_r + \sum_{j=1}^{N} \sqrt{Q_j} g_j s_j' + n_{da} + n_{dd} \tag{8.186}$$

where $a = \frac{1}{\sqrt{P_s |h|^2 + 2\sigma_{ra}^2 + 2\sigma_{rc}^2}}$ is the amplification factor, P_r, g, and y_r are defined as before, Q_j is the transmission power of the ith interferer, g_j is the channel gain from the ith interferer to the destination, s_j' is its transmitted symbol, and n_{da} is the antenna noise and n_{dd} is the baseband noise at the destination with means zero and variances $2\sigma_a^2$ and $2\sigma_d^2$, respectively. Note that the amplification factor is determined by the noise power and the signal power only. The reason that it does not include the interference power is because, in practice, the relay will only have knowledge of the channel gain and the noise power through estimation processes but will not be able to estimate the interference power due to its variability.

Using (8.186), the end-to-end SNR in this case can be derived as

$$\gamma = \frac{\gamma_r \gamma_d}{\gamma_d + \frac{(1-\alpha)(P_s|h|^2 + 2\sigma_{ra}^2 + 2\sigma_{rd}^2)/(2\alpha\eta)}{(\sum_{i=1}^{N} P_i|h_i|^2 + 2\sigma_{ra}^2 + 2\sigma_{rd}^2)(P_s|h|^2 + \sum_{i=1}^{N} P_i|h_i|^2)}} \tag{8.187}$$

where

$$\gamma_r = \frac{P_s|h|^2}{\sum_{i=1}^{N} P_i|h_i|^2 + 2\sigma_{ra}^2 + 2\sigma_{rd}^2} \tag{8.188}$$

and

$$\gamma_d = \frac{|g|^2}{\sum_{j=1}^{N} Q_j|g_j|^2 + 2\sigma_{da}^2 + 2\sigma_{dd}^2} \tag{8.189}$$

are the hop SINR of the source-to-relay and relay-to-destination links, respectively. If one lets $N = 0$, the case without interference, as discussed before, can be obtained.

Using (8.187), the outage probability for Nakagami-m fading channels was derived in Chen (2016) as

$$
\begin{aligned}
P_o(\gamma_0) = 1 &- \left(\frac{m_{I1}}{P_{I1}\Omega_{I1}}\right)^{Nm_{I1}} \left(\frac{m_1}{P_s\Omega_1}\right)^{m_1} \left(\frac{m_{I2}}{P_{I2}\Omega_{I2}}\right)^{Nm_{I2}} \\
&\times \sum_{l=0}^{m_2-1} \sum_{l'=0}^{l} \frac{(m_2/\Omega_2)^l \binom{l'}{l} (2\sigma_{da}^2 + 2\sigma_{dd}^2)^{l-l'} (Nm_{I2} + l' - 1)!}{\Gamma(Nm_{I1})\Gamma(m_1)\Gamma(Nm_{I2})l!} \\
&\times \int_0^\infty \int_{\gamma_0 z + 2\gamma_0\sigma_{ra}^2 + 2\gamma_0\sigma_{rd}^2}^\infty e^{-\frac{m_2(2\sigma_{da}^2 + 2\sigma_{dd}^2)}{\Omega_2} X(y,z) - \frac{m_1}{P_s\Omega_1} y - \frac{m_{I1}}{P_{I1}\Omega_{I1}} z} \\
&\times \frac{[X(y,z)]^{l'} y^{m_1-1} z^{Nm_{I1}-1}}{\left[\frac{m_{I2}}{Q_{I2}\Omega_{I2}} + \frac{m_2 X(y,z)}{\Omega_2}\right]^{Nm_{I2}+l'}} dy dz
\end{aligned} \tag{8.190}
$$

where $X(y,z) = \frac{1-\alpha}{2\alpha\eta} \frac{\gamma_0(y + 2\sigma_{ra}^2 + 2\sigma_{rd}^2)}{(y+z)(y-\gamma_0 z - 2\gamma_0\sigma_{ra}^2 - 2\gamma_0\sigma_{rd}^2)}$, m_{I1} and Ω_{I1} are the m parameter and the average fading power of the interferer at the relay, m_{I2} and Ω_{I2} are the m parameter and the average fading power of the interferer at the destination, m_1 and Ω_1 are the m parameter and the average fading power of h, and m_2 and Ω_2 are the m parameter and the average fading power of g. Details of this derivation can be found in Chen (2016).

Using (8.190), the achievable rate for delay-constrained applications can be shown as

$$R = \frac{(1-\alpha)R_0}{2}[1 - P_o(2^{2R_0} - 1)]. \tag{8.191}$$

Figures 8.32 and 8.33 show R for TS versus α in a Nakagami-m fading channel with Rayleigh interferer under different conditions, when $2\sigma_{ra}^2 = 2\sigma_{rd}^2 = 2\sigma_{da}^2 = 2\sigma_{dd}^2 = 1$ and $P_s = P_{I1} = Q_{I2} = \Omega_{I1} = \Omega_{I2} = 1$, while Ω_1 and Ω_2 vary with the average SINRs $\Delta_1 = \frac{\Omega_1}{NP_{I1}\Omega_{I1} + 2\sigma_{ra}^2 + 2\sigma_{rd}^2}$ and $\Delta_2 = \frac{\Omega_2}{NQ_{I2}\Omega_{I2} + 2\sigma_{ra}^2 + 2\sigma_{rd}^2}$, respectively. One can see that the rate increases when η increases, N increases, R_0 increases, or the m parameter increases,

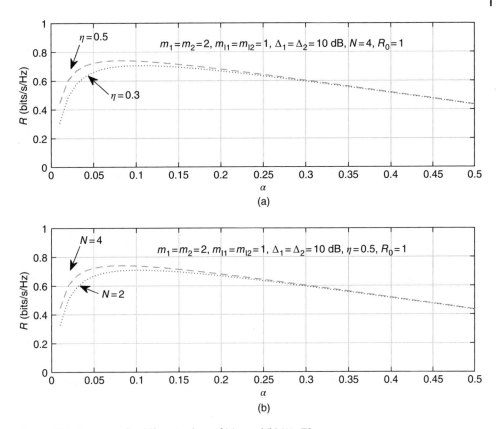

Figure 8.32 R versus α for different values of (a) η and (b) N in TS.

as more energy can be harvested with larger η or N or when channel conditions are better. Also, similar to before, an optimum α exists in all cases, and the optimum value changes with η, N, R_0, or the m parameter. The observation that the rate increases with N implies that, in these cases, the interference is beneficial, or the benefit of harvesting the interference outweighs the degradation caused by the interference.

8.7.1.2 Power Splitting

If PS is used, the source transmits the signal to the relay for $\frac{T}{2}$ seconds, followed by $\frac{T}{2}$ seconds during which the relay forwards the information to the destination. In the first $\frac{T}{2}$ seconds, the signal will be split into two parts, one part for information decoding and the other part for energy harvesting. The received signal at the relay for information decoding is given by

$$y_r = \sqrt{(1-\rho)P_s}hs + \sqrt{1-\rho}\sum_{i=1}^{N}\sqrt{P_i}h_i s_i + \sqrt{1-\rho}n_{ra} + n_{rd} \qquad (8.192)$$

where ρ is the PS factor and all other symbols are defined as before. In this case, the antenna noise n_{ra} is split, because PS is performed at the RF part, while the baseband noise n_{rd} is not split.

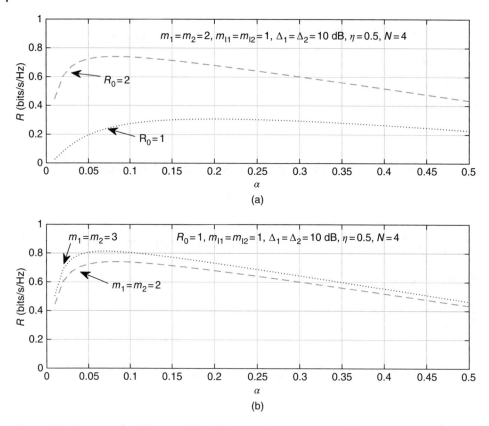

Figure 8.33 R versus α for different values of (a) R_0 and (b) m parameters in TS.

Then, the harvested energy using the other part of the signal is given by

$$E_r = \eta\rho\left(P_s|h|^2 + \sum_{i=1}^{N} P_i|h_i|^2\right)\frac{1}{2}. \tag{8.193}$$

Again, $T = 1$ and $E\{|s|^2\} = E\{|s_i|^2\} = 1$ have been used for simplicity. The extra energy from the interference is also evident in (8.193). The transmission power of the relay is thus given by

$$P_r = \frac{E_r}{1/2} = \eta\rho\left(P_s|h|^2 + \sum_{i=1}^{N} P_i|h_i|^2\right). \tag{8.194}$$

Using the harvested energy, the received signal at the destination is given as

$$y_d = \sqrt{P_r}agy_r + \sum_{j=1}^{N} \sqrt{Q_j}g_js'_j + n_{da} + n_{dd} \tag{8.195}$$

where $a = \dfrac{1}{\sqrt{(1-\rho)P_s|h|^2+2(1-\rho)\sigma_{ra}^2+2\sigma_{rd}^2}}$ is the amplification factor in this case, and all other symbols are the same as before.

Using (8.195), the end-to-end SINR in this case can be derived as

$$
\gamma = \cfrac{\gamma_r' \gamma_d}{\gamma_d + \cfrac{\left(P_s|h|^2 + 2\sigma_{ra}^2 + \frac{2\sigma_{rd}^2}{1-\rho}\right)/(\eta\rho)}{\left[\sum_{i=1}^{N} P_i|h_i|^2 + 2\sigma_{ra}^2 + \frac{2\sigma_{rd}^2}{1-\rho}\right](P_s|h|^2 + \sum_{i=1}^{N} P_i|h_i|^2)}}
\tag{8.196}
$$

where $\gamma_r' = \dfrac{P_s|h|^2}{\sum_{i=1}^{N} P_i|h_i|^2 + 2\sigma_{ra}^2 + 2\sigma_{rd}^2/(1-\rho)}$ is the hop SINR of the source-to-relay link for PS and
γ_d is the same as before. Again, if one lets $N = 0$, the case without interference can be
obtained. Otherwise, when N increases, more interference is incurred.

Using (8.196), the outage probability for PS in Nakagami-m fading channels was
derived in Chen (2016) as

$$
P_o(\gamma_0) = 1 - \left(\frac{m_{I1}}{P_{I1}\Omega_{I1}}\right)^{Nm_{I1}} \left(\frac{m_1}{P_s\Omega_1}\right)^{m_1} \left(\frac{m_{I2}}{Q_{I2}\Omega_{I2}}\right)^{Nm_{I2}}
$$

$$
\times \sum_{l=0}^{m_2-1} \sum_{l'=0}^{l} \frac{(m_2/\Omega_2)^l \binom{l'}{l} (2\sigma_{da}^2 + 2\sigma_{dd}^2)^{l-l'} (Nm_{I2} + l' - 1)!}{\Gamma(Nm_{I1})\Gamma(m_1)\Gamma(Nm_{I2})l!}
$$

$$
\times \int_0^\infty \int_{\gamma_0 z + 2\gamma_0 \sigma_{ra}^2 + 2\gamma_0 \sigma_{rd}^2/(1-\rho)}^{\infty} e^{-\frac{m_2(2\sigma_{da}^2 + 2\sigma_{dd}^2)}{\Omega_2} X(y,z) - \frac{m_1}{P_s\Omega_1} y - \frac{m_{I1}}{P_{I1}\Omega_{I1}} z}
$$

$$
\times \frac{[X(y,z)]^l y^{m_1-1} z^{Nm_{I1}-1}}{\left[\frac{m_{I2}}{Q_{I2}\Omega_{I2}} + \frac{m_2 X(y,z)}{\Omega_2}\right]^{Nm_{I2}+l'}} \, dy \, dz
\tag{8.197}
$$

where $X(y,z) = \dfrac{\gamma_0}{\eta\rho} \dfrac{y + 2\sigma_{ra}^2 + 2\sigma_{rd}^2/(1-\rho)}{(y+z)[y - \gamma_0 z - 2\gamma_0 \sigma_{ra}^2 - 2\gamma_0 \sigma_{rd}^2/(1-\rho)]}$ now and other symbols are defined as
before. More details on (8.197) are available in Chen (2016). Using (8.197), the
achievable rate for delay-constrained applications can be calculated using (8.191).

Figures 8.34 and 8.35 show R for PS versus ρ in a Nakagami-m fading channel with
Rayleigh interferer under different conditions. The same parameter settings are used as
those in Figures 8.32 and 8.33. Similar observations can be made. Also, the curves for ρ
in PS are generally flatter than those in TS, indicating that there are more choices of ρ
that give near-optimal performances.

8.7.2 Multi-Hop

We have been studying energy harvesting dual-hop relaying systems so far. However,
in some applications, range extension is quite important to provide services for remote
nodes. In these applications, multiple relays can be used to form a multi-hop relaying
system that can extend the range considerably. Next, we will consider a multi-hop energy
harvesting AF relaying system and study how the energy changes with hops. Some of the
discussion and derivation can be found in Mao et al. 2015.

Consider a relaying system with one relaying link from the source node to the destination node via several hops. There is no direct link between the source node and
the destination node, which is the aim of network coverage extension. All nodes are
half-duplex and have a single antenna. The signal from the source node is relayed by
multiple nodes in several hops, one node in each hop, until it arrives at the destination

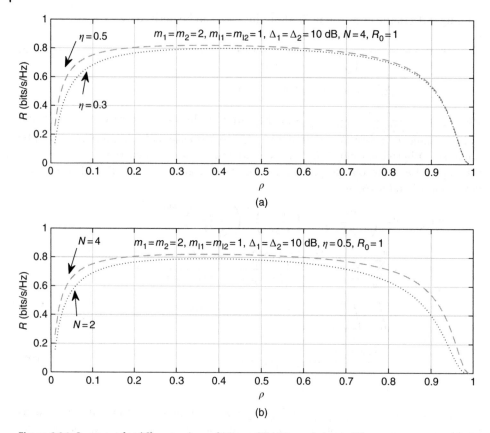

Figure 8.34 R versus ρ for different values of (a) η and (b) N parameters in PS.

node. In each hop, the relaying node first harvests the energy of the signal from the previous hop and the harvested energy is then used to forward the signal to the next hop. Assume that in each hop it takes T seconds to transmit the signal and that the initial transmission power of the source is P_0. In this section, the largest number of hops will be calculated, that is, the hop when the signal dies down due to insufficient energy for forwarding. This calculation depends on the energy harvesting strategies used by the relays. Thus, it is necessary to discuss the TS and PS methods separately.

8.7.2.1 Time Switching

In the TS method, the relay uses a portion of the relaying time to harvest the signal from the source node or the previous hop. Denote α_m as the portion at the mth relay in the mth hop, with $0 \le \alpha_m \le 1$ and $m = 1, 2, \cdots$. Thus, among the relaying time T in the mth hop, $\alpha_m T$ will be used to harvest the energy and $(1 - \alpha_m)T$ will be used for forwarding the received signal.

Consider AF first. In the AF protocol, the received signal from the previous hop is amplified and forwarded to the next hop directly. Thus, the received signals at the first relay, the second relay, and the mth relay are given by

$$y_1 = \sqrt{P_0}a_1 s + n_{1a} + n_{1c} \tag{8.198}$$

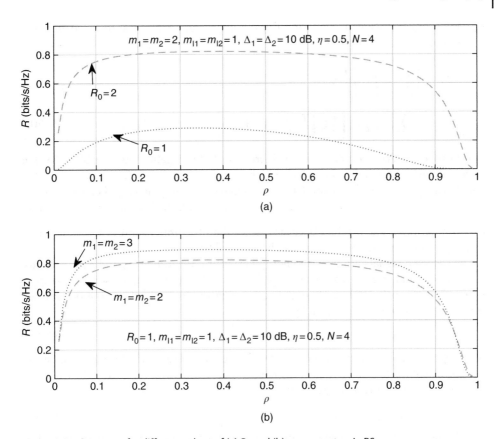

Figure 8.35 R versus ρ for different values of (a) R_0 and (b) m parameters in PS.

$$y_2 = \sqrt{P_1} g_1 a_2 y_1 + n_{2a} + n_{2c} \tag{8.199}$$
$$\vdots$$
$$y_m = \sqrt{P_{m-1}} g_{m-1} a_m y_{m-1} + n_{ma} + n_{mc}, \tag{8.200}$$

respectively, where P_0 is the initial transmission power of the source node, $P_1, \cdots, P_{m-1}$ are the transmission powers of the first until the $(m-1)$th relays that have been harvested from $y_1, \cdots, y_{m-1}$, respectively, $a_1, a_2, \cdots, a_m$ are the channel gains of the first, second, $\cdots$, mth hops, respectively, $g_1, \cdots, g_{m-1}$ are the amplification factors at the first until the $(m-1)$th relays, respectively, s is the transmitted BPSK signal with equal *a priori* probabilities for $s = 1$ and $s = -1$, $n_{1a}, n_{2a}, \cdots, n_{ma}$ are the AWGN introduced by the RF fronts at the first relay, the second relay, and the mth relay, respectively, and $n_{1c}, n_{2c}, \cdots, n_{mc}$ are the AWGN introduced in the RF to baseband conversions at the first relay, the second relay, and the mth relay, respectively. Note that we have discussed before that there are two types of noise in energy harvesting relaying, one from the antenna and one from the baseband. Furthermore, assume that the channel gains $a_1, a_2, \cdots, a_m$ are known. Also, assume that $n_{1c}, n_{2c}, \cdots, n_{mc}$ are Gaussian random variables with mean zero and variance $\sigma_{1c}^2, \sigma_{2c}^2, \cdots, \sigma_{mc}^2$, respectively, and $n_{1a}, n_{2a}, \cdots, n_{ma}$ are Gaussian random variables with mean zero and variance

$\sigma_{1a}^2, \sigma_{2a}^2, \cdots, \sigma_{ma}^2$, respectively. Using the iterative relationship, the received signal at the mth relay can be rewritten as

$$y_m = \prod_{i=1}^{m} a_i \prod_{i=1}^{m-1} g_i \prod_{i=0}^{m-1} \sqrt{P_i} s$$

$$+ \sum_{i=1}^{m} \left[\prod_{j=i+1}^{m} a_j \prod_{j=i}^{m} g_j \prod_{j=i}^{m-1} \sqrt{P_j} (n_{ia} + n_{ic}) \right]. \tag{8.201}$$

Two types of AF protocol are considered next: fixed-gain; and variable-gain. For fixed-gain AF, the amplification factor is set to $g_i = 1$ for $i = 1, 2, \cdots, m-1$. Other constants can be considered similarly. For variable-gain AF, the amplification factor is set to $g_i = \dfrac{1}{\sqrt{P_{i-1}|a_i|^2 + \sigma_{ic}^2 + \sigma_{ia}^2}}$ for $i = 1, 2, \cdots, m-1$. Also, in (8.201), if the lower limit of the summation or the product becomes larger than the upper limit of the summation or the product, the sum or the product is assumed to be 1.

The received signals are used in energy harvesting for $\alpha_1 T$, $\alpha_2 T$, and $\alpha_m T$ seconds, respectively, and in relaying for $(1-\alpha_1)T$, $(1-\alpha_2)T$, and $(1-\alpha_m)T$ seconds, respectively. Thus, from (8.198), the energy harvested at the first relay from the transmitted signal of the source node is given by

$$E_{h1} = \eta P_0 a_1^2 \alpha_1 T \tag{8.202}$$

where η is the conversion efficiency of the energy harvester and all other symbols are defined as before. Similarly, the energy harvested at the second relay from the transmitted signal of the first relay is given by

$$E_{h2} = \eta P_1 g_1^2 a_2^2 P_0 a_1^2 \alpha_2 T. \tag{8.203}$$

Finally, the energy harvested at the mth relay from the transmitted signal of the $(m-1)$th relay is given by

$$E_{hm} = \eta \prod_{i=0}^{m-1} P_i \prod_{i=1}^{m-1} g_i^2 \prod_{i=1}^{m} a_i^2 \alpha_m T. \tag{8.204}$$

Since each hop has a relaying time of T, the transmitted power at the first relay, the second relay, and the m-th relay can be calculated as Mao et al. 2015

$$P_1 = \frac{E_{h1}}{T} = \eta P_0 a_1^2 \alpha_1 \tag{8.205a}$$

$$P_2 = \frac{E_{h2}}{T} = \eta P_1 g_1^2 a_2^2 P_0 a_1^2 \alpha_2 \tag{8.205b}$$

$$P_m = \frac{E_{hm}}{T} = \eta \prod_{i=0}^{m-1} P_i \prod_{i=1}^{m-1} g_i^2 \prod_{i=1}^{m} a_i^2 \alpha_m, \tag{8.205c}$$

respectively. Note that these relationships are in fact recurrence relationships. Using 8.205a–8.205c, non-recurrence expressions of the transmission powers can also be obtained by replacing the transmission powers on the right-hand sides of these equations.

One sees that, since $\eta < 1$, $\alpha_1 \leq 1$, and $a_1 < 1$ due to channel attenuation, $E_{h1} = \eta \alpha_1 a_1^2 E_0$ is only a portion of the initial amount of energy transmitted by the source node as $E_0 = P_0 T$. Also, $E_{h2} = \eta g_1^2 a_2^2 P_0 a_1^2 \alpha_2 E_{h1}$ is only a portion of the harvested energy at the first relay. Thus, the harvested energy at each hop decreases at an accelerated rate when the hop number increases. At some hop, the energy will not be sufficient for transmission. We use the throughput criterion such that

$$R_0 = \log_2(1 + \gamma_m)(1 - \alpha_m) \tag{8.206}$$

where

$$\gamma_m = \frac{\prod\limits_{i=1}^{m} a_i^2 \prod\limits_{i=1}^{m-1} g_i^2 \prod\limits_{i=1}^{m-1} P_i (1 - \alpha_m) T}{\sum\limits_{i=1}^{m} \left[\prod\limits_{j=i+1}^{m} a_j^2 \prod\limits_{j=i}^{m-1} g_j^2 \prod\limits_{j=i}^{m-1} P_j (\sigma_{ia}^2 + \sigma_{ic}^2) \right]} \tag{8.207}$$

$m = 1, 2, \cdots$ is the hop index, $(1 - \alpha_m)$ in (8.206) takes into account the throughput loss due to energy harvesting time, and $(1 - \alpha_m)$ in (8.207) takes into account the transmission time loss due to energy harvesting.

It is of great interest to find the first value of m for which $\log_2(1 + \gamma_m)(1 - \alpha_m) < R_0$. Before this happens, each hop has a throughput of R_0 such that the TS coefficient α_m can be calculated as Mao et al. 2015

$$\alpha_m = 1 - \frac{(2^{R_0/(1-\alpha_m)} - 1)W}{\prod\limits_{i=1}^{m} a_i^2 \prod\limits_{i=1}^{m-1} g_i^2 \prod\limits_{i=1}^{m-1} P_i T} \tag{8.208}$$

for the mth relay, where $W = \sum_{i=1}^{m} [\prod_{j=i+1}^{m} a_j^2 \prod_{j=i}^{m-1} g_j^2 \prod_{j=i}^{m-1} P_j (\sigma_{ia}^2 + \sigma_{ic}^2)]$. This equation does not lead to a closed-form expression of α_m but it can be easily solved using mathematical software.

In the DF protocol, the received signal from the previous hop is first decoded and then the decoded information is forwarded to the next hop. Thus, the received signals at the first relay, the second relay, and the mth relay are given by

$$y_1 = \sqrt{P_0} a_1 s + n_{1a} + n_{1c} \tag{8.209}$$

$$y_2 = \sqrt{P_1} a_2 s_1 + n_{2a} + n_{2c} \tag{8.210}$$

$$\vdots$$

$$y_m = \sqrt{P_{m-1}} a_m s_{m-1} + n_{ma} + n_{mc}, \tag{8.211}$$

respectively, where $s_1 = sgn\{y_1 a_1^*\}, \cdots, s_{m-1} = sgn\{y_{m-1} a_{m-1}^*\}$ are the decoded and forwarded information at the relays, $sgn\{x\} = 1$ when $x > 0$ and $sgn\{x\} = -1$ when $x < 0$ is the signum function, and all other symbols are defined as before. Unlike AF, since $E\{|s_{m-1}|^2\} = 1$, the performance of the mth hop in DF only depends on P_{m-1}, a_m, σ_{ma}^2, and σ_{mc}^2. Next, the largest number of hops, that is, the maximum value of m given an initial amount of energy $E_0 = P_0 T$, will be studied.

For DF, the harvested energy at the first relay can be derived from (8.209) as

$$E_{h1} = \eta P_0 a_1^2 \alpha_1 T. \tag{8.212}$$

Similarly, the harvested energy at the second relay can be derived from (8.210) as

$$E_{h2} = \eta P_1 a_2^2 \alpha_2 T \tag{8.213}$$

and the harvested energy at the mth relay can be derived from (8.211) as

$$E_{hm} = \eta P_{m-1} a_m^2 \alpha_m T. \tag{8.214}$$

Using them, the transmission powers at the first relay, the second relay, and the mth relay, respectively, can be derived as Mao et al. 2015

$$P_1 = \frac{E_{h1}}{T} = \eta P_0 a_1^2 \alpha_1 \tag{8.215a}$$

$$P_2 = \frac{E_{h2}}{T} = \eta P_1 a_2^2 \alpha_2 \tag{8.215b}$$

$$P_m = \frac{E_{hm}}{T} = \eta P_{m-1} a_m^2 \alpha_m. \tag{8.215c}$$

Again, these relationships are recurrent.

Comparing AF with DF, one sees that their energies harvested at the first relay are the same. For the second relay, $E_{h2} = \eta a_2^2 \alpha_2 E_{h1}$ for DF and $E_{h2} = g_1^2 P_0 a_1^2 \eta a_2^2 \alpha_2 E_{h1}$ for AF. Thus, for fixed-gain AF with $g_1 = 1$, the harvested energy for AF is $10\log_{10}(P_0 a_1^2)$ dB more than the harvested energy for DF, implying that fixed-gain AF in this case could support more hops than DF due to more harvested energy. On the other hand, for variable-gain AF with $g_1 = \frac{1}{\sqrt{P_0 a_1^2 + \sigma_{1c}^2 + \sigma_{1a}^2}}$, the harvested energies of AF and DF are almost identical when the SNR of the initial received signal is large such that $P_0 a_1^2 \gg \sigma_{1c}^2 + \sigma_{1a}^2$.

Similarly, using the throughput criterion, one has

$$R_0 = \log_2(1 + \gamma_m)(1 - \alpha_m) \tag{8.216}$$

where

$$\gamma_m = \frac{P_{m-1} a_m^2 (1 - \alpha_m) T}{\sigma_{ma}^2 + \sigma_{mc}^2} \tag{8.217}$$

and $m = 1, 2, \cdots$ is the hop index. The TS coefficient α_m can be calculated as Mao et al. 2015

$$\alpha_m = 1 - \frac{(2^{R_0/(1-\alpha_m)} - 1)(\sigma_{ma}^2 + \sigma_{mc}^2)}{P_{m-1} a_m^2 T} \tag{8.218}$$

for the mth relay, until the throughput drops below the threshold as $\log_2(1 + \gamma_m)(1 - \alpha_m) < R_0$, for which the value of m is determined as the largest number of hops supported by the initial energy of E_0.

8.7.2.2 Power Splitting

In the PS method, the relay splits a portion of the received signal power as harvested energy. Denote ρ_m as the splitting factor at the mth relay, with $0 \le \rho_m \le 1$ and $m = 1, 2, \cdots$.

Consider AF first. In this case, the parts of the received signals at the first relay, the second relay, and the mth relay for relaying are given by

$$y_1 = \sqrt{(1 - \rho_1)P_0} a_1 s + \sqrt{1 - \rho_1} n_{1a} + n_{1c} \tag{8.219}$$

$$y_2 = \sqrt{(1 - \rho_2)P_1} g_1 a_2 y_1 + \sqrt{1 - \rho_2} n_{2a} + n_{2c} \tag{8.220}$$

$$\vdots$$

$$y_m = \sqrt{(1 - \rho_m)P_{m-1}} g_{m-1} a_m y_{m-1} + \sqrt{1 - \rho_m} n_{ma} + n_{mc}, \tag{8.221}$$

respectively, for T seconds, where all the symbols are defined as before. Using them, the signal at the mth relay to be relayed is derived as

$$
y_m = \prod_{i=0}^{m-1} \sqrt{(1 - \rho_{i+1})P_i} \prod_{i=1}^{m} a_i \prod_{i=1}^{m-1} g_i s
$$
$$
+ \sum_{i=1}^{m} \left[\prod_{j=i}^{m-1} \sqrt{(1 - \rho_{j+1})P_j} \right.
$$
$$
\left. \cdot \prod_{j=i+1}^{m} a_i \prod_{j=i}^{m-1} g_j ((1 - \rho_i) n_{ia} + n_{ic}) \right].
$$

(8.222)

The same received signals are also used for energy harvesting. Using the harvested energies, the transmission powers at the first relay, the second relay, and the mth relay, respectively, can be calculated as Mao et al. 2015

$$
P_1 = \eta P_0 a_1^2 \rho_1
$$

(8.223a)

$$
P_2 = \eta P_0 P_1 g_1^2 a_1^2 a_2^2 (1 - \rho_1) \rho_2
$$

(8.223b)

$$
P_m = \eta \prod_{i=0}^{m-1} P_i \prod_{i=1}^{m-1} g_i^2 \prod_{i=1}^{m} a_i^2 \prod_{i=1}^{m-1} (1 - \rho_i) \rho_m.
$$

(8.223c)

Comparing these results with those for the TS method, one sees that the harvested power in PS has an additional term of $\prod_{i=1}^{m-1}(1 - \rho_i)$. Since this term is smaller than 1 and decreases quickly as m increases, the harvested power using the PS method is much smaller than that using the TS method, under similar conditions. As a result, the largest number of hops using the PS method is much smaller than that using the TS method, which is not desirable for network coverage extension.

Again, applying the throughput per hop restriction, one has

$$
R_0 = \log_2(1 + \gamma_m)
$$

(8.224)

where

$$
\gamma_m = \frac{\prod_{i=0}^{m-1}(1 - \rho_{i+1})P_i \prod_{i=1}^{m} a_i^2 \prod_{i=1}^{m-1} g_i^2 T}{U}
$$

(8.225)

where $U = \sum_{i=1}^{m}[\prod_{j=i}^{m-1} \sqrt{(1 - \rho_{j+1})P_j} \prod_{j=i+1}^{m} a_i^2 \prod_{j=i}^{m-1} g_j^2 ((1 - \rho_i)\sigma_{ia}^2 + \sigma_{ic}^2)]$. The PS splitting factor ρ_m can be calculated as Mao et al. 2015

$$
\rho_m = 1 - \frac{(2^{R_0} - 1)\sigma_{mc}^2}{\prod_{i=0}^{m-1}(1 - \rho_{i+1})P_i \prod_{i=1}^{m} a_i^2 \prod_{i=1}^{m-1} g_i^2 T - W}
$$

(8.226)

where

$$
W = \left\{ \sum_{i=1}^{m-1} \left[\prod_{j=i}^{m-1} \sqrt{(1 - \rho_{j+1})P_j} \prod_{j=i+1}^{m} a_i^2 \prod_{j=i}^{m-1} g_j^2 ((1 - \rho_i)\sigma_{ia}^2 + \sigma_{ic}^2) \right] + \sigma_{ma}^2 \right\} (2^{R_0} - 1).
$$

(8.227)

Unlike the TS coefficient α_m, the PS splitting factor ρ_m does have a closed-form expression. However, note that the calculation of ρ_m requires knowledge of $\rho_{m-1}, \cdots, \rho_1$. Thus, this is essentially a recurrence relationship.

For DF, the received signals at the first relay, the second relay, and the mth relay for relaying can be expressed as

$$y_1 = \sqrt{(1-\rho_1)P_0}a_1 s + \sqrt{1-\rho_1}n_{1a} + n_{1c} \qquad (8.228)$$

$$y_2 = \sqrt{(1-\rho_2)P_1}a_2 s_1 + \sqrt{1-\rho_2}n_{2a} + n_{2c} \qquad (8.229)$$

$$\vdots$$

$$y_m = \sqrt{(1-\rho_m)P_{m-1}}a_m s_{m-1} + \sqrt{1-\rho_m}n_{ma} + n_{mc}, \qquad (8.230)$$

respectively, for the whole relaying period T seconds, where all the symbols are defined as before.

Parts of the received signals are also used for energy harvesting. Using the PS method, the transmission powers at the first relay, the second relay, and the mth relay, respectively, can be calculated as Mao et al. 2015

$$P_1 = \eta P_0 a_1^2 \rho_1 \qquad (8.231a)$$

$$P_2 = \eta \rho_2 P_1 a_2^2 \qquad (8.231b)$$

$$P_m = \eta \rho_m P_{m-1} a_m^2. \qquad (8.231c)$$

These values can also be calculated recursively. Then, applying the throughput restriction to each hop, one has

$$R_0 = \log_2(1 + \gamma_m) \qquad (8.232)$$

where

$$\gamma_m = \frac{(1-\rho_m)P_{m-1}a_m^2 T}{(1-\rho_m)\sigma_{ma}^2 + \sigma_{mc}^2} \qquad (8.233)$$

and $m = 1, 2, \cdots$ is the hop index. Using this restriction, the PS splitting factor ρ_m can be calculated as Mao et al. 2015

$$\rho_m = 1 - \frac{(2^{R_0} - 1)\sigma_{mc}^2}{P_{m-1}a_m^2 T - \sigma_{ma}^2(2^{R_0} - 1)}. \qquad (8.234)$$

To summarize, the calculation of the largest number of hops is performed as in the following. For TS, α_m is first calculated, and then P_m and the throughput are calculated for $m = 1, 2, \cdots$. Once the throughput is less than R_0, the calculation stops and the value of m in the previous iteration is decided as the largest number of hops. In the case of PS, ρ_m is first calculated, followed by P_m, both of which will be used to calculate the throughput to find the first value of m that makes the throughput below R_0, for $m = 1, 2, \cdots$.

8.7.2.3 Numerical Examples

In this section, numerical examples will be presented to show the dependence of the largest number of hops on different relaying protocols, different harvesting strategies as well as different system parameters. To do this, in the calculations, the parameters are set as $\sigma_{ma}^2 = \sigma_{mc}^2 = 0.01$ for $m = 1, 2, \cdots$, $a_m^2 = 0.1$ for $m = 1, 2, \cdots$, and $T = 1$. Also, let $\gamma_0 = \frac{a_1^2 P_0 T}{\sigma_{1c}^2 + \sigma_{1a}^2}$ be the initial SNR in the first relay, which is directly related to the initial amount of energy $P_0 T$ from the source node. The effects of R_0, η, and γ_0 are examined.

Tables 8.2–8.6 show the obtained largest numbers of hops in different conditions for different relaying protocols and harvesting strategies. Several observations can be made from these tables.

First, under the same conditions, the value of the largest number of hops generally increases when the initial SNR γ_0 and the conversion efficiency η increase or when the throughput requirement R_0 decreases. This means that one may extend the network coverage by either increasing the amount of energy transferred from the source node and subsequently harvested by all relaying nodes, or improving the efficiency of the energy harvester or reducing the performance requirement. This is expected. Among the three parameters examined, one can also see that the initial SNR γ_0 has the largest impact on the network coverage extension. For example, in Table 8.2, the largest number of hops could have an increase of 5 hops from 2 hops to 7 hops, when γ_0 increases from 10 dB to 30 dB, while for η and R_0, the maximum increase is between 0 and 2 hops when η increases or R_0 decreases. In all the cases, the change of η from 0.3 to 0.5 leads to an increase of 1 hop at most, when $\gamma_0 \leq 25$ dB, indicating that it may not be worth

Table 8.2 The largest number of hops for fixed-gain AF with TS.

(R_0,η)	10 dB	15 dB	20 dB	25 dB	30 dB
(1,0.3)	2	2	3	4	5
(1,0.5)	2	2	3	4	7
(2,0.3)	2	2	3	2	4
(2,0.5)	2	2	3	2	6

Table 8.3 The largest number of hops for variable-gain AF with TS.

(R_0,η)	10 dB	15 dB	20 dB	25 dB	30 dB
(1,0.3)	2	2	3	3	3
(1,0.5)	2	3	3	3	4
(2,0.3)	2	2	2	3	3
(2,0.5)	2	2	2	3	3

Table 8.4 The largest number of hops for DF with TS.

(R_0,η)	10 dB	15 dB	20 dB	25 dB	30 dB
(1,0.3)	2	2	3	3	3
(1,0.5)	2	3	3	3	4
(2,0.3)	2	2	2	3	3
(2,0.5)	2	2	3	3	3

Table 8.5 The largest number of hops for variable-gain or fixed-gain AF with PS.

(R_0 , η)	10 dB	15 dB	20 dB	25 dB	30 dB
			γ_0		
(1,0.3)	1	1	1	1	1
(1,0.5)	1	1	1	1	1
(2,0.3)	1	1	1	1	1
(2,0.5)	1	1	1	1	1

Table 8.6 The largest number of hops for DF with PS.

(R_0 , η)	10 dB	15 dB	20 dB	25 dB	30 dB
			γ_0		
(1,0.3)	1	1	2	2	2
(1,0.5)	1	2	2	2	3
(2,0.3)	1	1	1	2	2
(2,0.5)	1	1	2	2	2

improving the design of the energy harvester to extend network coverage when the initial amount of energy is low to medium. This is important, as improvement of the energy harvester often requires a considerable amount of time and effort.

Secondly, comparing different relaying protocols, one sees that fixed-gain AF relaying has the largest numbers of hops when the TS method is used. In this case, variable-gain AF relaying and DF have very similar performances. The reason for this is that fixed-gain AF relaying leads to more harvested energies at the relays and therefore can be transferred for more times or more hops while variable-gain AF and DF have similar harvested energies for large SNRs. When the PS method is used, DF relaying has the largest numbers of hops, while variable-gain relaying and fixed-gain AF relaying have similar performances. This can be explained by comparing the harvested power of AF and DF, where one sees that the harvested power of AF is much smaller than that of DF due to the extra term of $\prod_{i=1}^{m-1}(1 - \rho_i)$ in the product and thus, the energy transferred from the source node can be exhausted quickly in AF relaying.

Thirdly, comparing the TS method with the PS method, one sees that the TS method can achieve many more hops than the PS method, under similar conditions. This is especially true for AF relaying, where the largest number of hops drops from 7 in Table 8.2 to 1 in Table 8.5, under the same other conditions. This is explained as follows: for the TS method in AF relaying, the powers of the signals for harvesting and for relaying are the same, and only the relaying time is switched. Thus, the first relay can choose to take a small amount of time and therefore harvest a small amount of energy (power is fixed during this), conserving the majority of the energy for later use by the following relays. However, for the PS method in AF relaying, the powers of the signals for harvesting and relaying are different. Even if the first relay chooses to split a small portion of power for relaying and therefore harvest a small amount of energy (time is fixed during this), with the good intention of passing on the majority of energy for later use by the following relays, according to AF relaying, the first relay is in fact only using a large transmission power (conserved for the following relays) to transmit a very weak signal (due to a small portion of power split) to the next relay. This is as undesirable as using a small transmission power (harvesting most energy from the source node at the first relay) to transmit a strong signal (due to a large portion of power split for the first hop). In both situations, the received power at the next relay will be very small, wasting a huge amount of conserved energy. Thus, the PS method is very ineffective for network coverage extension based on harvesting. If one has to use the PS method due to the hardware requirement, from these tables, DF relaying will be a better option than AF relaying.

In summary, the best way of extending the network coverage using multi-hop relaying with energy harvesting is to use fixed-gain AF relaying combined with the TS method, followed by using variable-gain AF or DF relaying with the TS method, and DF relaying with the PS method. One should avoid using AF with PS to extend the network coverage.

To see how the TS coefficient α_m and the PS splitting factor ρ_m vary with the hop number Figure 8.36 shows them for the case when $\gamma_0 = 30$ dB, $R_0 = 2$, and $\eta = 0.3$. For the PS method, one sees that the splitting factor ρ_1 of the AF relaying is very close to 1, in fact around 0.99, trying to save as much energy as possible for later use in the following hops. However, due to the small splitting factor, the received power at the second relay is scaled down by $1 - \rho_1 \approx 0.01$, leading to a huge loss of energy in effect. Thus, AF relaying combined with PS is not suitable for multi-hop relaying with energy

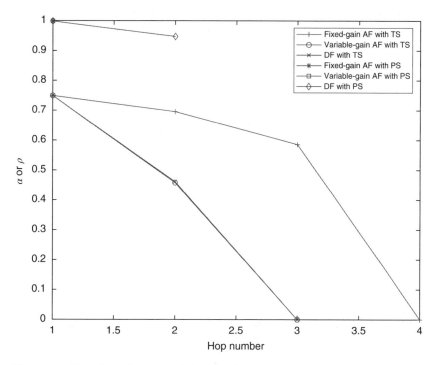

Figure 8.36 The values of α or ρ versus hop number when $\gamma_0 = 30$ dB, $R_0 = 2$, and $\eta = 0.3$.

harvesting. For variable-gain AF relaying and DF relaying, their TS coefficients at the three hops, α_1, α_2 and α_3, are almost identical, which agrees with observations from the tables. For fixed-gain AF relaying, α_1 is around 0.75, α_2 is around 0.7, α_3 is around 0.6, and α_4 is around 0, which decrease at an accelerated rate. In all the TS curves, the TS coefficient goes to 0 when the hop number increases, implying the last hop where all energies should be used for relaying without any energy harvesting to meet the throughput requirement at this hop.

The transmitted power at the first relay (the one after the source node) P_1 is an important indicator of how power decays during energy harvesting relaying, by comparing it with the initial transmitted power of $P_0 = \gamma_0 * (\sigma_{1c}^2 + \sigma_{1a}^2)/a_1^2$ at the source node. Figures 8.37–8.39 show the calculated P_1 versus γ_0 for different values of η and R_0 using the TS method. One sees that P_1 for fixed-gain and variable-gain AF and P_1 for DF using the TS method are identical. This can also be seen from the expressions of those P_1. For each figure, the value of P_1 increases dramatically when γ_0 increases, and in fact γ_0 has the largest impact on P_1. In these figures, η also has a larger impact on P_1 than R_0. At $\gamma_0 = 30$ dB, the initial transmitted power is $P_0 = 200$ and the transmitted power at the first relay is $P_1 = 8.5$. Thus, power decays quickly during relaying and harvesting.

Figure 8.40 shows P_1 versus γ_0 for different values of η and R_0 using fixed-gain AF with PS. Again, P_1 for fixed-gain and variable-gain AF and P_1 for DF using the PS method are identical, as predicted by the derived expressions of P_1, so that P_1 for variable-gain AF and DF are not shown here. From Figure 8.40, the value of P_1 again increases dramatically when γ_0 increases, and in fact γ_0 has the largest impact on P_1. In this case, η also

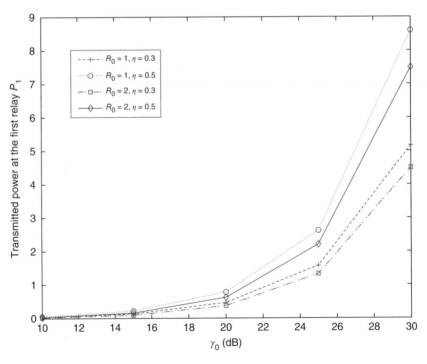

Figure 8.37 P_1 for different values of γ_0, R_0 and η for fixed-gain AF with TS.

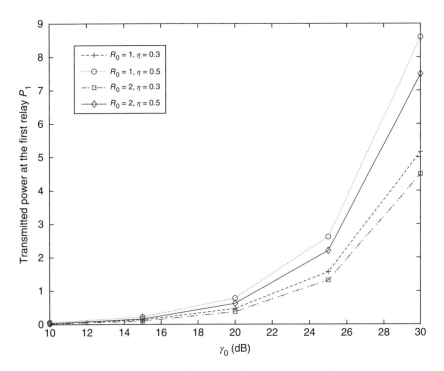

Figure 8.38 P_1 for different values of γ_0, R_0 and η for variable-gain AF with TS.

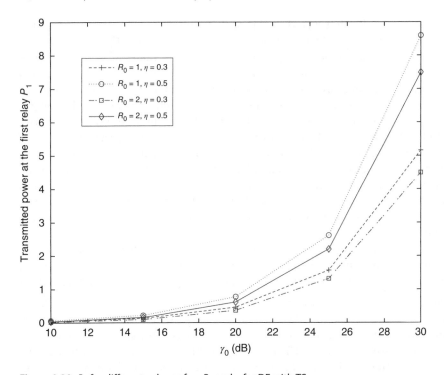

Figure 8.39 P_1 for different values of γ_0, R_0 and η for DF with TS.

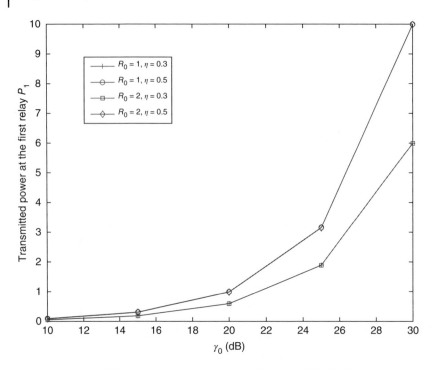

Figure 8.40 P_1 for different values of γ_0, R_0 and η for fixed-gain AF with PS.

has a large impact on P_1. However, the value of R_0 has little impact on P_1, implying that the throughput requirement does not change the received power in the first relay.

Figures 8.41 and 8.42 show α_1 and ρ_1 versus γ_0 for different values of R_0 and η, respectively, for fixed-gain AF. In general, the values of α_1 and ρ_1 increase when the value of γ_0 increases, indicating that more energy will be harvested and transferred to the next hop than is used at the first hop for relaying or decoding, as expected, as once the throughput requirement is met, the first relay should pass on all the remaining energy that increases with γ_0 to the next hop for extended coverage. Interestingly, in this case, the value of η has little impact on α_1 or ρ_1. Comparing Figure 8.41 with Figure 8.42, one sees that the PS factor ρ_1 is relatively high, ranging from 0.82 to nearly 1 in these cases, while the TS coefficient α_1 is relatively low, ranging from 0.32 and 0.86 in these cases. Thus, the PS method harvests more energy at the first relay than the TS method, in the cases considered. Similar graphs for variable-gain AF and DF can also be obtained. They are omitted here.

Figure 8.43 shows how the values of α_m and ρ_m change with m. They decay very quickly and become zero when $m = 3$. Figure 8.44 shows the largest number of hops versus γ_0, where $2\sigma_{ma}^2 = 2\sigma_{md}^2 = 0.01$ and $|g_m|^2 = 0.1$. Also, $\gamma_0 = \frac{|g_1|^2 P_s T}{2\sigma_{1d}^2 + 2\sigma_{1a}^2}$ is the initial SNR, which is directly related to the initial amount of energy $P_s T$ from the source node. For PS, this number is always 1, meaning that PS is not suitable for such multi-hop relaying. For TS, the number increases with γ_0, as expected, as more initial energy allows the signal to go further. This can be explained. For TS in AF, the power for harvesting and for forwarding is the same. Only the time is switched. Thus, the first relay can spend a small amount of

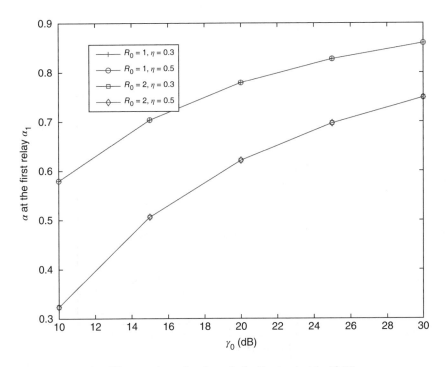

Figure 8.41 α_1 for different values of γ_0, R_0 and η for fixed-gain AF with TS.

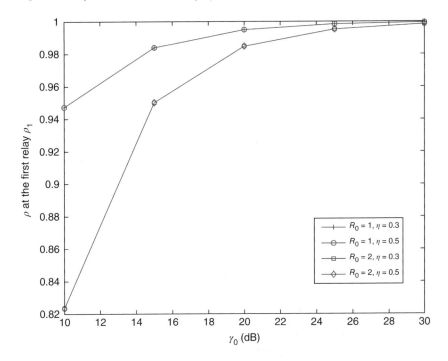

Figure 8.42 ρ_1 for different values of γ_0, R_0 and η for fixed-gain AF with PS.

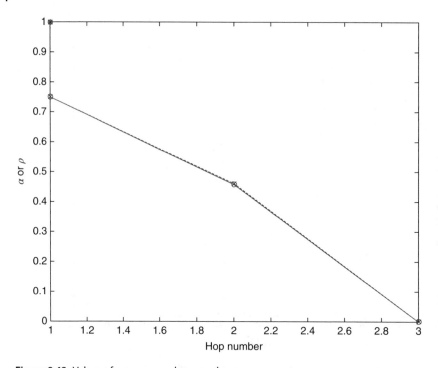

Figure 8.43 Values of α or ρ versus hop number.

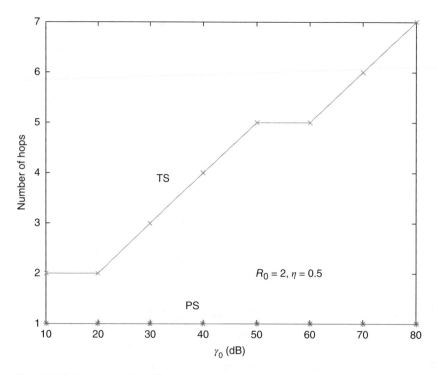

Figure 8.44 Largest number of hops versus γ_0.

time to harvest a small amount of energy (power is fixed in this case), leaving the rest of the energy for harvesting in the following hops. However, for PS in AF, the power for harvesting and forwarding are different. If the first relay splits a small portion of power to harvest a small amount of energy (time is fixed in this case), a strong signal will be forwarded but with weak power. If the first relay splits a large portion for harvesting, a weak signal will be forwarded but with strong power. In both cases, the received signal at the next relay will be very weak. Thus, PS is very ineffective for network coverage extension using energy harvesting.

8.7.3 Others

There are other important issues in energy harvesting relaying. For example, in the above, we have only considered the half-duplex mode. This is a typical operational mode for wireless devices. However, full-duplex has also received a lot of research interest recently. For example, in Zhong et al. (2014), full-duplex relaying was considered. In this study, the relay transmits the signal to the destination at the same time as the source transmits the signal to the relay. Thus, the received signal at the full-duplex relay contains the signal from the source as well as the interference caused by its own transmission. This interference can be partly canceled, as the relay knows its transmitted information. Zhong et al. (2014) then evaluated the performance of this full-duplex relaying system. In Zeng and Zhang (2015a) the source node transmits information to the relay in the first phase. In the second phase, the source transfers power to the relay while the relay forwards the signal to the destination and performs energy harvesting at the same time. Thus, the relay is able to harvest its own transmitted signal. This is called self-energy recycling. In all these radios, to enable transmission and reception at the same time, multiple antennas with different purposes are often used.

Also, we have only considered the use of multiple relays to form a multi-hop link for range extension. Another typical use of multiple relays is virtual antennas that can form multiple dual-hop links for diversity gain. In this case, relay selection can be performed for a simple structure, as discussed in Section 8.2.5. However, relay selection in energy harvesting relaying is slightly different from that in conventional relaying. In energy harvesting relaying, one may use the maximum harvested energy as the selection criterion to maximize the harvested energy at the relay, while conventional relaying normally maximizes the end-to-end or hop SNR. For example, in Michalopoulos et al. (2015), the maximum energy and the maximum SNR criteria were considered in a tradeoff to select the relay that gives both satisfactory capacity and energy transfer. Multiple antennas can also be used at different nodes to further increase the diversity gain (Ben Khelifa et al. 2016; Ben Khelifa and Alouini 2017b).

Finally, the relays considered so far always have fixed locations. Consequently, we have always included the distance-dependent path loss in the average fading power in the above discussion. Sometimes it is also necessary to express the path loss explicitly, for example, for relay placement designs. In this case, the average fading power can be replaced by $\frac{\Omega}{d^v}$, where Ω is the average fading power used before, d is the distance between two nodes, and v is the path loss exponent. In other cases, the nodes could be in motion so that their locations become random. The random location leads to random power loss in the relaying, which could degrade the relaying and harvesting performance. Ding et al. (2014) evaluated the effect of random location on the relaying performance. These studies require stochastic geometry theories.

8.8 Summary

In this chapter, energy harvesting relaying has been discussed in detail. First, background knowledge in conventional relaying has been introduced. Then, different types of energy harvesting relaying systems have been studied. In particular, we have focused on three important systems where the nodes harvest energy from the ambient environment, from the dedicated power transmitter, as well as systems where the relay harvests energy from the source. Following this, we have examined several important issues in energy harvesting relaying, including interference and multi-hop relaying.

Compared with conventional relaying, energy harvesting relaying is restricted by its energy supply, similar to the designs discussed in Chapter 6, Chapter 7, and other chapters. Thus, most design problems in energy harvesting relaying are related to the restricted energy. In systems where both source and relay harvest energy from the ambient environment, this is reflected by the energy arrival process or the energy causality. In systems where both source and relay harvest energy from the dedicated power transmitter or from each other in the system, this is reflected by the constrained transmission time or transmission power. In these systems, the optimum designs often maximize the achievable rate with respect to these energy, time or power parameters. Based on these designs, one can easily obtain solutions to similar problems where different nodes harvest energy from difference sources, as they can be considered as variants of the problems discussed in this chapter.

Compared with other energy harvesting communications systems, energy harvesting relaying requires at least three nodes and two transmission phases. The connection from one phase to the next phase can then be a problem, as the signal must appear in the first phase before it appears in the second phase. Hence, we have the signal or data causality. Also, the two-hop or multi-hop structures make the design problems more complicated than those in the conventional one-hop structure.

In conclusion, for energy harvesting relaying, the increased number of nodes and the increased number of hops lead to more design problems and make each of the design problems more complicated, but as a promising solution to the energy supply at the relaying node, energy harvesting relaying is seeing more and more applications.

References

G. Amarasuriya, C. Tellambura, and M. Ardakani. Asymptotically-exact performance bounds of AF multi-hop relaying over Nakagami fading. *IEEE Transactions on Communications*, 59:962–967, 2011.

M.A. Antepli, E. Uysal-Biyikoglu, and H. Erkal. Optimal packet scheduling on an energy harvesting broadcast link. *IEEE Journal on Selected Areas in Communications*, 29:1721–1731, 2011.

A. Aprem, C.R. Murthy, and N.B. Mehta. Transmit power control policies for energy harvesting sensors with retransmissions. *IEEE Journal of Selected Topics in Signal Processing*, 7:895–906, 2013.

S. Avallone and A. Banchs. A channel assignment and routing algorithm for energy harvesting multiradio wireless mesh networks. *IEEE Journal on Selected Areas in Communications*, 34:1463–1476, 2016.

F. Azmat, Y. Chen, and N. Stocks. Predictive modelling of RF energy for wireless powered communications. *IEEE Communications Letters*, 20:173–176, 2016.

F. Azmat, Y. Zhou, and Y. Chen. New cooperative strategy for cognitive radios with wireless powered primary users. *International Journal of Communication Systems*, 31:1–1, 2018.

B.C. Babu and S. Gurjar. A novel simplified two-diode model of photovoltaic (PV) module. *IEEE Journal of Photovoltaics*, 4:1156–1161, 2014.

Y.H. Bae and J.W. Baek. Achievable throughput analysis of opportunistic spectrum access in cognitive radio networks with energy harvesting. *IEEE Transactions on Communications*, 64:1399–1400, 2015.

S. Beeby and N. White. *Energy Harvesting for Autonomous Systems*. Artech House, 2010.

F. Ben Khelifa and M.-S. Alouini. Prioritizing data/energy thresholding-based antenna switching for SWIPT-enabled secondary receiver in cognitive radio networks. *IEEE Transactions on Cognitive Communications and Networking*, 3:782–800, 2017a.

F. Ben Khelifa and M.-S. Alouini. Precoding design of MIMO amplify-and-forward communication system with an energy harvesting relay and possibly imperfect CSI. *IEEE Access*, 5:578–594, 2017b.

F. Ben Khelifa, A. Sultan, and M.-S. Alouini. Sum-rate enhancement in multiuser MIMO decode-and-forward relay broadcasting channel with energy harvesting relays. *IEEE Journal on Selected Areas in Communications*, 34:3675–3684, 2016.

F. Ben Khelifa, K. Tourki, and M.-S. Alouini. Proactive spectrum sharing for SWIPT in MIMO cognitive radio systems using antenna switching technique. *IEEE Transactions on Green Communication and Networking*, 1:204–222, 2017.

Energy Harvesting Communications: Principles and Theories, First Edition. Yunfei Chen.
© 2019 John Wiley & Sons Ltd. Published 2019 by John Wiley & Sons Ltd.

C. Bergozini, D. Brunelli, and L. Benini. Comparison of energy intake prediction algorithms for systems powered by photovoltaic harvesters. *Microelectronics Journal*, 41:766–777, 2010.

H.J. Bergveld, W.S. Kruijt, and P.H.L. Notten. Electronic-network modeling of rechargeable NiCd cells and its application to the design of battery management systems. *Journal of Power Sources*, 77:143–158, 1999.

E. Bjornson, J. Hoydis, M. Kountouris, and M. Debbah. Massive MIMO systems with non-ideal hardware: energy efficiency, estimation, and capacity limits. *IEEE Transactions on Information Theory*, 60:7112–7139, 2014.

E. Boshkovska, D.W.K. Ng, N. Zlatanov, and R. Schober. Practical non-linear energy harvesting model and resource allocation for SWIPT systems. *IEEE Communications Letters*, 19:2082–2085, 2015.

M. Buettner, G. Yee, E. Anderson, and R. Han. X-MAC: a short preamble MAC protocol for duty-cycled wireless sensor networks. *ACM SenSys 2006*, 1:307–320, 2006.

S. Buzzi, C.-L. I, T.E. Klein, H.V. Poor, C. Yang, and A. Zappone. A survey of energy-efficient techniques in 5G networks and challenges ahead. *IEEE Journal on Selected Areas in Communications*, 34:697–709, 2016.

Y. Cao, X.-Y. Liu, M.-Y. Wu, and M.K. Khan. EHR: routing protocol for energy harvesting wireless sensor networks. *IEEE International Conference on Parallel and Distributed Systems*, 1:56–63, 2016.

B.K. Chalise, H.A. Surweera, G. Zheng, and G.K. Karagiannidis. Beamforming optimization for full-duplex wireless-powered MIMO systems. *IEEE Transactions on Communications*, 65:3750–3764, 2017.

H. Chen, Y. Li, J.L. Rebelatto, B.F. Uchoa-Filho, and B. Vucetic. Harvest-then-cooperate: wireless-powered cooperative communications. *IEEE Transactions on Signal Processing*, 63:1700–1711, 2015.

Y. Chen. Energy-harvesting af relaying in the presence of interference and Nakagami-m fading. *IEEE Transactions on Wireless Communications*, 15:1008–1017, 2016.

Y. Chen and N.C. Beaulieu. Optimum pilot symbol assisted modulation. *IEEE Transactions on Communications*, 55:1536–1546, 2007.

Y. Chen and N.C. Beaulieu. Performance of collaborative spectrum sensing for cognitive radio in the presence of Gaussian channel estimation errors. *IEEE Transactions on Communications*, 57:1944–1947, 2009.

Y. Chen, D.B. da Costa, and H. Ding. Effect of CCI on WPC with time-division energy and information transmission. *IEEE Wireless Communications Letters*, 5:168–171, 2016a.

Y. Chen, D.B. da Costa, and H. Ding. Interference analysis in wireless power transfer. *IEEE Communications Letters*, 21:2318–2321, 2017a.

Y. Chen, W. Feng, R. Shi, and N. Ge. Pilot-based channel estimation for AF relaying using energy harvesting. *IEEE Transactions on Vehicular Technology*, 66:6877–6886, 2017b.

Y. Chen, G.K. Karagidinis, H. Lu, and N. Cao. Novel approximations to the statistics of products of independent random variables and their applications in wireless communications. *IEEE Tranactions on Vehicular Technology*, 61: 443–454, 2012.

Y. Chen, J.A. Nossek, and A. Mezghani. Circuit-aware cognitive radios for efficient communications. *IEEE Wireless Communications Letters*, 2:323–326, 2013.

Y. Chen and H.-S. Oh. A survey of measurement-based spectrum occupancy modeling for cognitive radios. *IEEE Communications Surveys and Tutorials*, 18:848–859, 2016a.

Y. Chen and H.-S. Oh. Spectrum measurements modelling and prediction based on wavelets. *IET Communications*, 10:2192–2198, 2016b.

Y. Chen, K.T Sabnis, and R.A. Abd-Alhameed. New formula for conversion efficiency of RF EH and its wireless applications. *IEEE Transactions on Vehicular Technology*, 65:9410–9414, 2016b.

Y. Chen, R. Shi, W. Feng, and N. Ge. AF relaying with energy harvesting source and relay. *IEEE Transactions on Vehicular Technology*, 66:874–879, 2017c.

Y. Chen, C.-X. Wang, H. Xiao, and D. Yuan. Novel partial relay selection for cooperative diversity in Nakagami-m fading channels. *IEEE Transactions on Vehicular Technology*, 60:3497–3503, 2011a.

Y. Chen, C. Wang, and B. Zhao. Performance comparison of feature-based detectors for spectrum sensing in the presence of primary user traffic. *IEEE Signal Processing Letters*, 18:291–294, 2011b.

Y. Chen, N. Zhao, and M.-S. Alouini. Wireless energy harvesting using signals from multiple fading channels. *IEEE Transactions on Communications*, 65: 5027–5039, 2017d.

Y. Cheng, P. Fu, Y. Ding, B. Li, and X. Yuan. Proportional fairness in cognitive wireless powered communication networks. *IEEE Communications Letters*, 21:1397–1400, 2017.

Y. Chen, Z. Xie, and N. Zhao. Energy analysis of co-channel harvesting in wireless networks. *IEEE Communications Letters*, 22:530–533, 2018.

W. Chung, S. Park, S. Lim, and D. Hong. Spectrum sensing optimization for energy-harvesting cognitive radio systems. *IEEE Transactions on Wireless Communications*, 13:2601–2613, 2014.

B. Clerckx and E. Bayguzina. Waveform design for wireless power transfer. *IEEE Transactions on Signal Processing*, 64:6313–6328, 2016.

A. Collado and A. Georgiadis. Optimal waveforms for efficient wireless power transmission. *IEEE Microwave and Wireless Components Letters*, 24:354–356, 2014.

D.R. Cox. Prediction by exponentially weighted moving averages and related methods. *Journal of the Royal Statistical Society: Series B (Methodological)*, 23:414–422, 1961.

T.V. Dam and K. Langendoen. An adaptive energy-efficient MAC protocol for wireless sensor networks. *ACM SenSys 2003*, 1:171–180, 2003.

P.D. Diamantoulakis, K.N. Pappi, G.K. Karagiannidis, H. Xing, and A. Nallanathan. Joint downlink/uplink design for wireless powered networks with interference. *IEEE Access*, 5:1534–1547, 2017.

Z. Ding, I. Krikidis, B. Sharif, and H.V. Poor. Wireless information and power transfer in cooperative networks with spatially random delays. *IEEE Transactions on Wireless Communications*, 13:4440–4453, 2014.

Z. Ding, X. Lei, G.K. Karagiannidis, R. Schober, J. Yuan, and V.K. Bhargava. A survey on non-orthogonal multiple access for 5G networks: research challenges and future trends. *IEEE Journal on Selected Areas in Communications*, 35: 2181–2195, 2017a.

Z. Ding, Y. Liu, J. Choi, Q. Sun, M. Elkashlan, C.-L. I, and H.V. Poor. Application of non-orthogonal multiple access in LTE and 5G networks. *IEEE Communications Magazine*, 55:185–191, 2017b.

R. Doose, K.R. Chowdhury, and M.D. Felice. Routing and link layer protocol design for sensor networks with wireless energy transfer. *Proceedings of IEEE GLOBECOM*, 1:1–5, 2010.

M. Doyle, T.F. Fuller, and J. Newman. Modeling of galvanostatic charge and discharge of the lithium/polymer/insertion cell. *Journal of the Electrochemical Society*, 140:1526–1533, 1993.

A. El Shafie, N. Al-Dhahir, and R. Hamila. A sparsity-aware cooperative protocol for cognitive radio networks with energy-harvesting primary user. *IEEE Transactions on Communications*, 63:3118–3131, 2015.

M. Erol-Kantarci and H.T. Touftah. Suresense: sustainable wireless rechargeable sensor networks for the smart grid. *IEEE Wireless Communications*, 19:30–36, 2012.

Z.A. Eu, H.-P. Tan, and W.K.G. Seah. Design and performance analysis of MAC schemes for wireless sensor networks powered by ambient energy harvesting. *Ad Hoc Networks*, 9:300–323, 2011.

X. Fafoutis and N. Dragoni. ODMAC: an on-demand MAC protocol for energy harvesting wireless sensor networks. *Proceedings of ACM PE-WASUN*, 1:49–56, 2011.

FCC. Notice of proposed rule making. *ET Docket No. 04-113*, 2004.

FCC Spectrum Policy Task Force. Report of the spectrum efficiency working group. *Technical Report*, 2002.

I. Flint, X. Lu, N. Privault, D. Niyato, and P. Wang. Performance analysis of ambient RF energy harvesting: a stochastic geometry approach. *IEEE Global Telecommunications Conference 2014*, 1:1448–1453, 2014.

Y. Gao, Y. Chen, and A. Bekkali. Performance of passive UHF RFID in cascaded correlated generalized Rician fading. *IEEE Communications Letters*, 20:660–663, 2016a.

Y. Gao, Y. Chen, and A. Hu. Throughput and BER of wireless powered DF relaying in Nakagami-m fading. *Science China Information Sciences*, 60, 2017.

Y. Gao, Y. Chen, Z. Xie, and G. Hu. Wireless energy-harvesting cognitive radio with feature detectors. *KSII Transactions on Internet and Information Systems*, 10:4625–4641, 2016b.

Y. Gao, Y. Chen, Y. Zhou, and N. Cao. BER and achievable rate analysis of wireless powered communications with correlated links. *IET Communications*, 12:310–316, 2018.

J. Gong, S. Zhang, X. Wang, S. Zhou, and Z. Niu. Supporting quality of service in energy harvesting wireless links: the effective capacity analysis. *IEEE International Conference on Communications Workshops*, 1:901–906, 2014.

J.A. Gow and C.D. Manning. Development of a photovoltaic array model for use in power-electronics simulation studies. *IEEE Proceedings – Electrical Power Applications*, 146:193–200, 1999.

I.S. Gradshteyn and I.M. Ryzhik. *Tables of Integrals, Series and Products*. Academic Press, 2000.

P. Grover and A. Sahai. Shannon meets Tesla: wireless information and power transfer. *Proceedings of the IEEE International Symposium on Information Theory*, 1:2363–2367, 2010.

T. Ha, J. Kim, and J.-M. Chung. HE-MAC: harvest-then-transmit based modified EDCF MAC protocol for wireless powered sensor networks. *IEEE Transactions on Wireless Communications*, 17:3–16, 2018.

M.A. Hasan and S.K. Parida. An overview of solar photovoltaic panel modeling based on analytical and experimental viewpoint. *Renewable and Sustainable Energy Reviews*, 60:75–83, 2016.

D. Hasenfratz, A. Meier, C. Moser, J.-J. Chen, and L. Thiele. Analysis, comparison, and optimization of routing protocols for energy harvesting wireless sensor networks. *IEEE International Conference on Sensor Networks, Ubiquitous, and Trustworthy Computing*, 1:19–26, 2010.

M.O. Hasna and M-S. Alouini. Outage probability of multihop transmission over Nakagami fading channels. *IEEE Communications Letters*, 7:216–218, 2003.

M.O. Hasna and M.-S. Alouini. Harmonic mean and end-to-end performance of transmission systems with relays. *IEEE Transactions on Communications*, 52:130–135, 2004.

S. Haykin. Cognitive radio: brain-empowered wireless communications. *IEEE Journal on Selected Areas in Communications*, 23:201–220, 2005.

C. He, A. Arora, M.E. Kiziroglou, D.C. Yates, D. OHare, and E.M. Yeatman. MEMS energy harvesting powered wireless biometric sensor. *Proceedings of the International Workshop on Wearable and Implantable Body Sensor Networks 2009*, 1:207–212, 2009.

P. He, L. Zhao, S. Zhou, and Z. Niu. A dual band rectenna using broadband Yagi antenna array for ambient RF power harvesting. *IEEE Transactions on Vehicular Technology*, 64:4525–4536, 2015.

C.K. Ho and R. Zhang. Optimal energy allocation for wireless communications with energy harvesting constraints. *IEEE Transactions on Signal Processing*, 60:4808–4818, 2012.

E. Hossain and M. Hasan. 5G cellular: key enabling technologies and research challenges. *IEEE Instrumentation and Measurement Magazine*, 18:11–21, 2015.

C. Hoymann, W. Chen, J. Montojo, A. Golitschek, C. Koutsimanis, and X. Shen. Relaying operation in 3GPP LTE: challenges and solutions. *IEEE Communications Magazine*, 50:156–162, 2012.

K.-Y. Hsieh, F.-S. Tseng, and T.-L. Ku. A spectrum and energy cooperation strategy in hierarchical cognitive radio cellular networks. *IEEE Wireless Communications Letters*, 5:252–255, 2016.

J. Hsu, S. Zahedi, A. Kansal, M. Srivastava, and V. Raghunathan. Adaptive duty cycling for energy harvesting systems. *Proceedings of IEEE ISLPED 2006*, 1:180–185, 2006.

C. Huang, R. Zhang, and S. Cui. Throughput maximization for the Gaussian relay channel with energy harvesting constraints. *IEEE Journal on Selected Areas in Communications*, 31:1469–1479, 2013.

K. Huang and V.K.N. Lau. Enabling wireless power transfer in cellular networks: architecture, modeling and deployment. *IEEE Transactions on Wireless Communications*, 13:902–912, 2014.

D. Hwang, K.C. Hwang, D.I. Kim, and T.-J. Lee. Self-energy recycling for RF powered multi-antenna relay channels. *IEEE Transactions on Wireless Communications*, 16:812–824, 2017.

F. Iannello, O. Simeone, and U. Spagnolini. Medium access control protocols for wireless sensor networks with energy harvesting. *IEEE Transactions on Communications*, 60:1381–1389, 2012.

R. Jain, A. Durresi, and G. Babie. Throughput fairness index: an explanation. *ACM Forum Contribution 99-0045*, 1999.

X. Ji, J. Xu, Y.L. Che, Z. Fei, and R. Zhang. Adaptive mode switching for cognitive wireless powered communications systems. *IEEE Wireless Communications Letters*, 6:386–389, 2017.

L. Jiang, H. Tian, S. Gjessing, and Y. Zhang. Secure beamforming in wireless-powered cooperative cognitive radio networks. *IEEE Communications Letters*, 20:522–525, 2016.

Y. Jing and H. Jafarkhani. Single and multiple relay selection schemes and their achievable diversity orders. *IEEE Transactions on Wireless Communications*, 8:1414–1423, 2009.

N.L. Johnson, S. Kotz, and N. Balakrishnan. *Continuous Univariate Distributions - I*. Wiley, 1994.

H. Ju and R. Zhang. Throughput maximization in wireless powered communication networks. *IEEE Transactions on Wireless Communications*, 13:418–428, 2014.

M. Ju, K. Kang, K. Hwang, and C. Jeong. Maximum transmission rate of PSR/TSR protocols in wireless energy harvesting DF-based relay networks. *IEEE Journal on Selected Areas in Communications*, 33:2701–2717, 2015.

P. Kamalinejad, C. Mahapatra, Z. Sheng, S. Mirabbasi, V.C.M. Leung, and Y.L. Guan. Wireless energy harvesting for the Internet of Things. *IEEE Communications Magazine*, 53:102–108, 2015.

A. Kansal, J. Hsu, S. Zahedi, and M.B. Srivastava. Power management in energy harvesting sensor networks. *ACM Transactions on Embedded Computing Systems*, 6:1–35, 2007.

M. Kashef and A. Ephremides. Optimal partial relaying for energy-harvesting wireless networks. *IEEE/ACM Transactions on Networking*, 24:113–122, 2016.

K. Kawashima and F. Sato. A routing protocol based on the power generation pattern of sensor nodes in energy harvesting wireless sensor networks. *International Conference on Network-Based Information Systems*, 1:470–475, 2013.

S.M. Kay. *Fundamentals of Statistical Processing, Volume I: Estimation Theory*. Prentice Hall, 1993.

S.M. Kay. *Fundamentals of Statistical Processing, Volume II: Detection Theory*. Prentice Hall, 1998.

F.H. Khan, Y. Chen, and M.-S. Alouini. Novel receivers for AF relaying with distributed STBC using cascaded and disintegrated channel estimation. *IEEE Transactions on Wireless Communications*, 11:1370–1379, 2012.

T.A. Khan, A. Alkhateeb, and R.W. Heath. Millimeter wave energy harvesting. *IEEE Transactions on Wireless Communications*, 15:6048–6062, 2016.

J. Kim, and J.-W. Lee. Energy adaptive MAC protocol for wireless sensor networks with RF energy transfer. *IEEE ICUFN 2011*, 1:89–94, 2011.

J. Kim and J.-W. Lee. Performance analysis of the energy adaptive MAC protocol for wireless sensor networks with RF energy transfer. *IEEE ICTC 2011*, 1:14–19, 2011.

S. Kisseleff, I.F. Akyildiz, and W.H. Gerstacker. Magnetic induction-based simultaneous wireless information and power transfer for single information and multiple power receivers. *IEEE Transactions on Communications*, 65: 1396–1410, 2017.

K. Kotani and T. Ito. High efficiency CMOS rectifier circuit with self Vth cancellation and power regulation functions for UHF RFIDs. *IEEE ASSCC 2007*, 1:119–122, 2007.

K. Kotani, A. Sasaki, and T. Ito. High-efficiency differential-drive CMOS rectifier for UHF RFIDs. *IEEE Journal of Solid-State Circuits*, 44:3011–3018, 2009.

I. Krikidis, G. Zheng, and B. Ottersten. Harvest-use cooperative networks with half/full-duplex relaying. *IEEE WCNC 2013*, 1:4256–4260, 2013.

M.L. Ku, Y. Chen, and K.J.R. Liu. Data-driven stochastic models and policies for energy harvesting sensor communications. *IEEE Journal on Selected Areas in Communications*, 33:1505–1520, 2015.

M.-L. Ku, Y. Han, H.-Q. Lai, Y. Chen, and K.J.R. Liu. Power waveforming: wireless power transfer beyond time reversal. *IEEE Transactions on Signal Processing*, 64:5819–5834, 2016.

J.N. Laneman, D.N.C. Tse, and G.W. Wornell. Cooperative diversity in wireless networks: efficient protocols and outage behavior. *IEEE Transactions on Information Theory*, 50:3062–3080, 2004.

T. Le, K. Mayaram, and T. Fiez. Efficient far-field radio frequency energy harvesting for passively powered sensor networks. *IEEE Journal of Solid-State Circuits*, 43:1287–1302, 2008.

P. Lee, Z.A. Eu, M. Han, and H.-P. Tan. Empirical modeling of a solar-powered energy harvesting wireless sensor node for time-slotted operation. *IEEE WCNC 2011*, 1:179–184, 2011.

S. Lee and R. Zhang. Cognitive wireless powered network: spectrum sharing models and throughput maximization. *IEEE Transactions on Cognitive Communications and Networks*, 1:335–346, 2015.

S. Lee, R. Zhang, and K. Huang. Opportunistic wireless energy harvesting in cognitive radio networks. *IEEE Transactions on Wireless Communications*, 12:4788–4799, 2013.

V. Leonov. Thermoelectric energy harvesting of human body heat for wearable sensors. *IEEE Sensors Journal*, 13:2284–2291, 2013.

Z. Li, Y. Peng, W. Zhang, and D. Qiao. J-RoC: a joint routing and charging scheme to prolong sensor network lifetime. *IEEE International Conference on Network Protocols*, 1:373–382, 2011.

H. Liang, C. Zhong, H.A. Suraweera, G. Zheng, and Z. Zhang. Optimization and analysis of wireless powered multi-antenna cooperative systems. *IEEE Transactions on Wireless Communications*, 16:3267–3281, 2017.

Y.C. Liang, Y. Zeng, E.C.Y. Peh, and A.T. Hoang. Sensing-throughput tradeoff for cognitive radio networks. *IEEE Transactions on Wireless Communications*, 7:1326–1337, 2008.

L. Liu, R. Zhang, and K.-C. Chua. Wireless information transfer with opportunistic energy harvesting. *IEEE Transactions on Wireless Communications*, 12:288–300, 2013.

L. Liu, R. Zhang, and K.-C. Chua. Secrecy wireless information and power transfer with MISO beamforming. *IEEE Transactions on Signal Processing*, 62:1850–1863, 2014a.

L. Liu, R. Zhang, and K.-C. Chua. Multi-antenna wireless powered communication with energy beamforming. *IEEE Transactions on Communications*, 62:4349–4361, 2014b.

P. Liu, S. Gazor, I.-M. Kim, and D.I. Kim. Noncoherent relaying in energy harvesting communication systems. *IEEE Transactions on Wireless Communications*, 14:6940–6954, 2015a.

R. Liu and W. Trappe. *Securing Wireless Communications at the Physical Layer*. Springer, 2009.

Y. Liu, Z. Yang, R. Yu, Y. Xiang, and S. Xie. An efficient MAC protocol with adaptive energy harvesting for machine-to-machine networks. *IEEE Access*, 3:358–369, 2015b.

Y. Liu, Y. Zhang, R. Yu, and S. Xie. Integrated energy and spectrum harvesting for 5G wireless communications. *IEEE Network*, 29:75–81, 2015c.

X. Lu, P. Wang, D. Niyato, D.I. Kim, and Z. Han. Wireless networks with RF energy harvesting: a contemporary survey. *IEEE Communications Surveys and Tutorials*, 17:757–789, 2015.

Y. Luo, J. Zhang, and K.B. Letaief. Optimal scheduling and power allocation for two-hop energy harvesting communication systems. *IEEE Transactions on Wireless Communications*, 12:4729–4741, 2013.

Y. Ma, H. Chen, Z. Lin, Y. Li, and Vucetic B. Distributed and optimal resource allocation for power beacon-assisted wireless-powered communications. *IEEE Transactions on Communications*, 63:3569–3583, 2015.

Y. Mao, G. Yu, and C. Zhong. Energy consumption analysis of energy harvesting systems with power grid. *IEEE Wireless Communications Letters*, 2:611–614, 2013.

M. Mao, N. Cao, Y. Chen, and Y. Zhou. Multi-hop relaying using energy harvesting. *IEEE Wireless Communications Letters*, 4:565–568, 2015.

G. Martinez, S. Li, and C. Zhou. Wastage-aware routing in energy-harvesting wireless sensor networks. *IEEE Sensors Journal*, 14:2967–2974, 2014.

J. Masuch, M. Delgado-Restituto, D. Milosevic, and P. Baltus. An RF-to-DC energy harvester for co-integration in a low power 2.4GHz transceiver frontend. *IEEE International Symposium on Circuits and Systems 2012*, 1: 680–683, 2012.

B. Medepally and N.B. Mehta. Voluntary energy harvesting relays and selection in cooperative wireless networks. *IEEE Transactions on Wireless Communications*, 9:3543–3553, 2010.

D. Michalopoulos, H.A. Suraweera, and R. Schober. Relay selection for simultaneous information transmission and wireless energy transfer: a tradeoff perspective. *IEEE Journal on Selected Areas in Communications*, 33: 1578–1594, 2015.

N. Michelusi, K. Stamatiou, and M. Zorzi. Transmission policies for energy harvesting sensors with time-correlated energy supply. *IEEE Transactions on Communications*, 61:2988–3001, 2013.

A. Minasian, S. ShahbazPanahi, and R.S. Adve. Energy harvesting cooperative communication systems. *IEEE Transactions on Wireless Communications*, 13: 6118–6131, 2014.

P.D. Mitcheson, E.M. Yeatman, G.K. Rao, A.S. Holmes, and B.S. Shariff. Energy harvesting from human and machine motion for wireless electronic devices. *Proceedings of the IEEE*, 96:1457–1486, 2008.

R. Moghe, Y. Yang, F. Lambert, and D. Divan. A scoping study of electric and magnetic field energy harvesting for wireless sensor networks in power system applications. *Proceedings of IEEE ECCE 2009*, 1:3550–3557, 2009.

M. Moradian and F. Ashtiani. Optimal relaying in a slotted aloha wireless network with energy harvesting nodes. *IEEE Journal on Selected Areas in Communications*, 33:1680–1692, 2015.

M.Y. Naderi, P. Nintanavongsa, and K.R. Chowdhury. RF-MAC: a medium access control protocol for re-chargeable sensor networks powered by wireless energy harvesting. *IEEE Transactions on Wireless Communications*, 13:3926–3937, 2014.

A.A. Nasir, X. Zhou, S. Durrani, and R.A. Kennedy. Relaying protocols for wireless energy harvesting and information processing. *IEEE Transactions on Wireless Communications*, 12:3622–3636, 2013.

A.A. Nasir, X. Zhou, S. Durrani, and R.A. Kennedy. Throughput and ergodic capacity of wireless energy harvesting based DF relaying networks. *Proceedings of IEEE ICC 2014*, 1:4066–4071, 2014.

D.W.K. Ng, E.S. Lo, and R. Schober. Multiobjective resource allocation for secure communication in cognitive radio networks with wireless information and power transfer. *IEEE Transactions on Communications*, 65:3166–3184, 2016.

K. Nguyen, V.-H. Nguyen, D.-D. Le, Y. Ji, D.A. Duong, and S. Yamada. ERI-MAC: an energy-harvested receiver-initiated MAC protocol for wireless sensor networks. *International Journal of Distributed Sensor Networks*, 10:1–8, 2014.

P. Nintanavongsa, U. Muncuk, D.R. Lewis, and K.R. Chowdhury. Design optimisation and implementation for RF energy harvesting circuits. *IEEE Journal of Emerging and Selected Topics in Circuits and Systems*, 2:24–33, 2012.

K. Nishioka, N. Sakitani, Y. Uraoka, and T. Fuyuki. Analysis of multicrystalline silicon solar cells by modified 3-diode equivalent circuit model taking leakage current through periphery into consideration. *Solar Energy Materials Solar Cells*, 91:1222–1227, 2007.

O. Orhan and E. Erkip. Energy harvesting two-hop communication networks. *IEEE Journal on Selected Areas in Communications*, 33:2658–2670, 2015.

O. Ozel, K. Tutuncuoglu, J. Yang, S. Ulukus, and A. Yener. Transmission with energy harvesting nodes in fading wireless channels: optimal policies. *IEEE Journal on Selected Areas in Communications*, 29:1732–1743, 2011.

O. Ozel and S. Ulukus. Achieving AWGN capacity under stochastic energy harvesting. *IEEE Transactions on Information Theory*, 58:6471–6483, 2012.

G. Pan, H. Lei, Y. Deng, L. Fan, J. Yang, Y. Chen, and Z. Ding. On secrecy performance of MISO SWIPT systems with TAS and imperfect CSI. *IEEE Transactions on Communications*, 64:3831–3843, 2016.

J.A. Paradiso and T. Starner. Energy scavenging for mobile and wireless electronics. *IEEE Pervasive Computing*, 4:18–27, 2005.

S. Park and D. Hong. Optimal spectrum access for energy harvesting cognitive radio networks. *IEEE Transactions on Wireless Communications*, 12:6166–6179, 2013.

S. Park and D. Hong. Achievable throughput of energy harvesting cognitive radio networks. *IEEE Transactions on Wireless Communications*, 13:1010–1022, 2014.

S. Park, H. Kim, and D. Hong. Cognitive radio networks with energy harvesting. *IEEE Transactions on Wireless Communications*, 12:1386–1397, 2013.

M. Peng, Y. Liu, D. Wei, W. Wang, and H.-H. Chen. Hierarchical cooperative relay based heterogeneous networks. *IEEE Wireless Communications*, 18: 48–56, 2011.

Y. Peng, Z. Li, W. Zhang, and D. Qiao. Prolonging sensor network lifetime through wireless charging. *Proceedings of IEEE RTSS*, 1:129–139, 2010.

M. Pinuela, P.D. Mitcheson, and S. Lucyszyn. Ambient RF energy harvesting in urban and semi-urban environments. *IEEE Transactions on Microwave Theory and Techniques*, 61:2715–2726, 2013.

J.R. Piorno, C. Bergonzini, D. Atienza, and T.S. Rosing. Prediction and management in energy harvested wireless sensor nodes. *Proceedings of Wireless VITAE 2009*, 1:6–10, 2009.

J. Polastre, J. Hill, and D. Culler. Versatile low power media access for wireless sensor networks. *ACM SenSys 2004*, 1:95–107, 2004.

J. Pradha J, S.S. Kalamkar, and A. Banerjee. Energy harvesting cognitive radio with channel-aware sensing strategy. *IEEE Communications Letters*, 18:1171–1174, 2014.

M. Pratibha, K.H. Li, and K.C. Teh. Channel selection in multichannel cognitive radio systems employing RF energy harvesting. *IEEE Transactions on Vehicular Technology*, 65:457–462, 2016.

M. Pratibha, K.H. Li, and K.C. Teh. Optimal spectrum access and energy supply for cognitive radio systems with opportunistic RF energy harvesting. *IEEE Transactions on Vehicular Technology*, 66:7114–7122, 2017.

J.G. Proakis. *Digital Communications*. McGraw-Hill, 2001.

L.P. Qian, G. Feng, and V.C.M. Leung. Optimal transmission policies for relay communication networks with ambient energy harvesting relays. *IEEE Journal on Selected Areas in Communications*, 34:3754–3768, 2016.

QinetiQ. Cognitive radio technology: a study for Ofcom. *Technical Report*, 2007.

T. Rakia, Y.-C. Yang, F. Gebali, and M.-S. Alouini. Optimal design of dual-hop VLC/RF communication system with energy harvesting. *IEEE Communications Letters*, 20:1979–1982, 2016.

R. Ramanathan and R. Hain. Toplogy control of multihop wireless networks using transmit power adjustment. *IEEE INFOCOM 2000*, 2:404–413, 2000.

D. Ramasur and G.P. Hancke. A wind energy harvester for low power wireless sensor networks. *Proceedings of the IEEE International Instrumentation and Measurement Technology Conference*, 1:2623–2627, 2012.

P. Ramezani and A. Jamalipour. Throughput maximization in dual-hop wireless powered communication networks. *IEEE Transactions on Vehicular Technologies*, 66:9304–9312, 2017.

R. Rao, S. Vrudhula, and D.N. Rakhmatov. Battery modeling for energy aware system design. *Computer Magazine*, 36:77–87, 2003.

U. Raza, P. Lukkarni, and M. Sooriyabandara. Low power wide area networks: an overview. *IEEE Communications Surveys and Tutorials*, 19:855–873, 2017.

C. Renner, J. Jessen, and V. Turau. Lifetime prediction for supercapacitor powered wireless sensor nodes. *Proceedings of FGSN 2009*, 1:55–58, 2009.

S. Roundy, P.K. Wright, and J. Rabaey. A study of low level vibrations as a power source for wireless sensor nodes. *Computer Communications*, 26:1131–1144, 2003.

S. Scorcioni, L. Larcher, and A. Bertacchini. A 868MHz CMOS RF-DC power converter with −17dBm input power sensitivity and efficiency higher than 40% over 14dB input range. *2012 ESSCIRC*, 1:109–112, 2012a.

S. Scorcioni, L. Larcher, and A. Bertacchini. Optimised CMOS RF-DC converters for remote wireless powering of RFID applications. *2012 IEEE International Conference on RFID*, 1:47–53, 2012b.

S. Scorcioni, L. Larcher, and A. Bertacchini. A reconfigurable differential CMOS RF energy scavenger with 60% peak efficiency and −21 dBm sensitivity. *IEEE Microwave and Wireless Components Letters*, 23:155–157, 2013.

H. Shafieirad, R.S. Adve, and S.S. Panahi. Max-SNR opportunistic routing for large-scale energy harvesting sensor networks. *IEEE Transactions on Green Communications and Networking*, PP:1–1, 2018.

C. Shen, W.-C. Li, and T.-H. Chang. Wireless information and energy transfer in multi-antenna interference channel. *IEEE Transactions Signal Processing*, 62:6249–6264, 2014.

M. Sherman, A.N. Mody, R. Martinez, and C. Rodrigues. IEEE standards supporting cognitive radio and networks, dynamic spectrum access, and coexistence. *IEEE Communications Magazine*, 46:72–79, 2008.

Q. Shi, W. Xu, T.-H. Chang, Y. Wang, and E. Song. Joint beamforming and power splitting for MISO interference channel with SWIPT: An SOCP relaxation and decentralized algorithm. *IEEE Transactions on Signal Processing*, 62: 6194–6208, 2014.

M.K. Simon and M.-S. Alouini. *Digital Communications Over Fading Channels*. Wiley, 2005.

A. Singh, M.R. Bhatnagar, and R.K. Malik. Secrecy outage of a simultaneous wireless information and power transfer cognitive radio system. *IEEE Wireless Communications Letters*, 5:288–291, 2016.

P. Sitka, P. Corke, L. Overs, P. Valencia, and T. Wark. Fleck – a platform for real-world outdoor sensor networks. *Proceedings of ISSNIP 2007*, 1:709–714, 2004.

P.C. Sofotasios, S. Muhaidat, G.K. Karagiannidis, and B.S. Shariff. Solutions to integrals involving the Marcum Q-function and applications. *IEEE Signal Processing Letters*, 22:1752–1756, 2015.

C. Song, Y. Huang, P. Carter, J. Zhou, S. Yuan, Q Xu, and M. Kod. A novel six-band dual CP rectenna using improved impedance matching technique for ambient RF energy harvesting. *IEEE Transactions on Antennas and Propagation*, 64:3160–3171, 2016.

C. Song, Y. Huang, J. Zhou, and P. Carter. Improved ultrawideband rectennas using hybrid resistance compression technique. *IEEE Transactions on Antennas and Propagation*, 65:2057–2062, 2017.

M. Stoopman, S. Keyrouz, H.J. Visser, K. Philips, and W.A. Serdijn. A self-calibrating RF energy harvester generating 1V at −26.3 dBm. *2013 Symposium on VLSI Circuits (VLSIC)*, 1:C226–C227, 2013.

M. Stoopman, S. Keyrouz, H.J. Visser, K. Philips, and W.A. Serdijn. Co-design of a CMOS rectifier and small loop antenna for highly sensitive RF energy harvesters. *IEEE Journal of Solid-State Circuits*, 49:622–634, 2014.

G. Stuber. *Principles of Mobile Communication*. Kluwer Academic Publishing, 2001.

S. Sudevalayam and P. Kulkarni. Energy harvesting sensor nodes: survey and implications. *IEEE Communications Surveys and Tutorials*, 13:443–461, 2011.

H. Sun, Y. Guo, M. He, and Z. Zhong. A dual band rectenna using broadband Yagi antenna array for ambient RF power harvesting. *IEEE Antennas and Wireless Propagation Letters*, 12:918–921, 2013.

Y. Sun, O. Gurewitz, and D. Johnson. RI-MAC: a receiver-initiated asynchronous duty cycle MAC protocol for dynamic traffic loads in wireless sensor networks. *ACM SenSys 2008*, 1:1–14, 2008.

K.C. Syracuse and W.D.K. Clark. A statistical approach to domain performance modeling for oxyhalide primary lithium batteries. *Proceedings of the 12th Annual Battery Conference on Applications and Advances*, 1:14–17, 1997.

R. Tandra and A. Sahai. SNR walls for signal detection. *IEEE Journal of Selected Topics in Signal Processing*, 2:4–16, 2008.

B. Tong, Z. Li, G. Wang, and W. Zhang. How wireless power charging technology affects sensor network deployment and routing. *IEEE International Conference on Distributed Computing Systems*, 1:438–447, 2010.

C. Tracy and H. Widom. On orthogonal and symplectic matrix ensembles. *Communications in Mathematical Physics*, 177:727–754, 1996.

T.A. Tsiftsis, G.K. Karagiannidis, P.T. Mathiopoulos, and S.A. Kotsopoulos. Nonregenerative dual-hop cooperative links with selection diversity. *EURASIP Journal of Wireless Communications and Networking*, 2006:1–8, 2006.

H. Urkowitz. Energy detection of unknown deterministic signals. *Proceedings of the IEEE*, 55:523–531, 1967.

M. Usman and I. Koo. Access strategy for hybrid underlay-overlay cognitive radios with energy harvesting. *IEEE Sensors Journal*, 14:3164–3173, 2014.

L.R. Varshney. Transporting information and energy simultaneously. *Proceedings of the IEEE International Symposium on Information Theory*, 1612–1616, 2008.

M.G. Villalva, J.F. Gazoli, and E.R. Filho. Comprehensive approach to modeling and simulation of photovoltaic arrays. *IEEE Transactions on Power Electronics*, 24:1198–1208, 2009.

E. Visotsky, S. Kuffner, and R. Peterson. On collaborative detection of TV transmissions in support of dynamic spectrum sharing. *Proceedings of the IEEE 1st Symposium on Dynamic Spectrum Access Networks*, 1:338–345, 2005.

B. Wang, D. Tsonev, S. Videv, and H. Haas. On the design of a solar-panel receiver for optical wireless communications with simultaneous energy harvesting. *IEEE Journal on Selected Areas in Communications*, 33:1612–1623, 2015.

K. Wang, Y. Chen, and M.-S. Alouini. BER and optimal power allocation for amplify-and-forward relaying using pilot-aided maximum likelihood estimation. *IEEE Transactions on Communications*, 62:3462–3475, 2014.

T. Wang, A. Cano, G.B. Giannakis, and J.N. Laneman. High-performance cooperative demodulation with decode-and-forward relays. *IEEE Transactions on Communications*, 55:1427–1438, 2007.

W.S. Wang, T. O'Donnell, N. Wang, M. Hayes, B. O'Flynn, and C. O'Mathuna. Design considerations of sub-mW indoor light energy harvesting for wireless sensor systems. *ACM Journal on Emerging Technologies in Computing Systems*, 6:4424–4435, 2010.

D. Wu and R. Negi. Effective capacity: a wireless link model for support of quality of service. *IEEE Transactions on Wireless Communications*, 2:630–643, 2003.

W. Xiao, W.G. Dunford, and A. Capel. A novel modeling method for photovoltaic cells. *IEEE Power Electronics Specialists Conference 2004*, 1:1950–1956, 2004.

W. Xiao, F.F. Edwin, G. Spagnuolo, and J. Jatskevich. Efficient approaches for modeling and simulating photovoltaic power systems. *IEEE Journal of Photovoltaics*, 3:500–508, 2013.

L. Xie, Y. Shi, Y.T. Hou, and W. Lou. Wireless power transfer and applications to sensor networks. *IEEE Wireless Communications*, 20:140–145, 2013.

Z. Xie, Y. Chen, Y. Gao, Y. Wang, and Y. Su. Wireless powered communication networks using peer harvesting. *IEEE Access*, 5:3454–3464, 2017.

D. Xu and Q. Li. Joint power control and time allocation for wireless powered underlay cognitive radio networks. *IEEE Wireless Communications Letters*, 6:294–297, 2017.

G. Yang, C.K. Ho, and Y.L. Guan. Dynamic channel estimation and power allocation for wireless power beamforming. *Proceedings of the IEEE International Conference on Communications 2014*, 1:1–5, 2014.

G. Yang, G.-Y. Lin, and H.-Y. Wei. Markov chain performance model for IEEE 802.11 devices with energy harvesting source. *IEEE GLOBECOM*, 1:5212–5217, 2012.

J. Yang and S. Ulukus. Optimal packet scheduling in an energy harvesting communication system. *IEEE Transactions on Communications*, 60:220–230, 2012.

W. Ye, J. Heidemann, and D. Estrin. An energy-efficient MAC protocol for wireless sensor networks. *IEEE INFOCOM 2002*, 3:1567–1576, 2002.

S. Yin, Z. Qu, and S. Li. Achievable throughput optimization in energy harvesting cognitive radio systems. *IEEE Journal on Selected Areas in Communications*, 33:407–422, 2015.

S. Yin, E. Zhang, Z. Qu, L. Yin, and S. Li. Optimal cooperation strategy in cognitive radio systems with energy harvesting. *IEEE Transactions on Wireless Communications*, 13:4693–4707, 2014.

H. Yoo, M. Shim, and D. Kim. Dynamic duty-cycle scheduling schemes for energy-harvesting wireless sensor networks. *IEEE Communications Letters*, 16:202–204, 2012.

W. Yu, L. Musavian, and N. Qiang. Tradeoff analysis and joint optimization of link-layer energy efficiency and effective capacity toward green communications. *IEEE Transactions on Wireless Communications*, 15:3339–3353, 2016.

M.-R. Zenaidi, Z. Rezki, and M.-S. Alouini. Performance limits of online energy harvesting communications with noisy channel state information at the transmitter. *IEEE Access*, 5:1239–1249, 2017.

Y. Zeng, C.L. Koh, and Y.-C. Liang. Maximum eigenvalue detection: theory and application. *Proceedings of the IEEE International Conference on Communications (ICC'08)*, 1:4160–4164, 2008.

Y. Zeng and Y.-C. Liang. Eigenvalue-based spectrum sensing algorithms for cognitive radio. *IEEE Transactions on Communications*, 57:1784–1793, 2009a.

Y. Zeng and Y.-C. Liang. Spectrum-sensing algorithms for cognitive radio based on statistical covariances. *IEEE Transactions on Vehicular Technology*, 58: 1804–1815, 2009b.

Y. Zeng and R. Zhang. Full-duplex wireless-powered relay with self-energy recycling. *IEEE Wireless Communications Letters*, 4:201–204, 2015a.

Y. Zeng and R. Zhang. Optimized training design for multi-antenna wireless energy transfer in frequency-selective channel. *Proceedings of the IEEE International Conference on Communications 2015*, 1–5, 2015b.

C. Zhai, J. Liu, and L. Zheng. Cooperative spectrum sharing with wireless energy harvesting in cognitive radio networks. *IEEE Transactions on Vehicular Technology*, 65:5303–5316, 2016a.

C. Zhai, J. Liu, and L. Zheng. Relay-based spectrum sharing with secondary users powered by wireless energy harvesting. *IEEE Transactions on Communications*, 64:1875–1887, 2016b.

D. Zhang, Z. Chen, M.K. Awad, N. Zhang, H. Zhou, and X. Shen. Utility-optimal resource management and allocation algorithm for energy harvesting cognitive radio sensor networks. *IEEE Journal on Selected Areas in Communications*, 34:3552–3565, 2016.

R. Zhang and C.K. Ho. MIMO broadcasting for simultaneous wireless information and power transfer. *IEEE Transactions on Wireless Communications*, 12:1989–2001, 2013.

Y. Zhang, W. Han, D. Li, P. Zhang, and S. Cui. Power versus spectrum 2-D sensing in energy harvesting cognitive radio networks. *IEEE Transactions on Signal Processing*, 63:6200–6212, 2015.

N. Zhao, Y. Cao, R. Yu, Y. Chen, M. Jin, and V.C.M. Leung. Artificial noise assisted secure interference networks with wireless power transfer. *IEEE Transactions on Vehicular Technology*, PP:1–1, 2017a.

N. Zhao, S. Zhang, F.R. Yu, Y. Chen, A. Nallanathan, and V.C.M. Leung. Exploiting interference for energy harvesting: a survey, research issues, and challenges. *IEEE Access*, 5:10403–10421, 2017b.

G. Zheng, Z. Ho, E.A. Jorswieck, and Ottersten B. Information and energy cooperation in cognitive radio networks. *IEEE Transactions on Signal Processing*, 62:2290–2303, 2014.

C. Zhong, X. Chen, Z. Zhang, and G.K. Karagiannidis. Wireless-powered communications: Performance analysis and optimization. *IEEE Transactions on Communications*, 63:5178–5190, 2015a.

C. Zhong, H.A. Suraweera, G. Zheng, I. Krikidis, and Z. Zhang. Wireless information and power transfer with full duplex relaying. *IEEE Transactions on Communications*, 62:3447–3461, 2014.

C. Zhong, G. Zheng, Z. Zhang, and G.K. Karagiannidis. Optimum wirelessly powered relaying. *IEEE Signal Processing Letters*, 22:1728–1732, 2015b.

B. Zhou, H. Hu, S.-Q. Huang, and H.-H. Chen. Intracluster device-to-device relay algorithm with optimal resource allocation. *IEEE Transactions on Vehicular Technology*, 62:2315–2326, 2013a.

X. Zhou. Training-based SWIPT: optimal power splitting at the receiver. *IEEE Transactions on Vehicular Technology*, 64:4377–4382, 2015.

X. Zhou, R. Zhang, and C.K. Ho. Wireless information and power transfer: architecture design and rate-energy tradeoff. *IEEE Transactions on Communications*, 61:4754–4767, 2013b.

T. Zhu, Z. Zhong, Y. Gu, T. He, and Z.-L. Zhang. Leakage-aware energy synchronization for wireless sensor networks. *Proceedings of ACM MobiSys 2009*, 1:319–332, 2009.

Index

Energy Harvesting Communications: Principles and Theories, First Edition. Yunfei Chen.
© 2019 John Wiley & Sons Ltd. Published 2019 by John Wiley & Sons Ltd.